This book espouses an innovative theory of scientific realism in which due weight is given to mathematics and logic. The authors argue that mathematics can be understood realistically if it is seen to be the study of universals, of properties and relations, of patterns and structures, the sorts of things which can be in several places at once. Taking this kind of scientific Platonism as their point of departure, they show how the theory of universals can account for probability, laws of nature, causation, and explanation, and they explore the consequences in all these fields.

This will be an important book for all philosophers of science, logicians, and metaphysicians, and their graduate students. It will also appeal to those outside philosophy interested in the interrelationship of philosophy and science.

CAMBRIDGE STUDIES IN PHILOSOPHY

Science and Necessity

CAMBRIDGE STUDIES IN PHILOSOPHY

JAMES CARGILE *Paradoxes: a study in form and predication*
PAUL M. CHURCHLAND *Scientific realism and the plasticity of mind*
N. M. L. NATHAN *Evidence and assurance*
WILLIAM LYONS *Emotion*
PETER SMITH *Realism and the progress of science*
BRIAN LOAR *Mind and meaning*
J. F. ROSS *Portraying analogy*
DAVID HEYD *Supererogation*
PAUL HORWICH *Probability and evidence*
ELLERY EELLS *Rational decision and causality*
HOWARD ROBINSON *Matter and sense*
E. J. BOND *Reason and value*
D. M. ARMSTRONG *What is a law of nature?*
HENRY E. KYBURG, JR. *Theory and measurement*
MICHAEL H. ROBINS *Promising, intending and moral autonomy*
N. J. H. DENT *The moral psychology of the virtues*
R. A. DUFF *Trials and punishments*
F LINT SCHIER *Deeper into pictures*
ANTHONY APPIAH *Assertion and conditionals*
ROBERT BROWN *Analyzing love*
ROBERT M. GORDON *The structure of emotions*
FRANÇOIS RECANATI *Meaning and force*
WILLIAM G. LYCAN *Judgement and justification*
W. D. HART *The engines of the soul*
GERALD DWORKIN *The theory and practice of autonomy*
DAVID O. BRINK *Moral realism and the foundations of ethics*
PAUL MOSER *Knowledge and evidence*
D. M. ARMSTRONG *A combinatorial theory of possibility*
MICHAEL TYE *The metaphysics of mind*
CARL GINET *On action*
MARK RICHARD *Propositional attitudes*
JOHN BISHOP *Natural agency*
J. CHRISTOPHER MAHONEY *The mundane matter of the mental language*
GERALD GAUS *Value and justification*
MARK HELLER *The ontology of physical objects*

Science and Necessity

John Bigelow
La Trobe University

Robert Pargetter
Monash University

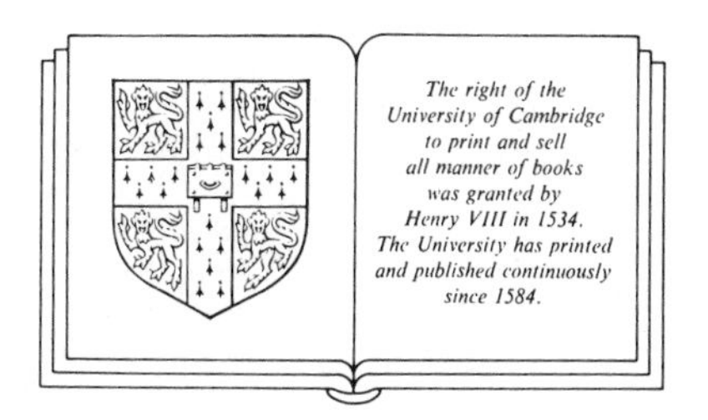
The right of the
University of Cambridge
to print and sell
all manner of books
was granted by
Henry VIII in 1534.
The University has printed
and published continuously
since 1584.

Cambridge University Press

Cambridge
New York Port Chester Melbourne Sydney

Published by the Press Syndicate of the University of Cambridge
The Pitt Building, Trumpington Street, Cambridge CB2 1RP
40 West 20th Street, New York, NY 10011, USA
10 Stamford Road, Oakleigh, Melbourne 3166, Australia

First published 1990

Printed in the United States of America

Library of Congress Cataloging-in-Publication Data

Bigelow, John, 1948–

Science and necessity / John Bigelow,
Robert Pargetter.

p. cm. – (Cambridge studies in philosophy)

Includes bibliographical references.

ISBN 0-521-39027-3

1. Science – Philosophy. 2. Logic. 3. Mathematics.
I. Pargetter, Robert. II. Title. III. Series.

Q175.B554 1991
501 – dc20 90-1700
CIP

British Library Cataloguing in Publication Data

Bigelow, John, 1948–

Science and necessity. – (Cambridge studies in
philosophy).

1. Scientific realism
I. Title II. Pargetter, Robert
501

ISBN 0-521-39027-3 hardback

Contents

Preface

In science, observations are important. They reveal to us the properties and relations of particular objects. Also important are unobserved, theoretical entities with their properties and relations. But logic is important too, and so is mathematics. Any scientific realism worth its salt must give due weight to mathematics and logic, as well as to observations.

In this book we argue for mathematical realism. Broadly speaking, we argue for a brand of Platonism, but a *scientific Platonism* rather than an otherworldly mysticism of the sort which is often conjured up when Plato is called to mind. Mathematics can be understood realistically, we argue, if it is seen to be the study of *universals,* of properties and relations, of patterns and structures, the kinds of things which can be in several places at once.

Logic too, we argue, can best be understood, in harmony with scientific realism, using the raw materials furnished by a scientific Platonism. Logic is concerned with the validity of arguments; and an argument is valid when the conclusion is a *necessary* consequence of the premisses, in other words when it is not *possible* for the conclusion to be false unless the premisses are mistaken. Consequently, any explanation of validity depends on the modal concepts of necessity and possibility. A scientific realist should be a modal realist, a realist about necessities and possibilities, as well as about the other things in science. The key to modal realism, we argue, is to be found in realism about universals.

Universals facilitate a realist understanding of mathematics, and of logic. They also generate realist accounts of probability, laws of nature, causation, and scientific explanation. Thus, they provide a unified theory of all the necessities and other modalities found in science. This book traces the consequences which follow, in all these fields, from our central theme of scientific Platonism.

Each of the topics addressed in this book is presented entirely from a realist perspective. We give our reasons for realism early in the book, and then we restrict our attention to realist theories. Thus, by the time we reach the topic of, for instance, causation, we have set aside a wide variety of fashionable philosophical perspectives. Our treatment of causation, therefore, does not survey the full range of plausible theories of causation. We address only those theories which contribute to the advancement of our central, scientific realism. The resultant narrowing of focus is cumulative. After discussing causation, for instance, we move on to the theory of scientific explanation, and by this time our focus has narrowed not only to realist theories, but to theories which grant a special status to causation. This book has the form of an extended argument, rather than an introductory survey.

We find ourselves most often contrasting our views with the views of those with whom we could be most naturally aligned. For instance, the scientific realism of Michael Devitt, the metaphysical realism of David Armstrong, and the modal realism of David Lewis are theories with which we most agree and on which we most rely. Nevertheless, they are the very theories which we argue against at greatest length.

Many of the chapters in this book have appeared as articles, but it would be a mistake to view the book as a collection of independent treatments of related topics. Each chapter grew, first of all, as a part of the book and only then, in some cases, was extracted as an article, as a means of testing and developing that component of the whole book.

Australasia at large, and Melbourne in particular, has been an exciting place in which to work. We feel a very special debt to David Armstrong, Frank Jackson, and David Lewis, but there are many others who over the years have helped shape our ideas. Among those who should be mentioned are Max Cresswell, Barbara Davidson, Michael Devitt, Brian Ellis, Peter Forrest, John Fox, Allen Hazen, Adrian Heathcote, George Hughes, Lloyd Humberstone, Elizabeth Prior, Denis Robinson, Kim Sterelny, Barry Taylor, and Pavel Tichý. We thank the secretarial staff at both La Trobe University and Monash University, particularly Gai Dunn and Lesley Whitelaw, for their assistance with the preparation of the text and our copy editor, Mary Nevader, for her painstaking and invaluable work. Our proof-readers, Alan Cooke and Ian Ravenscroft, not only worked diligently under a tight schedule to correct the final text, but also tried, in vain, to help us improve the quality of its content.

Finally, we appreciate a different kind of help and support from Faye and David, Elizabeth, Stephen, and Benjamin.

Symbols

λ	Categorial language, p. 100
$\wedge$	Conjunction, pp. 99, 102
d	Degrees of accessibility, p. 123
$\vee$	Disjunction, p. 102
$\mathcal{D}$	Domain, p. 120
ϕ	Empty set, pp. 142, 372–3
$\equiv$	Material equivalence, p. 102
$\supset$	Material implication, p. 102
α, β, γ	Metalinguistic variables, p. 99
$\Diamond\!\!\rightarrow$	Might counterfactual, p. 102
$\Box$	Necessity operator, p. 102
$\sim$	Negation, pp. 96, 102
Ψ	Open-sentence variables, pp. 369, 371
δ	Operator, p. 100
$\Diamond$	Possibility operator, p. 102
ψ	Predicate variables, p. 160
$\exists$, $\forall$	Quantifiers, pp. 43, 96, 102
Ω	Sample space, p. 121
$\mathcal{V}$	Semantic valuation function, pp. 99, 125–9
ω	Semantic value operator, p. 99
$\mathcal{E}$	Set of events or propositions, p. 121
ϵ	Set membership, pp. 144–5
$\mathcal{W}$	Set of possible worlds, p. 121
ω	Set of von Neumann numbers, pp. 372–3
$\Box\!\!\rightarrow$	Would counterfactual, p. 102

1

Realism and truth

1.1 SCIENTIFIC PLATONISM

Why is it that, in some parts of the world, cold weather is always followed, after a certain period of time, by warm weather and that, after another period of time, cold weather returns? Perhaps it is because Hades abducted Core and took her to be queen of the underworld. As a result, Core's mother Demeter, the goddess of corn, refused to allow anything to grow. Demeter threatened to leave the land barren until her daughter was returned to her. Zeus pleaded with her, but Demeter would not relent; so Zeus commanded Hades to return Core. However, because Core had eaten seven pomegranate seeds while a prisoner of Hades, the god of the underworld had a legitimate claim on her. As a compromise, Core spends four months each year in the underworld, and Demeter allows nothing to grow until her daughter is returned to her in the spring.

Compare this explanation of the seasons with the explanation given by Aristotle, and you will find a very marked contrast. Aristotle held that all of the heavenly bodies were attached to concentric spherical shells made of a heavenly substance called aether. The spherical earth was at the centre of the shells. Each of these shells rotated at a uniform speed around a fixed axis which was attached to the next outermost shell. Given the orientations of the axes and speeds of rotation, one could predict extremely accurately the motions of the sun through the seasons.

There may or may not be a fundamental difference in kind between science and other human pursuits, but there is at least a very large difference in degree. In degree of what? In degree of prominence given to *reasoning,* to *logic,* to *valid arguments.* Reason is not the only essential factor in science; nor is reason or rationality absent from other pursuits. But in science, logical argumentation is undeniably one of the most central features.

The distinctive feature of a valid argument is that it is *not possible* for its premisses to be true if its conclusion is false. Thus, at the core of science is the notion of *possibility* and its dual, *necessity*. These notions are crucial not only to argumentation, but also to other distinctive aspects of science.

Logical possibility and logical necessity are central to science; but there are other grades of possibility and necessity that are also important. For instance, in science we seek to find the bounds of both physical possibility and technical possibility. Laws of nature seem to require some sort of necessary connection among events. Plausibly, causation is not just a matter of accidental correlation between cause and effect. Explanation often seems to involve the application of reasoning, which converts something that was unexpected into something which can be seen to have been inevitable. And so on. Possibility and necessity pervade science.

This book aims to interpret the various roles which necessity and possibility play in science. The interpretation is from the standpoint of *realism*. In fact we defend three kinds of realism: *scientific, metaphysical,* and *modal*. Moreover, as we are committed to the correspondence theory of truth, we also defend *semantic* realism. This overall realist position can be aptly described as *scientific Platonism*.

1.2 SCIENTIFIC REALISM

Scientific realism is a cluster of claims which centre on the assertion that certain sorts of things exist and are related to one another in certain ways. Saying this much is easy, but it does not really illuminate the nature of scientific realism. This is because of the vagueness of 'certain sorts of things'. This phrase cannot be replaced simply by something like 'the things scientists mention in their theories'. Realists are not committed to the existence of *all* the entities that are invoked in useful theories of science. At times scientists have accepted and used theories which referred to valencies, orbitals, genes, and the like without being sure of or committed to the view that all such things exist. Currently, many quantum theorists do not hold a realist interpretation of quantum theory. We could truly say that scientific realists are realist about *most* of the things that *most* of their theories talk about; they generally adopt a realist interpretation of these theories; and they are committed to the existence of the most fundamental categories of an accepted science. Yet while true and perhaps helpful, this does not completely explicate the commitments of scientific realists. To say more, we shall take the indirect tack of comparing realism with the most common alternative view.

Realism is opposed by the doctrine of *instrumentalism*. Instrumentalists characteristically assert that no things of a certain sort exist, or at least that we have no adequate grounds for believing that they do exist. For instance, during the nineteenth century scientific realists asserted, while instrumentalists refused to assert, that there are such things as atoms.

Scientific realism and instrumentalism have two strands: claims about the *world* and claims about scientific *theories*. We shall focus at first on their claims about the world.[1] We believe, with realists, that the world is stocked with such things as plants and planets, frogs and photons, phenotypes and genotypes, oxygen and kinetic energy, magnetic fields and tectonic plates. But as will emerge more clearly later, we also make characteristically realist claims about scientific theories. We claim that certain scientific theories are *true,* and that they accurately represent what things exist in the world and how they are related to one another. The theories in question are, of course, theories about plants, planets, frogs, photons, and so forth. The belief that these theories are true is closely linked with the belief that there are such things as plants, planets, and so forth. Yet it is important not to prejudge important issues, as we would inevitably do if we were simply to conflate the two distinct strands of realism.

To begin with, let us focus on the more *extroverted* strand of realism, whereby, after thinking hard about, say, inheritance, and after observing and experimenting for years, we come to hold certain views about what there is in the world – genes, for instance. In doing this, we may spend very little time in *introverted* thought *about theories*. There may, in fact, be very few theories concerning our subject-matter at the beginning of our investigations. So we may spend little time studying the theories of others. And clearly, we cannot introspectively turn our attention to a study of our own theory until after we have developed a theory about the things

1 This is the kind of realism urged, e.g., by Devitt (1984) and Ellis (1987).

themselves. What we must do first is to decide whether there *are,* for instance, such things as dominant and recessive alleles, and so forth. Typically, a realist holds certain views about the world *and* certain views about theories. But we cannot hold views about theories until we have developed some theories to hold views about. The first step a realist takes is that of developing theories to the effect that there are certain things in the world, related in certain ways; one who develops such theories is a realist about those things.

An instrumentalist about those things is one who does not believe that there are any such things. Characteristically, however, an instrumentalist is not concerned only with whether such things exist. An instrumentalist rapidly switches the focus from things to theories, and turns the spotlight on realists' theories. Instrumentalists argue that realists' theories may be useful *instruments* for achieving certain practical results, but their usefulness should not be taken as grounds for believing that there really are such things as the entities which realists believe in and instrumentalists do not. They argue that, although such theories are useful, we should not infer that the claims they make are true or that the words or symbols they employ refer to anything in the world.[2]

This sets a trap for realists. It is tempting for them to fall entirely into the framework of instrumentalists and begin debating about the nature of theories. This may lead them to agree with the premiss that certain theories serve certain practical needs, and then to defend the view that we *can* infer, from the usefulness of the theories, that such theories are true and do refer to things in the world. Yet such an inference, from the usefulness of the theory to the existence of various things, is by its nature available to realists only *after* the theory has come into existence. Realists' primary inferences occur earlier than this – in the creation of the theory in the first place.

Let us offer an illustration from the history of science. The ancient Greeks looked at the way the stars moved slowly across the sky by night and the way the sun moved slowly across the sky by day. From various facts, about *where* the sun and stars were *when,* they inferred that the earth is a spherical body, at the centre of a spherical shell on which the stars are fixed. This shell rotates around

2 See Putnam (1978, 1981) and van Fraassen (1980).

the earth. Stars near the axis of rotation thus move more slowly than those farthest from the axis. Their motion is conferred on them by the shell in which they are embedded. If each moved independently of the others, it would be a miracle that they should so co-ordinate their speeds as to move *as if* they were fixed to a sphere. It would be more rational to ascribe their motion to that of a single shell which spins around the earth. Aristotle enlarged this theory by placing many more transparent spherical shells between the earth and the stars, to explain the wanderings of the planets. He was a *realist*. He believed that these shells existed and that they rotated in certain ways, transferring motion from one to another in such a way as to produce the motions we observe in the heavens, the motions of the stars, sun, moon, and planets.

If asked, no doubt Aristotle would have made affirmations not only about stars, spheres, and planets, but also about the words he used to describe stars, spheres, and planets. He would have asserted that his claims were *true* and that his Greek term for 'sphere' did *refer* to something which exists somewhere above us, which is made of some substance or other, and which moves in certain ways. He would have claimed that he had said of what is that it is, and of what is not that it is not.

From the time of Aristotle onwards, however, other astronomers became sceptical about whether such spheres existed. They began to employ a variety of theories in an attempt to, as is said, "save the phenomena". That is, they attempted to predict exactly where each heavenly body would be, at each time. If a theory served as a useful means of calculating and predicting, that was all that mattered to them. They did not believe that the things described in the theory really existed in the world.

This attitude developed concurrently with the creation of theories which were not only complex, but also in important ways arbitrary: each theory was mathematically equivalent to various rivals, and there were no reasons for preferring any of the rivals over the others. In consequence, it was difficult to believe in all the things referred to in the astronomical theories which followed Aristotle's. By the time of Ptolemy's astronomy, in the interests of accurate prediction, theories had been developed which replaced Aristotle's spheres by points which moved in circles around points, which were themselves moving in circles, and so forth. It is not impossible to take such theories realistically and to believe that there really are transparent marbles rolling around inside transparent hoops, and so forth. But it is not easy to sustain such beliefs.

Furthermore, a realist construal of Ptolemaic theories generates a variety of paradoxes, which instrumentalists can happily ignore. For instance, technical requirements in astronomy placed the earth away from the centre of the sun's orbit. (The number of days between solstices is not quite the same as the number of days between equinoxes.) This suggested that perhaps the earth might not be at the centre of the cosmos after all. Yet this raised a difficulty in understanding the phenomenon of falling bodies. The best theory of falling bodies available at the time (Aristotle's) required the earth to be at the exact centre of the universe. Hence, it seemed, we should preserve the earth's central position with respect to the outermost sphere of the stars. This forced upon astronomers a geometrically equivalent theory which placed the sun's orbit eccentric with respect to the centre of the universe. By keeping the earth at the centre of the universe, it was possible to restore the Aristotelian account of falling bodies on the earth itself, but we were then left with a new falling body – in fact a rising and falling body – the sun, with its eccentric motion. This is a severe problem for a realist.

Hence, there were strong motives for taking an instrumentalist view of the Ptolemaic astronomical theories. It may be wondered why such theories were concocted at all. It could not be because the inventors of such theories wanted to know *what was up there* and *why* the things up there moved as they did. It must have been, rather, because they wanted to predict observations very accurately. Why should they want to do that? Perhaps for navigational needs. And yet the effort invested and the precision achieved were far in excess of what was required for navigation. The notorious Ptolemaic epicycles were needed only for the planets, and the wandering planets are virtually useless for navigation. For navigation, a very simple theory, involving only the sun and the stars, was (and still is) all that was needed. Perhaps, then, accurate predictions were desirable for other motives. Perhaps accurate predictions of the movement of the stars and planets were erroneously believed to lead to accurate predictions of events of everyday life – by way of astrology. But, again, the supposed practical benefits could be obtained without the full Ptolemaic theoretical complexities. Perhaps the need for a calendar that did not very gradually change January from a winter month to an autumn, then a summer, then a spring month (and so on) provided a practical motive, though again, why should we worry so much about that? We would have centuries to adjust to the idea that January has crept back to a different season of the year. Or we could relabel the months from time to time. There was really no compelling practical reason why we needed to be able to predict, centuries in advance, exactly how much the calendar would have to be adjusted to keep January a winter month. Perhaps the making of accurate predictions had become an end in itself – a kind of intellectual Olympics, a heavenly Rubik's cube. Thereby, one astronomer could establish superiority over others, thus acquiring prestige. But whatever instrumentalists' motives for developing Ptolemaic theories, the motive which generated Aristotle's realism was clearly something quite different.

Copernicus, like Aristotle, was a realist. Copernicus believed that the earth orbited the sun, and that the earth thereby moved closer to a given star during one season and then farther from that star over the next two seasons. If asked to comment, not only about the heavens and earth, but also about the words in his book, Copernicus would no doubt have made characteristically realist comments, to the effect that his claims were *true* and that some of his terms *did refer* to things and movements which were really out there in the world.

Copernicus's book, describing the motions of the earth and planets, was finally published in the year he died. The Preface contained material which many believed to have been written, or at least endorsed, by Copernicus but which we now know to have been written by someone called Osiander. Osiander's Preface tells us that Copernicus's theory need not be taken realistically, but may be accepted from a purely instrumentalist standpoint: it facilitates computations, permits more accurate predictions to be made, and thereby yields practical benefits in such fields as calendar reform and navigation.

It is interesting, however, that the most eminent succeeding Copernicans, Kepler and Galileo, were clearly realists. Galileo in particular had remarkably little interest in the usefulness of the theory for making accurate astronomical predictions. He believed that things were, in broad terms, as Copernicus said they were: the earth does move around the sun. Such utility as the theory had, from an instrumental point of view, was of minor concern to Galileo.

These snippets of history reflect our conception of realism as growing in two stages.[3] In the first stage, a belief develops which is not about language or theories at all, but about what things there are in the world. This is *extroverted* realism. This belief is then threatened by instrumentalists, who deny the existence of those things and affirm only that speaking the way the realist has been speaking has instrumental value, so we may go on speaking that way as long as we cease to mean what we say. In the second stage of realism, these instrumentalist claims are countered. Realists become *introverted*. They then claim that what they have been saying *is true,* and not just in some instrumentalist or pragmatist sense of the term, in which 'true' just means 'useful'. Rather, realists claim, what they have been saying is true in the sense that there are certain things in the world, and the words they have been using in their theories do refer to these things, and the things in the world are related in the ways they are said to be related in realists' theories.

In summary, *extroverted* scientific realism is the assertion of the existence of a list of things which a realist believes in as a result of scientific endeavour and enquiry; *introverted* scientific realism is the assertion that certain scientific theories do refer to things in the world and truly describe those things. An introverted scientific realist, we claim, is committed to some version of what is known as the *correspondence theory of truth*. But this is a secondary position for a realist. The primary step for a realist is the formation of views about what the world contains; only after this primary step can a realist formulate a theory about how some things in the world (theories) correspond to other things in the world.

1.3 METAPHYSICAL REALISM

We defend realism not only in the sense in which its opponent is instrumentalism but also in the sense in which its opponent is *nominalism*. We defend metaphysical realism:[4] the theory that, in addition to the various individuals in the world, there are also various properties and relations – universals. Universals, unlike particulars, can be in many different places at the same time. Nominalists deny that there are such things. They claim that everything there is, is

3 Among the plethora of resumés of the history of astronomy, we highly recommend Dreyer (1953) and Dijksterhuis (1961).

4 We use the term 'metaphysical realism' in a manner quite distinct from Putnam (1978, 1981).

particular; everything there is, is in at most one place at any given time.

The debate between nominalism and metaphysical realism highlights the problem of priority between metaphysics and semantics. We urge that metaphysics precedes semantics[5] but that, once a realist metaphysics has generated a correspondence semantics, the semantic child can offer significant support to its metaphysical parent.

Traditionally, one of the major arguments for universals has been a semantic one. It has been noted that only a small number of terms in a natural language can be construed as referring to particular individuals which can be placed and dated. Proper names, demonstratives, pronouns, and definite descriptions plausibly refer to individuals: 'Gough', 'this', 'her', 'the Iron Lady', for instance. But many terms refer to no single individual in space-time. Yet we all know what they mean. Consider, for instance, terms like 'justice', 'beauty', 'circle', 'five'. It can plausibly be argued that such terms do refer to something real – to *ideals* towards which particulars may approach, or to *universals* which can be shared by many particulars, or to *abstractions* which can be studied by the intellect but not seen by the senses; or to something else. However we characterize them, there must be things of some sort or other corresponding to terms like 'justice' and so forth. So it may be argued.

We do not, however, rest our case for universals primarily on such a semantic argument. Rather, there are good reasons for believing in universals quite independently of semantics. Yet once we *have* argued that universals do exist, it is inevitable that we should welcome the fact that universals furnish us with rather smoother semantic theories than we could otherwise muster. In fact, this can be used as a good, extra argument in favour of universals. And so we cannot claim that the inferences all run one way, from metaphysics to semantics. There are arguments we should not ignore and which support metaphysical realism by drawing on semantics. It is therefore hard to keep semantics completely out of the picture, even temporarily.

Nevertheless, we argue that realism about universals does rest on

5 This methodology is in sympathy with Armstrong (1978) and Lewis (1983a). A further argument against the semantic source for metaphysical realism is found in Bigelow (1981), where it is argued that semantics could be formulated without reference to universals.

other good reasons, independent of semantics; and these independent reasons are the more fundamental ones. Metaphysical realism, like scientific realism, begins in an extroverted form. In fact, metaphysical realism constitutes just one aspect of general scientific realism. In attempting to understand the world we live in, we are led to believe not only in individuals, but also in their properties and relations. We are led to study these properties and relations. Furthermore, we are led to wonder what properties those properties and relations may have, and what relations hold among those properties and relations. Thereby we move into what are known as *second-degree* properties and relations, that is, the properties and relations instantiated by properties and relations. This also leads us to *second-order* properties and relations, that is, properties and relations instantiated by individuals in virtue of the existence of second-degree properties and relations. (For instance, 'being a colour' is a property of a property, whereas 'being coloured' is a property of individuals.) The distinction between second-degree and second-order universals will be explained more fully in Chapter 2; we think both sorts of universals hold the key to many puzzles in philosophy and science. And by moving to the study of properties and relations of properties and relations, we believe we launch ourselves into mathematics (as we shall see in more detail later). Metaphysical realism could, in fact, appropriately be relabelled *mathematical realism*.

In pursuing science in a thoroughly extroverted way, realists will find themselves believing not only in individuals, but also in properties, like mass and velocity. And they will notice that some masses stand in various *proportions* to one another. They will begin to argue, for instance, about whether *there is* a proportion meeting such-and-such conditions. Thereby they will have become participants in the theory of real numbers. And thereby extroverted realists will become what are known as Platonists.[6] For present purposes, we construe Platonism to be the doctrine that all these universals have genuine existence, be they first-order properties and relations of individuals, second-degree properties and relations of universals, or second-order properties and relations of individuals.

The central argument for univerals, then, is derived from mathe-

6 Bigelow (1988a) gives a detailed account of numbers in terms of a theory of universals.

matics, and from science more generally. And yet very swiftly, the debate shifts onto a semantic plane. The extroverted realist begins *speaking as though* "there are" mathematical entities of various sorts, and physical properties and relations of various sorts, and so on. Yet protests swiftly arise to the effect that we must not misconstrue such talk. We must not "reify" universals. We are being misled by a picture, embedded in forms of speech, which encourages Platonists to think of universals as though they "exist" in the same sense as individuals. Rather (the protest continues), we should recognize that the words used by extroverted realists should *not* be taken to refer to universals. The use of these words is merely a convenient shorthand expression of things which can be said, albeit in a more roundabout way, without ever imputing existence to universals.

Thus, Platonists, just like scientific realists more generally, are quickly drawn into an *introverted* realism in which they attempt to counter the nominalist and other instrumentalist arguments against taking their extroverted realism at face value. Realists thereby find themselves embroiled in arguing that science and mathematics *cannot* be reexpressed in ways which eschew all reference to universals. They argue that reference to universals cannot be paraphrased away without significant losses to mathematics and scientific theory in general.

The debate thereby becomes more and more abstruse and technical. Investigation focusses on locating the precise demarcation line in mathematics between those parts which can be "rescued" by paraphrasing universals away and those parts for which such paraphrases cannot be found. Then debates ensue over whether science really needs the "unrescuable" parts of mathematics and similarly whether there is any role in science for properties and relations which is not served by just talking about some sum of individuals.

The overall dialectic is clearly expressed by Quine. (In fact, Quine agonizes over the existence of *sets* rather than *universals,* or, as we would prefer to say, rather than *other* universals – see Sections 2.2, 8.4, and 8.5 for more on sets as universals. But if we let sets stand proxy for universals in general, the structure of the debate is clearly demarcated by Quine.) In his *Set Theory and Its Logic,* Quine explores what he calls 'virtual' set theory.[7] That is to say, he ex-

7 Quine (1969b). The criterion of ontological commitment is discussed in Quine (1961).

plores the extent to which when we are talking *as though* sets exist what we say can be reexpressed (in a more prolix manner) in ways which do not, even prima facie, ascribe existence to sets, but only to individuals.

In exploring virtual set theory, Quine is employing his famous *criterion of ontological commitment.* This is, in effect, a criterion for determining whether a proposed paraphrase is, or is not, successful in removing all reference to universals. That is a serious problem. Quine's solution runs, in outline, as follows.

First, we must paraphrase our theory in a language which has expressions which *explicitly* claim that certain things exist – so-called *quantifiers,* expressions meaning there are things which are thus and so. Then we collect all the sentences of our paraphrased theory which explicitly say, 'There are these things', 'There are those things', and so on. The things which our theory thus explicitly says there are, then, are according to Quine the only things which we are ontologically committed to: the only things which we should believe to exist.

In particular, Quine assumes that *predicates* in our theory do not incur any ontological commitment at all. When we say, for instance, that

There are things which are round and red,

this commits us to individuals, like apples, which are quantified over by the expression 'there are'. But it does *not* commit us to the existence of the two sets (the set of round things and the set of red things) or to the existence of the two qualities of roundness and redness. In presupposing this, Quine is taking a stand on a deep and important semantic question, on which we will say more at various places in this book. The relationship between predicates, like 'round' or 'red', and sets or universals is about as ticklish as any that we meet in this book. For the moment, let us enter a word of caution.

Consider the following case: We begin with the belief that everyone alive now has (or had) parents. Sometimes we say that two women are *sisters,* and we do so because we believe that *there are* two people who are parents of one and also parents of the other. We are ontologically committed to those parents. Yet suppose we were to adopt a theory in which we never refer to those parents at all. We use the predicate 'siblings', which describes their children, but we never *say* ‘there are’ parents in virtue of which they are siblings. In fact, suppose we found ways of paraphrasing every statement about the parents by framing a statement about the children. Instead of saying the mother gave birth to one sister before the other, we call one sister the ‘*elder* female sibling’. Instead of saying the father is a minister of religion, we describe the sisters as ‘minister’s daughters’ (and refuse to unpack this predicate further). Instead of saying the mother is a figure skater, we say the sisters are ‘daughters of a figure skater’ (and refuse to unpack this further). And so on.

It may or may not be possible to paraphrase parents away. Yet the question of whether there are parents is, surely, independent of the question of whether they can be purely verbally paraphrased away. Similarly, we hold that the question of whether there are universals is distinct from that of whether (as we realists would see it) all universals can be parcelled up into predicates in such a way that we never explicitly state that ‘there are’ any such things. There are, of course, important disanalogies between universals and parents. But there is nevertheless one crucial common factor: that the mere fact (if it is a fact) that they can be paraphrased away is not by itself enough to prove that they do not exist. Thus, we are not overly concerned with the question whether reference to universals can be avoided. Even if the paraphrasing trick can be successfully pulled on us, we will complain that the cards have been slipped up the nominalists’ sleeves.

Nevertheless, if we are to convince nominalists (who believe cards have ceased to be as soon as they have been slipped up someone’s sleeve), we must enter into their paraphrasing endeavours. Roughly speaking, the conclusion Quine and others come to is that sets (universals) can be paraphrased away from arithmetic and from chunks of the theory of rational and real numbers. But some parts of the theory of real numbers resist all rescue attempts by “hygienic” paraphrase. Some uses of real numbers in analysis – in the calculus and so on – resist reduction. This is troublesome, since the calculus was invented by Newton and Leibniz in order to explain such things as velocity and acceleration. And such things are central to science – they are not to be dismissed lightly. Nominalists cannot just ignore them. Thus, Quine reluctantly concedes that not all sets can be paraphrased away; therefore, we must admit the existence of sets. And so he abandons nominalism for Platonism.

The debate continues. Hartry Field, for instance, in *Science With-*

out Numbers, argues that those chunks of calculus which we need for physics can, after all, be paraphrased nominalistically.[8] We resist being embroiled in such technical debates. We cannot deny that it would be nice to know whether physics really needs various specific set-theoretical axioms – say, the power set axiom or the axiom of choice. It would be nice to know whether those axioms can be paraphrased nominalistically. And it would be nice to know just how elegant, or inelegant, those paraphrases can be. However, to think that the whole issue of nominalism versus realism reduces to technical questions of that sort is to lose one's sense of perspective.

Extroverted realists should not capitulate entirely to the nominalists' presumption that they are the only ones who are entitled to call the tune. It is a question of the burden of proof. Nominalists assume that we should avoid believing in universals unless there is demonstrably *no* option. So all our energy must be expended on exhaustive examinations of every possible strategy for avoiding universals. Extroverted realists need not allow themselves to be bullied into accepting those terms of debate. We urge that more time be spent exploring realist theories, without waiting for it to be demonstrated that we have no viable alternative.

Thus, our primary case for universals is the extroverted one rather than the negative, defensive reason, that realist language is both inescapable and semantically inexplicable without universals. For this reason, our case for universals is difficult to summarize. The reasons we believe in some specific universals may be quite different from the reasons we believe in others – just as the reasons for believing in genes may be quite different from the reasons for believing in neutrons.

Nevertheless, there is a common thread that is worth highlighting. When we study the world, not only do we notice the individuals in it; we also notice that these individuals have various properties and relations. This is the first step towards metaphysical realism: the recognition of *One Over Many,* of the recurrence of the same thing in many places. But the second step is taken when we advance from merely ascribing universals to particulars and begin talking about the universals themselves. Initially we talk, for instance, about whether one thing is nearer than another, or one event is earlier than another, and so on. But soon scientists and philosophers also begin to talk

8 Field (1980).

about, for instance, whether the 'nearer than' relation is isomorphic to the 'earlier than' relation (it isn't – but could it have been?). We wonder whether both relations are transitive, whether both are asymmetric and irreflexive, whether both are connected, continuous, and so forth. In a similar manner, we compare a relation like 'ancestor of' with a relation like 'parent of', and we wonder exactly how the one relation is related to the other. (The definition of the ancestral of the numerical successor relation is, in fact, central to arithmetic and a great stumbling block for nominalists.) Realism about universals really begins to have a severe bite precisely when it is these questions that are considered – that is, when we move from first-order to second-order universals.

And then, of course, another factor which intensifies the push to realism is the inevitable shift from extroverted realism about first- and second-order universals to introverted consideration of the semantics of sentences we use in expressing our theories about those universals. This semantic support for metaphysical realism is, we have argued, parasitic on the more basic, extroverted arguments for universals, yet it does offer powerful reinforcement. If universals exist, we can explain the semantics of various theories we accept. And this is manifestly strong support for the claim that universals do exist.

1.4 MODAL REALISM

Both scientific and metaphysical (or mathematical) realisms begin from a powerful extroverted base. Theorizing about the world leads very naturally and forcefully to the opinions that *there are* such things as bosons, irrational numbers, and so forth. Introverted, metatheorizing about such theories leads to a correspondence theory of truth. Such a theory of truth, a form of semantic realism, then reinforces the initial scientific and mathematical realisms.

With this framework to build upon, let us look at the most reviled brand of realism on the market: modal realism. Does modal realism have the same status as scientific or mathematical realism? Does it arise in the same way? Does it stand in the same relationship to the correspondence theory of truth? We argue that the answers to all these questions should be yes.

Modal realism, like its cousins, begins in an extroverted form. When it becomes self-conscious, it leads to a correspondence theory of truth for modal language. And the successes of this semantic theory then reinforce the extroverted realism from which it sprang. If there is any difference between modal realism and the others, it is only one of degree. Modal realism has a less secure extroverted base from which to launch itself. Consequently, it has to rely more heavily on reinforcement from semantics. This difference in degree is significant. It helps to explain why modal realism is so frequently discounted.

Modal realism, as we construe it, is a doctrine about what things there are, and is part of realism in the sense that we have used this term. It is the doctrine that *there are* such things as possible worlds, and possible outcomes of experiments, and so forth – "possibilia" for short.[9] As such, modal realism is not essentially a doctrine about symbols at all. It is, of course, stated using symbols, as every doctrine is; but these symbols are used to talk about things and not just about themselves or about other symbols. We leave it open at this stage what sorts of things possibilia will turn out to be – whether they are abstract or concrete, for instance. On our usage, modal realism stretches to cover both Lewis's celebrated "full-blooded" realism and what he calls "ersatz" modal realism. As we shall show, we take *realism* to be a distinct issue from *reductionism*. For now, we defend realism while remaining agnostic about reductionism.

Modal realists begin by speaking about possibilia, and not about words. Nevertheless, modal realists do follow up their claim about possibilia with a semantic claim *about their own claim,* to the effect that they really do mean what they are saying. When they say there are possible worlds, this 'there are' is to be explained semantically just like the familiar, everyday uses of these words in such sentences as 'There are radishes' and 'There are seals in Antarctica'. And when the details of these semantics are worked through, the doctrine of modal realism becomes more sharply defined. Possibilia, even more than universals, will gain in plausibility with supporting arguments from semantics. But first, let us consider the extroverted, nonsemantic sources of modal realism.

One presemantic source of modal realism lies in our everyday thoughts about media – about fictional characters, for instance. It is very natural to form the view that *there are,* say, five crew members

9 A clear presentation of such a doctrine is given in Lewis (1986a). For further discussion of the doctrine see Lycan (1979).

who regularly appear in *Love Boat*. Or that it only appeared as though there was a bullet that killed JR in *Dallas*, when in fact there was no such bullet and the whole episode was just one of the characters' dreams. On the face of things, it does seem obvious that *there are* fictional characters. Leave aside for the present all questions of whether these fictional characters are abstract or concrete, in this world or another, and so forth. Various views could be entertained, as we have already pointed out with reference to the distinction between realism and reductionism. What is crucial, however, is just the plausibility of the idea that they do exist.

Sceptical philosophers immediately launch a programme of paraphrase to free us from the need to believe in such entities as JR. Actors exist, as do reels of film, and patterns of light emitted by television screens, and so forth. But fictional characters do not exist at all. The standard way of paraphrasing fictional characters away is by insisting that *apparent* reference to such characters as in

There is someone who hates JR,

is always implicitly *prefixed* by a modifier which cancels ontological commitment, as in

In *Dallas*, there is someone who hates JR.

This kind of paraphrase cannot be applied directly to cases like 'I hate JR': this does not mean, 'In *Dallas*, I hate JR'. In fact, a great many hitches arise when we try to cancel ontological commitments of fictional talk by the simple expedient of adding a disclaiming prefix.

It is also worth noting that the attempt to clean up claims about fiction simply by prefixing them does conceal hard problems. Consider a prefixed claim like

In *Dallas*, no one shot JR.

The meaning of such a sentence is obviously determined by the meanings of its constituents – just as with any other sentence. In particular, it seems obvious that there is some semantic value attached to the operator

In *Dallas*

and some semantic value attached to the embedded sentence

No one shot JR.

The meaning of the whole, prefixed sentence is surely a result of putting the meaning of that operator together with the meaning of the embedded sentence. Yet if that is so, this commits us to giving some account of the meaning of the embedded sentence *before* we can determine the meaning of the prefixed sentence. And that means that we are caught in a vicious circle if we say that the embedded sentence is really just an abbreviation for the prefixed sentence.

Hence, the strategy of paraphrasing fictional characters away does meet with problems. It is a merit of modal realism that it makes it much easier to explain the meanings of prefixed sentences like

In *Dallas,* no one shot JR

in the usual way, as a function of the meaning of the embedded sentence, and the embedded sentence, in turn, as a function of the references of its components, including the fictional name 'JR'.[10]

If fiction were the only thing modal realism helped us to explain, the argument for modal realism would be much less than compelling for most philosophers. But modal realism also emerges, very naturally, from the inescapably important "fictions" that we survey whenever we reflectively attempt to make a rational choice among several alternatives. When deciding what it would be rational for us to do, we often think such things as 'There are *three* things I could do; if I did the first of these, then there would be *two* possible outcomes. One of these outcomes would be terrible, but it's very unlikely, whereas the other outcome is terrific and is highly probable. So the first possible action is a very good bet. The second possible action . . .' That is, in rational decision making, we consider, we talk about, we count, we ascribe existence to possible actions and possible outcomes.

In science, when we set up experiments, closely analogous considerations apply. We consider various possible initial conditions, we survey the range of possible outcomes given those initial conditions, and we estimate their probabilities. Then we compare our estimate of probabilities with the observed frequencies of such outcomes given these initial conditions. On the face of it, then, in

10 For this and many other intriguing complications in the semantics of media we are indebted to research by Jeffrey Ross.

science we do enumerate possibilities and ascribe probabilities to each. Modal realism takes this to commit us to the existence of possibilities. There may be more to say about the nature of such possibilities, exactly what they are made of, what status they have, and so forth; but a realist takes it that, whatever their constitution, we should allow that they do exist.

The importance of possibilia is reinforced if we look deeper into standard probability theory. The formal machinery of probability theory deals with a variety of things which, on examination, turn out to involve possibilia. In probability theory we assign probabilities to things called *events*. Yet many of the events which are discussed never occur. They are *possible* events. Furthermore, events are construed as *sets* of a special sort. Given two events, A and B say, there will be the event of either-A-or-B occurring. For instance, given the event of drawing a heart from a pack of cards and the event of drawing a diamond from a pack of cards, there will also be the event of drawing a heart *or* diamond – the event of drawing a red card. This is explained by taking the events A and B to be sets and the event A-or-B to be the set-theoretical *union* of sets A and B. The union of sets A, B, and all other "events" under consideration is called the *sample space*.

And yet what are the *members* of sets A and B? They are called *points*, or *elementary events*. Yet probability theorists tell us precious little about what the elementary events really are. For their purposes it does not matter what they are. The mathematics works smoothly no matter what set we take as the sample space (only as long as we choose a set with enough members). It is natural, however, for philosophers to ask what these elementary events might be. If we construe them as possible worlds, and take a realist stand on possible worlds, we can interpret standard probability theory entirely at face value, as literally true – without need for roundabout paraphrases or instrumentalist double-think. An event is something which could come about in a variety of possible maximally specific ways. So an event corresponds to a *set of possible worlds* otherwise known as a *proposition*. Probabilities are assigned to events or propositions – in other words, to sets of possible worlds.

Not only do mathematicians speak of the probabilities of events, they also talk about things called *random variables*. A random variable may be used to represent such a thing as, say, the height of a particular pea plant. Such a random variable determines more than just the actual height of the relevant plant; it represents not only that value but a range of possible values. What, then, is a random variable? It is a function which assigns a set of numbers to each elementary event in the sample space. A *function* is a kind of relation, and an *elementary event* is a possible world. So a random variable is, in fact, a relation which holds between a possible world and a set of numbers. The particular random variable we are considering is the relation which holds between any world and a set of numbers if and only if, in that world, that set of numbers includes the height of the designated pea plant. Thus, the mathematical description of a random variable amounts to an extroverted realism about possible worlds.

Thus, standard probability theory provides another justification for modal realism. The considerations which justified mathematicians in developing modern probability theory thereby support realism about possible worlds. And at least initially, they provide *nonsemantic* reasons for modal realism: justifications for extroverted modal realism.

Yet semantics cannot be postponed indefinitely. Introspection is, for a philosopher, inevitable. We must face the possibility that probability theory, decision theory, fiction, and so forth are all leading us into a false picture. It may be argued that, when we reflect on the semantics for probability theory, for instance, we will find that it does not refer to possible worlds after all. Its surface grammatical form is very close to that of genuinely referential language; yet its underlying logical form, its deep structure, is quite different – or so it may be argued. The modal realist must therefore face semantics as part of a defensive campaign against those who would paraphrase possibilia away.

However, when we turn to semantics, we will not find it to be exclusively a supplier of tools for the enemies of modal realism. It can also be a friend of modal realists. Semantics can provide a concrete framework which enables a modal realist to formulate a detailed theory of possibilia, to work out its consequences, and to test its adequacy. Modal realism without semantics is like an explorer who makes no maps of where he has been. Introverted modal realism does lead to a very powerful semantic framework. Not only does that help to articulate and clarify modal realism; it also provides another argument in its favour.

We urged that the justification for scientific realism is at first a justification for an extroverted realism, with support for introverted realism appearing at a second stage. With realism about universals – Platonic metaphysical realism – again the justification proceeded in two stages. When we consider modal realism we should not be surprised to find the justification again operating in two stages.

Nor should we be surprised or dismayed to find that it is here that a justification should rest even more heavily on introverted, semantic arguments. By the time we turn to modal matters, the general realist framework and the correspondence theory of truth have substantial support. In projecting the same framework and the same theory of truth to modal sentences, we find a strong justification for modal realism – the best explanation of the truth of modal sentences requires the real existence of possible worlds and other possibilia.[11] To seek an alternative framework and an alternative theory of truth at this stage in the development of both a unified and coherent metaphysics and a unified and coherent theory of semantics would be to fly against sound methodology.

In the case of modality, therefore, the argument for realism is much stronger than is generally recognized. There is initial support for extroverted modal realism. Yet by the time we reach modal realism, there is already a very strong case for a correspondence theory of truth. It is entirely reasonable then to extend an already well-supported theory from nonmodal to modal assertions. Modal realism is thus doubly supported: both as a direct outcome of investigation of the world and as a consequence of semantic theories which in turn rest on premodal realisms.

1.5 SEMANTIC REALISM

When something is true, in general there are *things* in the world which make it true, and there is *somehow* that these things are related to one another. If there were not those things, or if they were not thus related to one another, then it would not be true. It is true if and only if 'there are' certain things, and they are related 'thus'. That is to say, it is true if and only if it *corresponds to reality*. This is the core of the correspondence theory of truth;

11 It is instructive to remember that possible worlds were introduced into recent philosophical literature via semantic theory; see Kripke (1959, 1962) and Lewis (1968).

and it is obviously correct. It is the doctrine we call *semantic realism*.

A realist, we argue, should accept some sort of correspondence theory of truth. It would be a weird realist who told us that there are, say, atoms and then told us that she had not referred to atoms in her previous sentence, and that there are no relationships between her words and atoms that are necessary conditions for her to have been speaking truly.

Unfortunately, many view the affinity between realism and the correspondence theory of truth not as a reason for accepting the latter, but as a reason for rejecting the former. They think that the correspondence theory is out of date and problem ridden, a degenerating research programme, and hence that realism is in big trouble.

Objections to the correspondence theory draw some of their power from the conviction that a cluster of other theories are obviously right and that these other theories are rivals to the correspondence theory. Such objections are half right and half wrong. It is right to think that the other theories are correct, but wrong to see them as rivals to the correspondence theory. The alleged rival theories in question include "coherence" and "pragmatist" theories and the "redundancy" theory. In order to clarify the theory we *are* offering, we shall describe the bare bones of the theories which we are *not* offering.

Let us divide the world, first of all, into two sorts of things: those which can be true or false, *truthbearers* for short; and those which are not in any sense representations of anything at all and hence cannot be said to be in the relevant sense either true or false. There is a real problem, which we shall ignore here, about what things should be counted as truthbearers: sentences, assertions, propositions, beliefs, or (as we think) all of these and more besides. Without settling what a truthbearer *is,* however, we may ask what conditions it must meet if it is to count as *true*.

Coherence theories of truth say that a truthbearer counts as true just when it stands in specifiable relationships to *other truthbearers*.[12] Pragmatism broadens the range of admissible relations:[13] truth is said to depend on relations of a truthbearer not only to other

12 There are a great many presentations of the coherence theory, but for a traditional presentation and defense see Bradley (1914).

13 Again, presentations are legion; for a traditional exposition of pragmatism see James (1907).

truthbearers, but also to a context of human needs and goals, practical concerns, and social relations.

If coherence and pragmatist theories are to be genuine rivals to the correspondence theory, they must claim that truth depends *only* on interrelations among truthbearers and language users. It must not depend on relations to things other than language and language users. This means that such theories must claim that a truth *would still be true* whether or not there were things in the world, related thus and so, provided only that truthbearers had the same interrelations and utility to their users. For instance, consider the truth that Marie Curie was born in Warsaw. Rivals to the correspondence theory must say that this would still be true, even if Curie had not existed or had not been born in Warsaw, provided only that truthbearers had still been interrelated in the same ways that they are for us now.

This is manifestly false. It might be protested that, if Curie had not been born in Warsaw, then the relevant interrelations *could not have held* among truthbearers. Granting that, we could then infer that the interrelations in question entail that Curie was born in Warsaw. This means that those interrelations are not purely among truthbearers and practical projects of their users. Rather, the interrelations in fact include (entail) relations between truthbearers and Curie and Warsaw. Hence, the supposed rivals to a correspondence theory in fact collapse into their intended rival. The difference between correspondence and coherence evaporates.

Construed as rivals to the correspondence theory, the coherence and pragmatist theories must be rejected. Nevertheless, there are important facts which coherence and pragmatist theories are describing. There are, indeed, important relations among truthbearers and between truthbearers and their users.

Such interrelations are important for a variety of reasons. They are important for assessing the *rationality* of a person or a theory. And insofar as rationality is required for belief to count as knowledge, we may conclude that interrelations among truthbearers are important for epistemology. Yet in agreeing that we *cannot* ignore relations among truthbearers, we are very far from agreeing that we *can* ignore relations between truthbearers and other things. The importance of "holistic" interrelations for epistemology is no reason for opposing a correspondence theory of truth.

Another sphere in which holistic interrelations play a crucial role is in characterizing the nature of *understanding*. It is not possible for a person to understand a truthbearer without an awareness of its relationships to other truthbearers and to human activities. Insofar as "meaning" is to be counted as what people know when they understand something, we must conclude that holistic interrelations are important for semantics, as well as epistemology. This was one of the insights which led Wittgenstein from his earlier "picture theory" to his later slogan that "meaning is use".[14] Yet by admitting the importance of intralinguistic interrelations we do not threaten the correspondence theory of truth. In the first place, it is not plausible to think that an adequate theory of understanding requires *only* intralinguistic relations. Furthermore, even if we were to suppose that correspondence relations are unnecessary for the theory of understanding, why should we infer, from that basis, the falsity of the correspondence theory of truth? Such a conclusion would be warranted only if the correspondence theory had no justification other than as a contribution to a theory of understanding. That assumption is mistaken. A correspondence theory of truth may contribute towards, or supplement, a theory of understanding. But it is not merely a theory of understanding. It is not a theory of meaning, in the sense in which a theory of meaning is simply a theory of understanding. Correspondence (i.e., truth and reference) presupposes meaning; but it goes beyond meaning, in the sense of 'meaning' which is limited to what people must know if they are to understand something.

Truthbearers are related to one another and to human activities. They are also related to other things. A realist who talks about nonhuman, nonlinguistic things, say viruses, may admit that her words are related to other words and to social networks and human activities. Yet such a realist must also admit that her words are related to viruses. Her words would not be true if there were no viruses.

Another question worth raising is that of *translation*. Consider what must hold of someone's words if they are to count as a correct translation of the realist's talk about viruses. Clearly, relations among words and human activities will be relevant. Yet relations to things other than words or people are surely relevant too: relations to viruses, for instance. In determining the accuracy of translations

14 Wittgenstein (1922, 1953).

of a sentence, we must appeal not only to relations between that sentence and others, but also to relations between those sentences and nonlinguistic things.

Redundancy theories of truth are intended to be rivals to both the coherence theory and the correspondence theory.[15] If they are construed as theories about the meaning of the word 'true', perhaps there is some genuine rivalry here. There have been extensive debates over the success of strategies which aim to paraphrase assertions involving the word 'true' in terms which avoid the word 'true' altogether. We will not, however, pursue those debates here. The core of our correspondence theory of truth really has little to do with the semantics of the word 'true' in a natural language. The central point is that redundancy theorists cannot plausibly deny that both intralinguistic and extralinguistic relations are relevant to the issue of *translation*. There is something about two sentences, in virtue of which one is a translation of the other. This "something" is what we are describing in our correspondence theory of truth. Redundancy theorists may protest that it is a misdescription to call this a theory of "truth"; but that protest reduces to mere verbal quibble. We may wish to speak of the correspondence theory of translation or the correspondence theory of representation or of content – and to shy away from the word 'truth'. But the name of the theory is less important than its content.

Do not give too much weight to any preconceptions about the meaning of the term 'correspondence'. We use the term for whatever relations there are to things other than truthbearers, which determine truth or falsity. If truth does not depend solely on intrinsic properties of a truthbearer, relations among truthbearers, and between truthbearers and users, then there *must* be truth-relevant relations to other things. These other, truth-relevant relations, whatever they may be, are what we are calling correspondence relations. It is both common sense and sensible science to suppose that such relations exist. But what is more important for our purposes is the connection between realism and the correspondence theory. Any realist who turns his attention to semantics must adopt a correspondence theory of truth. Realists must be semantic realists.

Suppose someone began as an extroverted realist, believing in,

15 Ramsey (1929, 1931). We have also been influenced by discussions with John Fox.

say, Aristotle's crystalline spheres. Up to a certain time, he gave almost no thought to the relationship between symbols and the world. But suppose that when he finally considered this issue, he came to the view that his uses of the term 'crystalline spheres' *did not refer to anything*. It is our contention that such a person would thereby cease to be a realist; or at least, he would have to develop a divided mind, vacillating between realism when thinking about the heavens and instrumentalism when thinking about language. While holding an instrumentalist view of his astronomical theory, he could not be a realist about crystalline spheres. There is a kind of quasi-inconsistency in believing that there are crystalline spheres, but that in saying so you do not refer to crystalline spheres. That is close to Moore's paradoxical sentence 'The cat is on the mat but I don't believe it'.

In recent times this quasi-inconsistency has been asserted and defended under the title of *semantic eliminativism*.[16] The term 'refer' has been construed as a term in "folk psychology" or "folk semantics", which in turn has been described as a theory. It has then been argued that, if that theory is largely false, then 'refer' does not refer.

This presupposes that the reference of a term like 'refer' will be whatever it is in the world, if anything, which makes true most of the assertions in which the term appears. We reject that presupposition. We support theories of reference along the lines initiated by Kripke's *Naming and Necessity*.[17] As Kripke argues, a word may often refer to a thing even though the people who use the word have radically mistaken beliefs about what it refers to. For instance, the ancient Greeks referred to fires, just as we do, even when they believed that they were referring to an element, a substance, fire, which was observable only when released from composite substances in which fire was mixed with earth and air or water. The same thing applies to beliefs: people have referred to beliefs from time immemorial, even when they have had wildly mistaken views about the nature of the mind or soul in which believing occurred. In semantics, too, terms like 'refers' should be taken to refer, even when people have mistaken theories about what they refer to. The incoherence of

16 Discussions of eliminativism can be found in Quine (1969a), Leeds (1978), Churchland (1979), Stich (1983), and Devitt and Sterelny (1987).

17 Kripke (1980).

is, in fact, a further reason for thinking that a Kripkean theory applies to semantics as well as to other subject-matters.

Not all objections to correspondence, truth, and reference rest on the alleged *falsity* of the folk theory in which they are embedded. There is another line of argument which aims to undermine the *need* for truth and reference in our theory of the world. It is argued, for instance, that behaviour can be fully explained by "narrow" psychology – that is, by properties of a person's internal states which supervene entirely on how things are inside that person (understood to exclude any relational properties linking internal states with anything external). We do not need anything else to explain behaviour. Roughly, what a person does is determined by what she thinks she is referring to, not by additional facts about what she really is referring to whether she knows it or not. Referential relations do not cause behaviour. So what *do* we need truth and reference for, if not to explain behaviour?

Consider a parallel case. Why do we need colours? We might explain why a car stopped by reference to the colour of a light. Yet do we *need* to refer to the colour of the light? We could restrict our explanation to wavelengths of light and their effects on the mind. And we might conclude from this that there is no need for colours in our theory of the external world.

Yet there are well-known strategies for avoiding an eliminativist conclusion concerning colours. It can be maintained, in effect, that we *do need* colours in our theory of the world. We do not need to use colour words, but we do need to postulate a nexus of properties and relations which, in fact, constitute colours. Instead of eliminating them, we may take them to be combinations of things we do "need" in our theory of the world.

A similar stance may be taken concerning semantic properties and relations, like truth and reference. Instead of eliminating them, we may take them to be combinations of other things we do "need" in our theory of the world. We claim that semantic properties and relations are *supervenient* on the details of the physical properties and relations of person and environment (together with nonphysical experiential qualities, or "qualia", if you wish). That is to say, there could not be a difference in semantic properties and relations unless there were a difference in the physical characteristics or context.

This means that the full details about the physical situation (and qualia) *entail* the semantic properties and relations in question.

Supervenience of facts about any one subject-matter on facts about some other will always ensure that the *complete* truth about the latter will *entail* the facts about the former. We can show this to be so by a *reductio*. Suppose that C is a complete description of facts about one subject-matter and A is a complete description of facts about another. Suppose that the facts described by C supervene on those described by A. Then, for the purpose of a *reductio,* suppose that A does not entail C. Any such failure of entailment must mean, by definition of entailment, that there is a possibility that A is true and yet C false. In the actual world, A is true and C is true; but there is a logical possibility of A being true while C is false. Compare the actual situation with that possible situation. These two situations are the same with respect to the truth of A, yet differ with respect to the truth of C. There is a difference with respect to C, even though there is no difference with respect to A. And yet, by the definition of supervenience, that means that the facts about the subject-matter of C *do not* supervene on the facts about the subject-matter of A, contrary to hypothesis. The supposition that A does not entail C leads to a contradiction of supervenience of C-facts on A-facts. Hence, the supervenience of C-facts on A-facts must ensure that A entails C.

Applying this general result to the semantic case, we may conclude that, if semantic facts do supervene on physical facts, then the *full* physical story must *entail* those semantic facts. If a complete theory of the world "needs" the full physical story, it also "needs" whatever that physical story entails. Hence, provided that semantics is supervenient on the physical, it follows that a complete theory of the world does "need" semantics, in the appropriate sense of 'need'. We may not need to use *words* like 'truth' or 'reference'. For some purposes, we need not explicitly state the semantic consequences of what we believe. But that in no way implies that we do not "need" to "postulate" such things as truth and reference.

So we take it that our realism commits us to a correspondence theory of truth, that is, to semantic realism. Any decent realism must be complemented by a correspondence theory of truth and reference which relates *truthbearers* to *truthmakers*. Yet it is important to see that we are construing the correspondence theory of truth as a corollary of realism, and not as an input into the definition of realism. Thus, we do not define realism in epistemic or semantic terms. A variety of such definitions of realism are plausible, and yet we urge that they misconstrue realism.[18]

Consider, for instance, Putnam's characterization of what he calls "metaphysical realism" (in a sense that is quite distinct from that in which we use this term). Putnam attributes to the ("metaphysical") realist the view that, roughly,

> No matter what relations hold among truthbearers, and between truthbearers and people, it is possible that those truthbearers are false.

That is, a realist is supposed to think that even an "ideal theory" could be false. An ideal theory is one which satisfies whatever desiderata we may wish to specify – such as consistency, simplicity, utility – provided that these desiderata do not beg the question concerning realism. (We must disallow, for instance, the desideratum 'An ideal theory must be one which corresponds to reality'.)

To a realist, it is exceedingly plausible to suggest that an ideal theory could be wrong; so it is tempting to respond to Putnam's characterization of realism by saying, 'Yes, that is what a realist thinks'. And yet this is a mistake. Many realists have never thought about whether an ideal theory could or could not be false. When the question is raised, they may take it to be a corollary of their realism that an ideal theory could be false. Yet this is at most a corollary of their realism, not part of its definition.

In fact, we are inclined to think that Putnam's characterization may not even be a corollary after all. We believe that a consistent realist might have reasons for believing that an ideal theory could not be false. Yet this would not undermine her realism. It is possible that an argument could be mounted, from entirely realist premisses, to support the conclusion that an animal with functional states which count as beliefs must always have a substantial proportion of beliefs which are true. Bennett, for instance, makes such a claim in his work on linguistic behaviour.[19] In the ideal case, in

18 We thus reject the characterization of realism given by Putnam (1978, 1981) and by Dummett (1982). Putnam inspired realists for years; but then he redescribed the realism he had been defending as "internal" realism and distinguished it from what he called "metaphysical" realism, which he rejected. We think "internal realism" is not realism at all and *our* realism is the kind Putnam is rejecting, although we do not define it in the semantic and epistemological way in which Putnam defines "metaphysical realism."

19 Bennett (1976).

which the animal's functional states are *optimal,* it could be argued that the animal's body of beliefs must be true. This conclusion, derived from realist premisses, would not constitute a *reductio* of realism – not even if we apply the same conclusion to ourselves, since, after all, we too are animals. Hence, a realist could believe that an ideal theory must be true.

Furthermore, we doubt whether the very idea of an ideal theory can be explicated in the way that Putnam requires – that is, without begging the question concerning realism. Such notions as utility and consistency may not be as innocent of realism as antirealists assume. However, we set aside doubts about ideal theories and doubts about whether a realist must agree that ideal theories could be wrong. Even if we were to grant Putnam that ideal theories could be wrong, we would nevertheless insist that this is only a consequence of realism together with some extra assumptions; it is not definitional.

Putnam's characterization of realism is only one of a number of plausible epistemic or semantic characterizations of realism. Consider another example. Devitt gives an account of realism which is very close to our conception of extroverted realism.[20] A realist about, for instance, frogs will believe that frogs exist. To believe that frogs exist is not in itself to believe that the word 'frog' refers or that sentences containing the word 'frog' have a certain semantic character, or anything semantic at all. So says Devitt, and we concur.

Yet Devitt feels compelled to add a qualification. He characterizes the realist as one who believes that frogs (for example) exist *independently of the mental.* The motive for mentioning independence from the mental is to exclude certain views which Devitt does not wish to allow as versions of realism. For instance, he refuses to call Berkeley a realist. Yet Berkeley did believe that frogs exist: he just believed that they are mere aggregates of ideas in our minds. He believed that they exist, but did not believe them to be independent of the mental.

Notice that there is a close parallel between Devitt's qualification, "independence from the mental", and Putnam's idea that realists think an ideal theory "could be wrong". And it is subject to the same objection. Consider an extroverted "frogger" who is insatiably curious about frogs but has never given even a fleeting thought

20 Devitt (1984).

to psychology. It may never have occurred to such a person even to wonder whether frogs are or are not independent of the mental. So Devitt would not construe such a frogger to be a realist about frogs. This, we think, would be a mistake.

In a critique of Devitt's realism, Barry Taylor[21] has probed Devitt's appeal to "independence from the mental". It is not just the existence of objects that should be construed as independent of the mental. Taylor urges that a realist must be concerned with the "independence" of *facts about* objects, and not just the independence of the *existence* of objects. He then mounts a persuasive argument to the effect that, if realism is construed as he urges, then a realist is committed to a correspondence theory of truth, or something of that ilk. This is so because we cannot define independence from the mental for facts without reference to whether they would be *true* even if people like us had never existed.

Realism begins, however, with extroverted investigations of the world of frogs, photons, continuous real-valued functions, and so forth. Devitt's characterization of realism should be construed in such a way as to reflect this thoroughly extroverted, grass-roots realism. When he says that a realist believes frogs and so forth exist independently of the mental, we should construe this as

> The realist believes that frogs and so forth exist, and does not believe them to be dependent on the mental.

This is to be distinguished from

> The realist believes that frogs and so forth exist, and believes them to be independent of the mental.

The latter characterization will fit many realists. No doubt it will fit almost any sensible realists about, say, frogs who think to enquire how frogs are related to mental states. But it is the former, not the latter, which defines realism. The latter is a corollary, not the core, of realism.

Suppose a realist enquires about the relationship of frogs to minds and comes to, say, Berkeley's conclusion that frogs are just aggregates of ideas in our minds. In that case, the realist ceases to be a realist. No doubt Berkeley began as a realist. But when he began introspecting – that is, when he began thinking about his ideas of frogs, as well as about frogs – he argued his way out of his initial

21 Taylor (1987).

realism. He began by believing in frogs, but ended up believing only in aggregates of ideas, and calling *them* 'frogs'. In the same way, an astronomer who begins as a realist may argue her way out of realism by becoming persuaded of an instrumentalist theory of astronomical theories. Such a person may begin by believing in, say, crystalline spheres, but end up believing only in theories instead and using the word 'spheres' merely to refer to symbols in such theories.

So Berkeley ceases to be a realist when he construes frogs to be aggregates of ideas. Yet this does *not* mean that, when he was a realist, what he really believed was that frogs existed *independently of the mental*. When he was a realist, he did not believe frogs were independent of the mental. Nor did he believe they were dependent on the mental. He didn't consider the matter.

Thus, we do not accept Taylor's characterization of realism. What he takes to be definitional for realism, we take to be (at most) a consequence of realism. But it is an extremely plausible consequence of realism. A realist about frogs need not have considered their relationship to minds; but let's face it, as soon as a realist frogger does think about the matter, it will be obvious that the existence and frogginess of frogs is independent of the mental. Frogs have two eyes, four legs, longer back legs than front legs, no fur or feathers, they all have the ancestors they have; and all of these things are so whether or not we notice them. Frogs would still have two eyes and so forth even if people did not exist.

Thus, realism leads naturally to a recognition of independence from the mental. Once we have recognized such independence, we have the premisses from which Taylor argues to a correspondence theory of truth. We endorse this stage of Taylor's argument. Devitt is right to construe realism as a nonsemantic and nonepistemic doctrine. But Taylor is right in insisting that a reflective realist should hold something like a correspondence theory of truth. Realism should include semantic realism.

What we recommend is a partial reversal of the dominant trend in postmedieval philosophy. Since Descartes, there has been a tendency to begin philosophy with epistemology, the theory of knowledge, and to draw inferences to the nature of the world we live in – that is, to draw conclusions concerning metaphysics. In other words, there has been a tendency to begin with premisses concerning the human agent and to proceed to conclusions concerning the world around the human agent.

In the twentieth century, this scenario has been given a new twist, a "linguistic turn". There has been a tendency to begin not so much with epistemology as with semantics. Philosophers begin with considerations about language. The advantage of this starting point has been that the facts it draws on are supposedly ones which all speakers know simply by virtue of *knowing their own language*. This gives philosophy a neatly *a priori* status. And it enables philosophers to draw inferences about the nature of the world without dirtying their hands with ordinary scientific method.

Lately there has been a tendency to reverse these priorities. There has been a tendency, for instance, to "naturalize" epistemology and semantics, treating propositions concerning these matters as inferable from propositions in other areas, rather than the other way around. We give a qualified endorsement to this reversal of priorities. Yet we also argue that the reversal should not be overdone. Inferences are not entirely one-way from epistemology or semantics to metaphysics; but neither are they entirely one-way from metaphysics to epistemology or semantics. The traffic is two-way.

1.6 REALISM MISCONSTRUED

Realism, including modal realism, cannot be defended adequately if it is misconstrued as a semantic theory. We must distinguish between realism and semantic theories like the correspondence theory of truth. And yet we cannot adequately defend realism, especially modal realism, without recognizing an intimate relationship between realism and a correspondence theory of truth. It is not easy to keep a clear image of the relationship between realism and semantics, but unless we get the balance right, realism, particularly modal realism, is doomed.

The first thing to note is that a realist *can* be *extroverted*. What we mean by this is that realists can talk about the things they are realists about without giving any attention to their own talk. They need not make any assertions about their own theories. They are realists, but not because they assert that their theories have such-and-such semantic status. They are realists not because of the semantic assertions they do make (since they make none), but rather, in part, because of the semantic assertions they do *not* make. Realists are realists in part because they do *not* give an antirealist gloss on their extroverted realist assertions.[22]

Realists can say there are certain things, and they can mean what they say without having to *say* that they mean what they say. In fact, if they try to say that they mean what they say, they can succeed only if they are taken to mean what they say when they make that semantic assertion. A retreat to semantics is no guarantee that they mean what they say. An antirealist may hear an extroverted realist's assertions that there are certain things with certain characteristics and relations. The antirealist may then say, 'Yes, I can agree with all that. I say just the same things myself. Yet that does not constitute realism, since when I go on to give a semantic theory for all these assertions, it can be seen that all these things can be said by an antirealist, without any inconsistency whatever'.

For instance, an antirealist may hear an extroverted realist say that there are just three physically possible outcomes for some given experiment. The antirealist may reply, 'Yes, I think so too. But now listen to my semantic theory for what I have just said . . .'

Realists beware. This is a trap. It is tempting for realists to respond by arguing for a rival semantic account of their earlier, extroverted assertions. (By doing so, such realists cease to be extroverted. They have become introverted, since now they are thinking about their own theorizing, and not just about the subject-matter of that earlier theorizing.) Yet this strategy is doomed to disappointment, as a response to the antirealist. We have supposed that the antirealist has a semantic theory which applies to any extroverted assertions of the form

> There are things of such-and-such a sort, which are related to one another in such-and-such ways.

Realists insist that this semantic theory misunderstands their positions. What they believe is that

> There are things of such-and-such a sort, which are related to one another in such-and-such ways *and* which are related to those *words* in such-and-such ways.

22 This recapitulation of our position has been influenced by the way David Lewis, in personal correspondence, has characterized our argument.

If the antirealist's deflationary semantics is worth its salt, if it works at all for the former, extroverted realist assertions, then it will work equally well for this latter realist assertion. If an antirealist can give "innocent" reinterpretations of assertions about "dubious" entities, there is no reason why it should be difficult to give equally "innocent" reinterpretations of assertions about how these entities relate to words.

Realists, then, are people who say there are certain things and for whom deflationary semantic theories *are* mistaken. They "mean what they say". But they need not be people who *say* that such semantic theories are mistaken. No matter what they say, a consistent antirealist will always apply the antirealist semantics in such a way as to misinterpret systematically what the realist is saying.

Thus, realists are distinguished, ultimately, by what they do not say, by their refusal to give any deflationary account of what they have been saying to date. Realism, therefore, is not a semantic theory. This applies to modal realism as much as to other realisms. Modal realism is not the doctrine that modal language is best characterized by a correspondence theory of truth.

Given, however, that one *is* a realist about certain things, there is no reason one should not go on to explore the ways in which those things are related to one's words. There is no reason that one has to bother discussing words at all (as opposed simply to using them). But there is no reason that one should not become introverted and wonder how the words that have been used in theorizing about things are related to those things. And when this issue is raised, there is no coherent option for a realist but to adopt some sort of correspondence theory of truth. It is absurd to imagine that there are *no* interesting correspondence relations between the things one is a realist about and the words one uses in talking about them, no such relations whose presence is required in order for one's assertions to be true.

Realism thus leads to a correspondence theory of truth. A modal realist, for instance, will explain the truth of modal assertions by appeal to reference relations between words and possibilia, as found, for instance, in the standard model-theoretical treatment of the probability calculus. And realists will refuse to give that semantic story any ontologically deflationary gloss. None of this will persuade an antirealist who is determined to misinterpret the realist's assertions by some other metasemantics. But it is useless for the realist to *assert* that these are misinterpretations: such as assertion is as easy to misinterpret as the initial, extroverted assertions were.

Nevertheless, the hopelessness of refuting antirealists by employing semantics does *not* entail that semantics gives *no* support for realism. On the contrary, it is perfectly rational for a realist to draw considerable support from semantic sources. The successes of a referential semantics do reflect well on the realism which led to those semantics. In particular, the successes of a referential semantics for modal language reflect favourably on modal realism – for a realist. The fact that such semantics do not refute antirealism, but can be construed by antirealists as being of merely instrumental value, is not relevant. The fact remains that a realist is rational to see such semantics, realistically construed, as a success for realism.

In the case of modal realism, the support from semantics is especially powerful, for a realist. There is already a strong case for a referential semantics for premodal, scientific assertions. And it must not be underestimated how desirable it is to have a *uniform* account of truth across subject-matters. In science, assertions occur which mix terms with purely "physical" reference ('gene', 'positive charge', etc.) with mathematical terms and modal language. Mathematical and modal terms may occur in the very same sentence as "physical" terms that we are antecedently realist about. We need some account of how to compute the semantic values of such sentences from the semantic values of their constituents. It is so much easier to do this if we take a referential stance on all the terms – if we are Platonists about mathematical terms and if we are modal realists.

Furthermore, in science, inferences occur in which some premisses are purely "physical", while other premisses concern mathematics or modality. Such arguments may be valid. Validity is very neatly explained in referential semantics, in terms of truth. So if we *are* realists about mathematics and modality, very neat explanations are forthcoming for the validity of such "mixed" argumentation. For a realist, this can rationally be taken as powerful support for realism across the board: for scientific Platonism.

Antirealists can construe the theories described as merely what Putnam calls "internal realism", that is, as subject to an across-the-board semantic theory which antirealists can accept. As realists, we think such an across-the-board semantics misrepresents what we mean by what we say. But we have not undertaken to prove that here, and indeed it is hard to see what could possibly convince someone who is *determined* not to take you as meaning what you say. This would be as hopeless as attempting to gain the trust of someone who is invincibly paranoid. Nor have we undertaken to refute those who employ a referential semantics selectively, allowing "physical" referents but not employing referential semantics for semantic, mathematical, or modal language. Proponents of such a mixed semantic theory have their work cut out for them, but of course they are well aware of that. Yet a realist does not need to refute all rivals in order to take the virtues of a uniform referential semantics to provide significant rational support for realism.

Realism is doggedly misunderstood. It is not a semantic doctrine, yet it obtains substantial support from semantic realism. This is especially true of modal realism. Any realism requires semantic realism, and this in turn supports modal realism. Thus, modal realism is much more solidly grounded than its opponents realize. In consequence, most of the objections raised against it by antirealists miss the point and beg the question against realism.[23]

23 We are grateful for useful comments and suggestions by David Lewis on various sections of this chapter. Much of the chapter is based on Bigelow and Pargetter (1989a).

2

Quantities

2.1 A HIERARCHY OF UNIVERSALS

We have argued for scientific realism, metaphysical realism, and modal realism. The last two should, in fact, be thought of as part of the overall commitment of scientific realism, since science does include mathematics and logic. Mathematics furnishes a powerful argument for metaphysical realism. And logic, construed to include probability, involves necessity and related issues concerning possibilia and thereby furnishes a powerful argument for modal realism. We characterize this position as scientific Platonism. In addition, we have argued for semantic realism. Anyone who accepts any of the other realisms should also believe in reference and truth.

It is important to remember that, in our usage, a realist can be neutral on questions of *reductionism*. In defending realism about physical objects, universals, possibilia, truth, and reference, we leave ourselves the task of investigating what these things are, what they are constructed of, whether they are reducible to entities of some specifiable sort, and so forth.

We begin by considering universals: what are they? One crucial feature of (many) universals is their ability to be in several places at once. They, or at least some of them, are *recurrences*. We should not, however, place too much weight on plural localization as a defining characteristic of universals. Not all universals are multiply located. Some are hard to locate, with any plausibility, anywhere at all. Admittedly, a property is plausibly locatable when it is instantiated by something spatial: it is located wherever its instances are. But relations are harder to locate. Furthermore, some universals raise another problem. Some universals are not hard to locate; on the contrary, they are all too easy to locate – just as easy to locate as the individuals which instantiate them. The trouble is that they are only in one single location at any one time. A universal such as this could never be multiply instantiated, so it could never be in two places at once. It might be what is called an *individual essence* – a property a thing must have and no other thing could possibly have.

It could be denied that there are any such universals – Armstrong, for instance, has defended the view that all universals are at least potentially multiply located.[1] This doctrine might be worth defending if it really did permit a neat definition of universals simply in terms of multiple locatability. However, that characterization is unsatisfactory in any case. Furthermore, there are useful jobs which can be done by "one-place-at-a-time" universals. It would be a shame to excise them from our theory from the very outset, just to facilitate a neat definition of universals.

The fundamental distinction between particulars and universals turns on the instantiation relation. We can begin to get a grip on the distinction by beginning with paradigm universals, the ones which are indeed multiply located, the "recurrences". These multiply located things stand in a special relation to the uniquely located individuals which instantiate them. The individual exists; and the universal exists; but that is not all there is to be said about the matter. There is *somehow* that the universal stands to the individual. Both individual and universal could have existed, even if that individual had not instantiated that universal. In addition to the existence of *something* which is the individual and *something* which is the universal, there is also *somehow* that the individual stands to the universal. This somehow is *instantiation*. We shall say more about instantiation later. For the moment, we content ourselves with an ostensive definition – we explain what we mean by directing attention to a prominent example of it.

Instantiation is an antisymmetric relation: if x instantiates y, then y does not instantiate x. Any such antisymmetric relation will generate a hierarchy. It may generate a linear ordering (as in the case of the antisymmetric temporal relation 'earlier than'). Or it may generate a tree structure (as in the case of the family trees generated by the antisymmetric kinship relation 'parent of'). Instantiation gener-

1 Armstrong (1978) construes a universal as essentially a "One over *Many*", and so he will not admit universals which are necessarily uniquely instantiated.

ates a hierarchy which is more like a family tree than a temporal ordering.[2]

Particulars, or individuals, instantiate but are never instantiated. That is what is distinctive about them. Let us use the symbol *o* to refer to the collection of all individuals. Properties of individuals are things which are instantiated by things which are in turn not instantiated by anything. We shall refer to the set of all these properties of individuals by the symbol (*o*). The set of two-place relations between individuals will be referred to by the symbol (*o*, *o*); three-place relations by (*o*, *o*, *o*); and so on.

In addition to the sets

o,
(*o*),
(*o*, *o*),
etc.,

there will be sets of properties and relations which are instantiated not by individuals, but by other properties and relations. There will be, for instance, properties of properties, which we shall denote by

((*o*)).

The property of *being a virtue* might be an example of such a property.

There will be relations between an individual and a property of individuals; the set of these relations will be denoted by

(*o*, (*o*)).

An example might be that of *admiration:* if Russell admires wisdom, there is a relation between Russell and wisdom. This relation is instantiated by a pair of things. The first member of this pair is an individual (something which is not instantiated by anything). The second member of the pair is a property of individuals (something which is instantiated by things which are, in turn, not instantiated by anything). So the relation between Russell and wisdom is a universal in the set (*o*, (*o*)).

This allocation of admiration would have to be revised, of course, if we were to construe wisdom not as a property of an individual, but as a relation: wisdom is the love of knowledge.

2 The hierarchy we describe is not idiosyncratic. It echoes the structure of standard set theories. We have been influenced primarily by Russell (1903), Whitehead and Russell (1910), Cresswell (1973), and Montague (1974).

(And what is knowledge?) Admiration, virtue, and so on serve only as illustrations. It is important to remember that we do not assume any simple mapping of words, or descriptions, or concepts onto universals. We assume, only for the sake of a simple illustration, that there are such universals as wisdom, virtue, admiration, and so forth.

Suppose Russell admires not a property of individuals, but a property of properties. Suppose that being a virtue is a property of properties, a universal in the set $((o))$. Then when Russell admires virtue, he stands in a relation not to something in the set (o), but to something in the set $((o))$. Hence, the relation in which he stands to virtue is not in the set of universals

$$(o,\ (o)),$$

but rather in the set

$$(o,\ ((o))).$$

When Russell admires wisdom (simplistically construed), the admiration belongs to the set $(o,\ (o))$. When he admires virtue (simplistically construed), the admiration belongs to the set $(o,\ ((o)))$. Does this mean that the one thing, admiration, belongs to both sets? Or is the admiration in the one set a different kind of admiration than that in the other set?

This question compels us to be more explicit about how we are defining the sets (o), $((o))$, $(o,\ (o))$, and so on. The set (o), for instance, is a set of things which may be instantiated by individuals; but is it a set of things *all* of whose instances are individuals or *some* of whose instances are individuals? Are there any universals having some individuals as instances and some nonindividuals as instances? If so, do we count them in both of two distinct *types,* or in neither?

If there are any such universals, we shall class them in both of two separate types, so types will not be a family of *disjoint* sets. We assume that any given universal will belong to at least one type; but some may belong to several. The term 'multigrade relation' is used for a relation, such as living together, which may hold among different numbers of individuals. The twelve apostles lived together, the Pointer Sisters may have lived together, perhaps twenty Rajneeshis may live together, and so on – the relation is the same, though the number of related individuals varies. Thus, plausibly, living together is a relation belonging to type $(o,\ o)$ and to type (*o*, *o*, *o*), and so on. We extend the term 'multigrade' to any universals which belong to more than one type. We assume, though we do not try to prove, that there are multigrade universals.

Thus, assigning a universal to a type is a way of saying what types *some* of its instances belong to. Given any *n*-tuple of types, we may define the set of universals, some of whose instances are of those *n* types.

In this manner, we may use the antisymmetric relation of instantiation to distinguish an unlimited number of classes of universals. The hierarchy is generated by a very simple recursion. There is a class of individuals, the class *o*, defined to be things which are not instantiated by anything. All the other classes are generated by the rule

If $t_1, t_2, \ldots, t_n$ are any types, then $(t_1, t_2, \ldots, t_n)$ is a type.

The generated type $(t_1, t_2, \ldots, t_n)$ is defined to be the set of universals which can at least sometimes be instantiated by *n*-tuples of things, the first of which is of type t_1, and so on down to the *n*th, which is of type t_n.

This defines an infinite hierarchy of types. There is a single class, which we call the *domain* or the *universe,* which is the union of all the types. Each type is a subset of the domain.[3] Note that some of these subsets of the domain may be empty. We are working within the broad tradition of Armstrong's *a posteriori* realism about universals. That is, we treat it as an open question which universals there are.

We shall, however, assume (and to some extent we will argue, later, for the assumption) that many of the basic types will be nonempty classes. In particular, our metaphysics for quantities, vectors, and real numbers will assume that there are universals in the following types:

(*o*)	properties of individuals
(*o*, *o*)	"determinate" relations between individuals
((*o*))	properties of properties
((*o*, *o*))	properties of relations
(*o*, (*o*, *o*))	relations between individuals and relations
((*o*, *o*), (*o*, *o*))	relations between relations
((*o*), (*o*))	relations between properties (including, notably, such things as ratios or proportions between properties like length)

3 This construction of a hierarchy of types is modelled especially on Cresswell (1973); Cresswell traces the sources of his formulation back to Adjukiewicz (1935/1967).

(The levels of the type hierarchy that we require to be nonempty will be comparable to the levels of the set-theoretical hierarchy that are used in constructing real numbers in Whitehead and Russell's *Principia Mathematica*.)

Universals, then, are distributed into a hierarchy of types. In one sense, then, there are "higher-order" universals. In some contexts it is useful to draw distinctions between first-order and second-order universals[4] – although in other contexts this terminology is too restrictive to mesh with the complexities of the type hierarchy. There is a danger, however, in using terminology which suggests that some universals are of higher order than others. It is important to beware of confusing this metaphysical hierarchy with a *logical* hierarchy, reflected in what is known as higher-order *quantification*, a procedure which occurs in higher-order languages. Universals, of all types, are construed as things which can be *named*, ranged over by first-order quantifiers, like 'something', '$\exists x$'. The whole hierarchy of types can be described in a *first-order language*. So there is an important sense in which universals of all *types* are really still *first-order* entities.[5]

The hierarchy of types is generated by the instantiation relation. We will see later, in Section 3.7, that *instantiation* is itself a higher-order universal in a very important sense, a sense in which none of the universals within the hierarchy are higher order. Instantiation is higher order in a sense which is reflected in higher-order languages, and not just by its place in a hierarchy which can be described entirely within a first-order language.

The problem of instantiation, and truly higher-order universals, will wait until we have worked our way well into semantics. Our task now is to apply the hierarchy of types to the metaphysics of science.

4 In certain contexts, we find it useful to conform to the terminology recommended in Putnam (1975). In this terminology, a distinction is drawn between second-degree universals (properties of *properties*) and second-order universals (properties which individuals have in virtue of properties of their properties). This distinction is important with respect to the problem of quantities discussed in Section 2.4. In a broader setting, however, the distinctions among types are too numerous for a distinction between 'order' and 'degree' to be useful.

5 In the terminology of Frege (e.g., Frege, 1884), *our* universals of all types would count as "objects" rather than "concepts". The idea of higher-order languages, to which we refer, traces back to Frege's *Begriffschrift* (1879). We shall return to this issue in Section 3.7. The perspective we adopt owes much to Boolos (1975).

The type hierarchy of universals is, formally, very closely analogous to the hierarchy of sets. What, if anything, do sets have to do with universals? Philosophically, set theory is an enigma. We will not undertake a full confrontation with it here. Yet we cannot bypass it altogether, not without casting darkness on the metaphysics we will be offering for science. We cannot avoid giving some indication of where we plan to locate our theory of universals with respect to that towering contemporary landmark, set theory.[6]

What we offer is a hypothesis for which we can offer no conclusive proof but which does have a number of explanatory advantages. The hypothesis is that sets are universals. The set-theoretical hierarchy is itself a pared-down subset of the hierarchy of universals. Not every universal is a set, but every set is a universal. Members of a set instantiate the universal which is, in fact, what that set amounts to.

Where did set theory come from? From mathematics. When? Very recently, after thousands of years of progress in mathematics – one of the oldest of the intellectual disciplines and a discipline which has probably made more cumulative progress than any other. So why, after thousands of years, did mathematics finally spawn the set-theoretical hierarchy?

Our explanation is as follows. Mathematics deals with universals: properties, relations, patterns, and structures, which can be instantiated in many places at once – universals of the paradigm sort, namely, recurrences. Sets are universals of an especially abstract and rarefied kind and are not, in themselves, especially important universals. What has been discovered recently is that these relatively recondite universals *instantiate* all the properties, relations, patterns,

6 Many expositions of set theory are available. Our discussion does not hinge on any of the differences which distinguish one theory from another. If pressed to specify, we would endorse Zermelo–Fraenkel set theory (ZF). A good survey of the range of set theories is given by Quine (1969b). A good exposition of a close cousin of ZF is given by Lemmon (1969). The theory expounded by Lemmon has much in common with von Neumann–Bernays–Gödel (NBG), but is slightly different; it is, in fact, Mostowski–Kelley–Morse (MKM). This we regard as a variant of ZF with certain expository advantages, but it is probably philosophically misguided. It fails to recognize the irreducibility of higher-order to first-order quantification, which we argue for in Section 3.7. There is somehow that all sets are: MKM misconstrues this "somehow" as a queer sort of *object,* called a "proper class". This we think is a mistake.

and structures which mathematicians have been studying for thousands of years. It is because these universals instantiate all the others that they have become so important. Applied mathematicians explore questions about which patterns are instantiated by which things in the world, whether those things are biological, economic, or whatever. Pure mathematicians, however, are concerned only with the patterns themselves, and not with what things instantiate them. They need only one good instance of whatever pattern they wish to study, an instance which, if possible, introduces a minimum of extraneous distractions and idiosyncratic detail. To find such an instance, they need look no further than set theory. Or so it seems, so far. There have been hints that something called category theory may be able to instantiate more structures than set theory can. But at least until recently, whenever mathematicians have wanted to study a structure, they have been able to find instantiations from set theory.

How, then, did sets emerge from mathematics? They emerged from reflections which took place relatively high up in the hierarchy of universals. Mathematicians had been dealing for a long time with real numbers; and as we shall argue, real numbers are themselves fairly high up in the type hierarchy. Mathematicians also dealt with functions which mapped real numbers to real numbers; and they dealt with families of such functions, structures which included such functions, and so forth. All this took them farther and farther up the type hierarchy.

Sometimes mathematicians found that structures which grew out of entirely different sources turned out to be, in an important sense, identical structures. Sometimes the different ways in which they were presented concealed the identity of underlying structure. Sometimes it was immensely surprising, and exciting, to find that two things which had seemed entirely different from one another were, in fact, exactly the same structure, underneath the divergences of symbolisms.

Consider a special case of this sort of discovery. Suppose mathematicians are dealing with what they take to be two different functions. They then discover that these two functions are "really" – in a sense – the same. In what sense? Well, a function is really a *relation* of a specific sort, a relation which links any given thing with just one other thing. The two apparently different functions are two relations. When they are discovered to be in a sense identical, what has been discovered is that any two things which stand in the one relation also stand in the other relation. The two relations are *coextensive*. This is something very important that they have in common. In fact, a mathematician is very likely to be interested *only* in what they have in common. Attention shifts from the two relations to that which they have in common. And that which they have in common is their *extension:* a set.

Sets, we propose, are what coextensive universals have in common. They appear relatively high up in the type hierarchy. They are, in a sense, "abstract" to a relatively high degree. That is why they emerged so late in the history of mathematics.

Objections may be raised to this proposed construal of sets. It may be objected that not every set can be construed as what coextensive universals have in common. *Random* sets, for instance, are often intended to be such that there is nothing which is common to all and only the members of that set. Such sets are intended not to be the extension of any genuine universal. Such sets, therefore, are not what several coextensive universals have in common.

That is a fair objection and should prompt open-mindedness about exactly which universals we should take sets to be. It should not feed despair, however, over the prospects for a theory of this general form. The first sets to be recognized, we urge, did come to the attention of mathematicians because they were, in some sense, what various universals had in common. Further investigations uncovered more things of the same sort. And, indeed, it is still a matter for investigation exactly where sets should be located in the type hierarchy of universals. We shall return to that issue later, when we have worked our way through some semantics, in Section 3.5, and then again in Sections 8.4 and 8.5.

At this point, a historical observation is in order. The metaphysical framework we are developing here is, broadly speaking, within the same tradition as Armstrong's *a posteriori* realism. There are significant points of difference, both in detail and in spirit and motivation; but there is much common ground. Yet over the status of set theory, we side with Goodman and Quine against Armstrong.[7]

Goodman and Quine see the metaphysical question of *nominalism*

7 Goodman and Quine (1947) establish the perspective which makes set theory the focus of the debate between nominalism and metaphysical realism. See also Goodman (1951) and Quine (1969b), esp. chap. 6. The rival construal of set theory as a kind of nominalism occurs in Armstrong (1978, chap. 4).

as boiling down to the question of whether it is justifiable to affirm the existence of sets. They are right. Realists believe in universals; nominalists do not. The history of mathematics justifies the view that, if we are to believe in universals, we should include sets among those universals. But once we believe in sets, we find that these universals seem to be capable of doing all the work that the mathematician needs done. If we believe in sets, it seems that we no longer need any of the other universals. And so Ockham's razor sweeps in, leaving realists with only the set-theoretical stubble remaining from the type hierarchy of universals. Realists are those who retain *at least* the set-theoretical stubble. Quine and Goodman were so instinctively Ockhamistic that they gave scarcely a thought to those who might retain *more* than the essential, set-theoretical minimum. And they were so sensitive to the history of mathematics that they gave scarcely a thought to those who might retain portions of the type hierarchy *other* than the set-theoretical "minimum" – those who, like Armstrong, might believe in some universals, but not in sets.

Goodman and Quine were too hasty in summoning the ghost of Ockham to scare off all the universals except the set-theoretical minimum. Pure mathematicians may be able to get by with nothing but sets; but science needs other universals as well as sets. Even granting that we should not multiply entities without necessity, we should recognize that there are necessities in science which do call for universals and for more full-bloodedly "physical" universals than just sets. Nevertheless, Goodman and Quine did have fundamentally sound instincts in seeing the debate over the reality of sets as the contemporary incarnation of the old debate between nominalism and Platonism. Sets emerged from the hierarchy of universals, and they are to be located in that hierarchy *somewhere,* if they are to be believed in at all. In defending metaphysical realism, we are defending the existence of universals *including sets*. Factoring off some important but mistaken aspects of the package deals which are normally labelled Platonism, what is left is metaphysical realism. Metaphysical realism, in the sense which stands as arch-rival to nominalism, is *Platonism*.

And Platonism is really, primarily, *mathematical realism*. Metaphysical realism asserts the existence of the things which form the subject-matter of mathematics. Admittedly, metaphysical realism is somewhat more general, at least in principle. A metaphysical realist may assert the existence of universals which lie outside the scope of "mathematics". Yet in practice, the important universals are precisely properties, relations, patterns, and structures which form the subject-matter of mathematics, as we shall show in the following sections. Thus, metaphysical realism is, above all, mathematical realism. And since mathematics is integral to science, metaphysical realism is an essential component of any thoroughgoing scientific realism.

2.3 SAME YET DIFFERENT

The world contains more than just individuals, together with properties and relations among them. Either individuals exist, or they do not; they do not come in degrees. Likewise, either properties and relations are instantiated by given things, or they are not; they cannot be barely instantiated or partially instantiated or almost fully instantiated and so on.

Yet some things in the world seem not to be quite like that. Individuals can be counted; but a *substance,* like plutonium, cannot be counted. We can say 'how much', but not 'how many'. Individuals are reflected in language by so-called *count nouns* like 'door' and 'key'; substances correspond to *mass terms* like 'bread' and 'water'. We can have 'two doors and one key' but not 'two bread and one water'.

Properties and relations, too, divide into two sorts. With some, we need only say whether given things do or do not possess the properties or relations in question. But there are others which seem to come in amounts, and with these the simple "on" or "off" of being instantiated or not being instantiated seems to leave something out. There is more to be said with such properties and relations than whether they are instantiated by a particular individual or individuals or not. Consider pleasure. Even after specifying what kind of pleasure is on offer, whether intellectual or sensual, or some other sort, we may want to know more. We may want to know how much. Consider also mass, charge, or any other such basic physical property. To specify the mass of something, it is not enough to say whether the thing does, or does not, have mass: we must also say how *much* mass it has. Similarly with a relation like relative velocity, it is not enough to say whether one object has velocity relative to another object: we also want to know how quickly the relative positions are changing. We use the term *quanti-*

ties for properties like mass, and relations like relative velocity, which seem to come in amounts.

In explaining the nature of quantities, we are drawn to the core of traditional metaphysics. The theory of universals (i.e., of properties and relations) arises from a recognition that, in some sense, two distinct things may be both the same and yet different at the same time. This is superficially a contradiction. The conflict evaporates, however, when we say that the things which are the same in one respect may be different in other respects. Different properties or relations constitute those different respects.

Quantities cause problems, because it seems as though two things may be both the same and different – *in the very same respect*. Two things may be the same, in that both *have* mass; yet they may be different, in that one has *more* mass than the other.

Not only are quantities possessed to a greater or lesser extent; it is also possible to specify how much. We may say that the mass of *a* is closer to that of *b* than it is to that of *c*. And, of course, in some cases we go on to measure this difference. *a* may be twice as massive as *b*, and with the specification of a scale with a standard unit of mass, we may say that *a* is 2.4 kg and *b* is 4.8 kg. These features of quantities do not fit, in any obvious way, into the metaphysical framework of a hierarchy of universals. We will argue that they do fit; but it is not immediately obvious how.

Plato, it seems, was aware of the problem of coping with quantities within the context of a theory of universals (see especially the *Philebus*). If a property is something an individual either has, or does not have, with no intermediate possibility, then it is hard to handle quantities.

Plato's own theory, in contrast, leaves much more space for quantities. There are things called Forms, which exist independently of individuals. An individual does not simply *have* or *not have* a Form; rather, an individual may resemble, or "participate in", a Form to different *degrees*. Or, to put it in a slightly different way, an individual can approach an Ideal more or less closely. For instance, degrees of goodness are analysed as degrees to which a thing approximates The Good.

We have no intention of being drawn into scholarly reflections on the interpretations of Plato. We shall nevertheless use the term 'theory of Forms' as a label for a class of related theories reminiscent of Plato. All these theories account for quantities by positing *one* entity and a *varying* relationship between this entity and individuals. The entity explains what various individuals have in common; and the relationship explains the degrees to which quantities are manifested by individuals.

When abstractly set out, this class of theories subsumes a form of nominalism championed by Berkeley.[8] In this form of nominalism, a Platonic, abstract Form is replaced by a selected individual called a *paradigm*. What a variety of individuals have in common is that they resemble the same paradigm. The idea is that an individual either does or does not bear sufficient resemblance to a *paradigm*. Properties give way to paradigms, together with a threshold of "sufficient resemblance". It is less often noted, however, that there is no need to restrict our attention to a single threshold. We may take note of degrees of resemblance to a paradigm. And these may play the role of the degrees of "participation" in the Platonic theory of Forms.

All theories of this Platonic, or Berkeleyan, kind share a noteworthy feature. They explain how a property can come in degrees, by appeal to a *relationship* that comes in degrees. It does not matter whether the relationship is called participation, in memory of Plato, or resemblance, in memory of Berkeley. The point remains that it is a relationship that comes in degrees.

This means that the central problem posed by quantities is as yet unresolved. The problem is to explain how things can be the same, and yet different, in the *same respect*. In the theory of Forms, we are faced with various pairs which are all the same (in that they are all instances of resemblance to a particular paradigm or participation in a particular Form) and yet at the same time different (in that one pair displays *more* resemblance, or participation, than another). Other quantities have been explained away in terms of a single unexplained quantity. The notion of degrees of a relationship cries out for analysis, and in the case of resemblance or participation, the theory offers none.

We do not claim that the theory of Forms is mistaken: only that it is incomplete. There may indeed *be* some single entity to which individuals are more or less closely related. We claim only that, by itself, such a theory fails to explain what we are setting out to explain – the nature of quantities.

8 Berkeley (1710/1965).

The strategy of the theory of Forms is to posit a single entity, together with various relationships of individuals to that entity. An alternative strategy posits *many* properties – one for each specific degree of a quantity. For instance, it is posited that the predicates 'having a mass of 2.0 kg', 'having a mass of 4π kg', and so on, correspond to *distinct nonoverlapping properties.*

Plato's strategy makes it easy to say what various objects, with distinct masses, have in common; it is not as easy to say what makes them different. The alternative strategy easily handles what Plato finds difficult. It easily explains what makes objects with distinct masses different: they differ because each has a property the other lacks. Yet Plato's strategy does explain something which the alternative strategy seems not to handle as well. Things with different masses do have something in common. To reflect this adequately, the alternative theory must add another property to the picture – one which two objects with distinct masses have in common. This property, which they share, is called a *determinable;* the two distinct properties not shared by the two objects are called *determinates.* The determinable is, in this example, 'having mass' (or simply 'mass'); the determinates are 'having *this* much mass' and 'having *that* much mass' or 'having mass of so-and-so kg' and 'having mass of such-and-such kg'.

Mass provides, of course, just one example of determinables and determinates. Colour is also a determinable; and red, green, and so on are determinates. Note also that, although red is a determin*ate* relative to the determinable of colour, nevertheless, red is also a determin*able* relative to the more specific determinate *shades* of red. And pains, too, display the same spectrum of problems. Pain is a determinable; throbbing pain is, relative to that determinable, a determinate. Pain in dolphins also counts as a determinate relative to that determinable, as does pain in humans. Furthermore, within the category of, say, burning pain in humans, there are determinate degrees or intensities.

Positing both determinates and determinables gives answers both to what is different and what is common. Objects having the one determinable can have different determinates. And yet this account is still incomplete. Either the relationship between determinable and determinates is objective or it is not. If there is some objective relationship between the determinable and the determinates, we need some explication of what it is. But if there were no objective connection, it would be stipulative to claim that the objects falling under various determinates share some determinable. The so-called determinable then collapses into merely a convenient common label rather than any genuine common property. That is the route to nominalism, and we will not be sidetracked into that territory at this juncture. Thus, the relationship between determinable and determinate is one which, we urge, should not be taken as primitive. So the theory of determinables and determinates is incomplete.

This account of determinables and determinates is modelled on a position set out by W. E. Johnson.[9] In Johnson's version, *both* the determinable and the determinate are properties *of individuals*. For instance, the object which is coloured (a determinable) is the very same object which is, say, blue (a determinate). It is possible, however, to approach the matter somewhat differently. We begin with a distinct property for each individual with a different determinate. For instance, we begin with a battery of properties, one for each precise shade of colour. Then we note that all these properties have something in common. Each distinct shade is a shade *of colour*. There is a property, that of being a colour, which each determinate shade shares with each other determinate shade. This common property is a property of properties – a *second-degree* property.

How does this theory fare over the central problem for quantities – the problem of explaining apparent sameness and difference of the same object in the same respect? It explains the difference between two objects of, say, distinct shades of blue. Each has a property the other lacks. But how do we explain what two blue objects have *in common* when they are not precisely the same shade of blue?

To answer this, we will have to draw on the fact that although two objects of different shades of blue have two distinct properties, these properties in turn have something in common. They are both, say, shades *of blue*. Therefore, the objects have in common the property of having a shade of blue. That is, what they share is having a property which has the property of being a shade of blue.

9 Johnson (1921). Discussion of this position is found in A. Prior (1949) and Armstrong (1978, pp. 111–15).

Similarly, objects of quite different colours, say one grey and the other orange, share the property of having a property which is a colour. *Being coloured* is a property of individuals, and it amounts to the property of *having a colour.*

Similar observations may be made about pains and other mental states. On one construal, functionalist theories of the mind identify the property of having a pain with the property of having whatever property constitutes pain for that individual at that time. This construes 'having a pain' as a second-order property. Even if the property which constitutes pain in dolphins is not the same property as the one which constitutes pain in people, nevertheless the dolphin property and the people property may have something in common, called their 'functional role'. And then, derivatively, pain sufferers in general, whether dolphins or people, do have something in common. They have in common the second-order property of *having* a property with that functional role.[10]

On this theory, then, there are *three* sets of properties to keep track of. There are the basic, *first-order* properties of individuals, on which all the others supervene. There are the properties of the first-order properties – these are *second-degree* properties. And finally there are the *second-order* properties of individuals, the properties of having some first-order properties which have such-and-such second-degree properties. The basic puzzle about quantities is then explained in the following way. Objects with different determinates are different because each has a first-order property the other lacks; they are the same because they have the same second-order property.

This theory does not conflict with the original theory of determinables and determinates. It merely says something in places where the original theory was silent. In particular, it says something about what a determinable is (namely, a second-order property) and how it relates to its determinates.

The original theory of determinates and determinables need neither affirm nor deny that determinables are second-order properties. Yet there is good reason for supplementing the theory with this construal of determinables. As Johnson pointed out, determinables and determinates stand in a very tight and characteristic pattern of logical relationships. If an object has a determinate, this entails that

10 See Jackson, Pargetter, and Prior (1982) for a discussion of mental states and second-order properties.

it has the corresponding determinable. But the reverse does not hold: possession of that determinable does not entail possession of *that* determinate. And yet possession of that determinable does entail possession of *one of* the determinates falling under its scope. For instance, possession of mass does entail possession of either this specific mass, or that specific mass, or one of the other possible specific masses. A thing cannot just have mass without having a *particular* mass.

This pattern of entailments cries out for explanation, and it was the inability of the original theory of determinates and determinables to provide an explanation that prompted us to point to its incompleteness. The second-order theory of determinables does furnish just such an explanation in a simple, straightforward way. If you believe in determinables and determinates, and you reject the second-order story, then you must supply an alternative explanation of the entailment pattern. No such alternative account of determinables is on the market. And besides, the explanation provided by the second-order theory seems very plausible. So we will take it that the theory of determinables and determinates should be construed in the second-order fashion sketched above.

We believe that this *augmented theory of determinables and determinates* is on the right track, but it is *still* incomplete. It is less incomplete than the Platonic theory of Forms and the original theory of determinates and determinables, but incomplete nevertheless. And in fact, the issue which the augmented theory of determinables and determinates overlooks is one which Platonic theories place on centre stage. This issue is that of explaining how quantities can come *in degrees*. It is not the case, simply, that two objects are the same (both being, for instance, blue) and yet different (being different shades of blue). It is also the case that we can specify *how* different they are. We can say that two blue things *differ more* than two other blue things *with respect to colour*. Similarly, two pains may be more alike than two other pains. And similarly also, a mass may be closer to one mass than to some other one. It is this feature that is so well exposed by the Platonic *degrees* of participation or resemblance.

These facts about quantities are not easy to cope with, using even the augmented model of determinables and determinates. The augmented theory of determinables and determinates is still incomplete. So we move on to explore another avenue.

Quantities generate relations among objects. At first sight, these relations seem to be just consequences of the intrinsic properties of the objects. It may seem that all such relations rest on properties. For instance, consider the relation 'taller than'. Goliath stands in this relation to David. It is plausible that Goliath has an intrinsic property, his height, which he would have whether or not David existed. Similarly, David has an intrinsic property quite independently of Goliath. Given the height of Goliath and the height of David, it follows automatically that Goliath is taller than David. There is no need to posit any intrinsic relation 'taller than' over and above the intrinsic properties of height. The relation supervenes on these properties.

Yet there is something unsatisfactory about this treatment of relations like 'taller than'. There is, to begin with, a worry about whether there really is any such thing as an *intrinsic* property of height. It is tempting to suppose that the property rests on relations rather than the other way around. Plausibly, being the height you are is nothing more than just being taller than certain things and shorter than others. However, we will not pursue this line of attack. We think that it is, in the end, possible to defend intrinsic properties of height against the relationist critique. Nevertheless, even if intrinsic properties of height do exist, there remains a question about how such properties can ground relations among objects.

Consider the analysis of spatial relations, like 'east of'. Suppose we try to ground this relation in intrinsic properties of location. *a* is east of *b*, because *a is here* and *b is there*. But why should *a*'s being here, and *b* there, entail *a*'s being east of *b*? Only because *here* is a position which is *east of* the position which constitutes *there*. We presuppose the existence of spatial relations between their positions. Another famous case of this sort is furnished by time. The relation of *earlier* to *later* is arguably not solely a product of the intrinsic characters of the related events. Relations are inescapable. They cannot be grounded in properties alone.

The same applies to a relation like 'twice as massive as'. You may try to ground this in intrinsic properties of determinate masses. But why should object *a*'s having one property and object *b*'s having another property entail *a*'s being twice as massive as *b*? We must presuppose a relation between the property of *a* and the property of *b*. The property of having *this* mass must stand in a relation of proportion to the property of having *that* mass. Otherwise there would be no explanation of why objects with the first property are *more* massive than those with the second. And, indeed, there would be no explanation of why they are *that* much more massive than those with the second property.

Let us, then, abandon the attempt to ground quantities in properties alone. Let us start again. In the theory of Forms, individuals stand in a variety of relationships to a single thing (a Form or paradigm). Let us, instead, present them as standing in a variety of relationships *to one another.* This one is twice as massive as that one; that one is two-thirds as massive as another one; and so on.

On this account, for an individual to have a particular determinate property is just for it to stand in a particular range of relationships to other individuals. For instance, to have *such-and-such* a mass is to stand in a certain set of relationships to other massive objects. For an object to be a particular shade of blue is to stand in a set of relationships to other blue objects. And so on.

Such a battery of relations will make it easy to characterize one of the important features of quantities. When individuals possess a given quantity to distinct degrees, there is an important way in which they *differ.* This is captured in the relational theory. Two such individuals differ because each possesses different relationships to the other. For instance, we may have that *a* is taller than *b*, but of course *b* is not taller than *b*. So *a* and *b* differ because each has a (relational) property the other lacks.

Not only does this theory explain the presence of *some* difference between two objects: it also characterizes the different *kinds* of difference there can be. It can explain *how* different they can be. We may have, for instance, *a* being twice as bright as *b*, and *c* three times as bright as *d*. Then, clearly, it is not enough to say just that *a* differs from *b*, and *c* differs from *d*. It is important to say that they differ differently. Individual *c* differs more from *d* than *a* does from *b*. By appealing to two relationships, one between *a* and *b*, and the other between *c* and *d*, it is easy to capture the fact that the two pairs differ differently. In fact, to capture fully the way that *c* differs *more* from *d* than *a* does from *b*, we must add an ingredient. We must add that the two relations 'twice as bright' and 'three times as bright' themselves stand in a certain ordering relation. This same relationship holds between 'three times as bright' and 'six times as bright'. A completely different relationship holds between, for instance, the two relations 'twice as bright' and 'one-third as bright'. Let us, however, leave this last complication for the moment.

The relational theory handles *differences* rather well; but does it adequately explain what individuals have *in common?* Things which have, for instance, different determinate masses do, nevertheless, have something in common. We could simply add a property to the theory, a property which is shared by all the things which stand in mass relations to one another. Call this property simply 'mass'. When things stand in mass relations to one another, this entails that they share the property of mass. And when two things share the property of mass, this entails that there is *some* mass relation between them.

This double-barrelled theory, containing many relations and a common property, is structurally similar to the Johnsonian theory of determinables and determinates. And, like that theory, it leaves much unexplained. It lays out a complex pattern of entailments between the one common property and the many relations. Yet nothing has yet been said to explain why these entailments should hold.

These unexplained elements fall into place if we supplement the relational theory with a specific account of the nature of the supposed common property. We need an account of the common property which displays a clear connection between it and the many relations which underpin it. The simplest way to deliver the required connections is by construing the common property in a second-order manner. Let us explain, in two stages. In the first stage, we shall examine *what* objects with different mass have in common; and in the second stage we shall face the question of how some can have *more* in common than others.

First, let us consider the many relations associated with a quantity like mass. These will be relations like 'more massive than', 'half as massive as', and so forth. It is plausible to postulate that each of these relations has a common property. Each of them is a relation *of mass*. Let us explore this idea further. The story begins with a profusion of different mass relations among individuals. To this we add a single property which all of these relations have in common. How does this property of *relations* explain what all massy *objects* have in common? By the fact that its existence entails that all massy objects share a certain second-order property. For any massy object, there are some mass relations it has to other things. (Indeed, there are mass relations it has to itself – e.g., 'same mass as'.) What all massy objects have in common is the property of *standing in mass relations*.

We may thus construe the common property of massiness as a second-order property – that of having mass relations to things. This construal has a large advantage over theories which give the common property no analysis. It explains the pattern of entailments between the common property of massiness and the various mass relations.

The relational theory is, nonetheless, *still* incomplete. What has not yet been explained is the way that some mass relations are in some sense "closer" to one another than to others.

All mass relations have something in common; and so massy objects have a second-order property in common. This is the first stage in constructing a theory of quantities. But there are complexities which are not fully captured by this first stage. It enables us to explain how two things can *differ* (having different masses) but yet also be the *same* (both having mass). However, it does not explain *how much* two things may differ. It does not explain how a thing may differ in mass more from a second thing than it does from a third thing. Nor does it explain how two things may differ in mass to just the same degree as they differ in, say, volume: one may be, for instance, twice the mass *and* twice the volume of the other.

To obtain an adequate theory of quantities, we must thus introduce a second stage of construction. It is not sufficient to note that all mass relations have something in common, a common property. We must also note that mass relations can be more and less *similar* to one another. The relations 'five times as long' and 'six times as long' are in some sense *closer* to one another than either is to, say, 'one hundred times as long'. Consider three pairs:

$$\langle a, b \rangle,$$
$$\langle c, d \rangle,$$
$$\langle e, f \rangle.$$

Suppose a is five times as long as b, and c is six times as long as d, and e is one hundred times as long as f. Then there is an important respect in which the first two pairs are more closely similar to one another than either is to the third. Hence, we must recognize relations among relations – relations of *proportion*. We shall explain the nature of proportions more fully in Chapter 8.

The theoretical structure we offer, which in part is inspired by Frege,[11] thus comprises three fundamental ingredients: (1) individuals, (2) determinate relationships among individuals, and (3) relations of proportion among those determinate relationships. Consider, for instance, three individuals *a*, *b*, and *c* as our population at level (1). There will be a class of relations at level (2) holding among *a*, *b*, and *c*. These level (2) relations may then be grouped according to whether they stand in level (3) *proportions* to one another. All the mass relations stand in proportions to one another; all the volume relations stand in proportions to one another; but the mass relations do not stand in proportions to the volume relations, or vice versa.

Thus, level (3) relations of proportion classify level (2) determinates into equivalence classes. Within each such equivalence class of level (2), level (3) proportions will also impose an ordering. And it is this ordering which explains how one thing can be closer to a second in, say, mass, than it is to a third.

Furthermore, level (3) proportions can introduce finer distinctions within equivalence classes at level (2). Level (3) proportions are universals, so they can be multiply instantiated. Thus, one and the same relation of proportion may hold between several distinct pairs of level (2) relations. For instance, several distinct pairs of mass relations may stand in the same level (3) proportion.

Indeed, using level (3) proportions, we can generate classifications which cut across the equivalence classes of mass relations, volume relations, velocity relations, and so forth. For instance, two level (2) mass relations will stand in some specific proportion; and two level (2) volume relations may stand in exactly the *same* proportion. This explains how it can be that, when an object is twice the mass of and also twice the volume of another, these two relations are at the same time both the *same* and *different*. These relations are different in that one is a mass relation, standing in proportions to other mass relations, while the other is not a mass relation and stands in no proportions to any mass relations. And yet there is also something these two

11 Frege (1893) developed a theory of real numbers which construes them as relations among relations. These relations among relations are construed as *proportions* holding among magnitudes (i.e., quantities in our sense) and 'exactly the same proportion of magnitudes which we find among line segments we should also have among duration, masses, light intensities, etc.' (p. 156).

different relations have in common, as is reflected in the way we describe them: "twice" the mass of – "twice" the volume of. This common factor, too, is explained by level (3) proportions. There is a proportion between 'twice the mass of' and 'the same mass as'; and the very same proportion also holds between 'twice the volume of' and 'the same volume as'. And that is why the same word 'twice' occurs in the descriptions of both.

The cross-category similarities between determinate quantities helps to explain the usefulness of numbers in dealing with quantities.[12] (We shall say more about the nature of numbers in Chapter 8.) Given relations of proportion holding among mass relations, for instance, we may generate a complex range of derivative properties and relations. For instance, an object may have a certain mass relation to some designated object which is arbitrarily chosen as our *unit of measurement*. Perhaps it has the relation 'same mass as' to the unit. Thereby, it has a second-order property: the property of having the 'same mass as' relation to the unit. When this is so, the object has a mass of one unit, say 1 kg.

Suppose some other object has some different mass relation to the unit: it is, as we say, twice as massive as the unit. Thereby it has the property of having a mass of 2 kg. The number 2 appears here as a signal that the mass relation between the object and an object of unit mass stands in the proportion of 2 to 1 to the 'same mass as' relation, and the term 'kg' specifies to which unit it stands in that proportion.

Earlier, when we discussed the theory of Forms, our objection was not that such a theory is mistaken, but only that it is incomplete. Objects do have relations to paradigms; and our theory not only allows, but also explains, how these relations arise. It is by way of such relations to paradigms that scales of measurement arise. Objects which specify a standard unit of a scale of measurement play the exact role of a paradigm in a theory of Forms.[13]

12 Frege's three-level theory of relational quantities and proportions holding among them was intended as a basis for a theory of real numbers. Such a theory of real numbers received formal elaboration by Whitehead and Russell and by Quine. See sec. 10 of 'Whitehead and the Rise of Modern Logic' in Quine (1941), reprinted in Quine (1966). A formal theory of this sort is given a realistic interpretation, in terms of universals, in Bigelow (1988a).

13 As Newton (1728) said, 'By number we understand not so much a multitude of Unities, as the abstracted Ratio of any Quantity, to another Quantity of the same kind, which we take for Unity'. This view became orthodoxy; thus, Leonard

It may be worth noting that the level (3) relations among determinates may be more complex and discriminating for some classes of determinates than for others. Determinate physical quantities like mass stand in such a rich pattern of proportions to one another that it forces us to draw upon the full resources of the real number system. In our terminology, however, pains and colours, too, count as quantities. It is highly likely that the level (3) relations among these quantities will manifest a variety of structures which are less linear, and less discriminating, and so which manifest structures other than that of the real number system.

Thus, quantities will subdivide into categories, according to the nature of the level (3) interrelations they manifest. These subdivisions correspond to the distinctions, familiar in measurement theory, which are drawn between, for instance, *interval* and *ratio* scales of measurement.

Different sets of relations can manifest different patterns of proportions among their members. And different scales of measurement must have mathematical structures which reflect these different patterns. In cases like that of mass, the structure is in one sense very rich and discriminating; but it is also very regimented and linear. In cases such as pain, the proportions will be in certain respects much more complex. They may allow for nonlinear partial orderings, incomparabilities across the orderings of different kinds of pains, and so forth. Dolphin pains may stand in the same sort of intensity orderings that human pains stand in; but it may not be possible to rank the intensity of a dolphin pain against a human pain. Similarly, it may not be possible to rank some aches against some burning pains – or even one person's pains against another's.

Given the three levels – individuals, determinate relations, and proportions – we can account for the complex patterns of sameness and difference which characterize quantities. The level (3) relations generate a rich network of second-order properties and relations

Euler (1822) echoes Newton: 'Mathematics, in general, is the *science of quantity*. . . . Now we cannot measure or determine any quantity, except by considering some other quantity of the time as known, and pointing out their mutual relation . . . a number is nothing but the proportion of one magnitude to another arbitrarily assumed as the unit'. The recent set-theoretical transformation of mathematics has obscured the relationship between number and physical quantities, and hence also the nature of quantities themselves. Our theory may be regarded as retrieving and updating lost wisdom, buried under many years of positivism and nominalism, and "new math".

among objects. And not only do they generate all these second-order properties and relations, but they also explain the pattern of entailment relations which hold among them.

We emphasize that the theory is not one without a justificatory path. But not only do we hope that this justificatory path is persuasive, we hold out the challenge to find an alternative theory with as much explanatory power.

Nominalists will, of course, seek ways of avoiding the ontological cost of accepting not only relations but also relations among relations. We do not pretend to have proved that realism is the only tenable position. But we do claim that, if you are to be a realist at all, our three-tiered realism is to be recommended, because it does justice to quantities to a greater extent than other realist alternatives. Furthermore, the three-tiered structure flows naturally from the spirit of realism. Realism has its source in the recognition that things may be the same and yet different. If such recognition justifies universals at all, close attention to the complexities of such samenesses and differences leads straight to our three-tiered realism and to our three-level theory of quantities.

Finally, even if nominalism remains tempting to some, we claim that our realist theory has unifying and explanatory virtues which are very unlikely to be matched by nominalism (or for that matter by a realist who will accept only first-order universals). Almost inevitably, nominalists (or first-orderists) will be compelled to take a panoply of relational predicates as primitive. Worse, there will be complex patterns of entailment among these predicates, which will also have to be taken as primitive. In three-tiered realism, such predicates will be construed as second-order predicates, containing implicit quantification over both properties and proportions. And this construal will enable us to explain those entailments in a simple and natural manner. The three-level theory gives a much more explanatory account of quantities.[14]

2.6 VECTORS AND CHANGE

One pleasing feature of the three-level theory of quantities is its openness to a range of variations and generalizations. One of these

14 Our discussion of quantities is based on Bigelow and Pargetter (1988). Armstrong (1988) urges a variant on our theory which takes proportions to be relations among *properties* rather than relations among relations. In our treatment of vectors, in the following section, we move in Armstrong's direction on this issue.

modifications enables us to give a theory of vectors. And vectors have been a big problem for a very long time. Furthermore, vectors are ubiquitous in current physics: it would be rather gratifying if we could obtain a clearer understanding of the physical reality which underlies the mathematical machinations of vector spaces.

Vectors emerge from the problem of change. A very important special case is the problem of motion, motion being conceived as change of place. Change has been a problem since the time of ancient Greek science. The trouble arose because the very notion of change seems to involve a contradiction. Suppose a discus changes from round to square. This entails that the discus is round; and it entails that the discus is square – and isn't that contradictory? No, not if we recognize that instantiation is tensed. A thing can be round at one time and square at another. But is this all there is to the process of change, that a thing has one property at one time and another property at another time?

In the late Middle Ages a debate developed over this issue. Ockham and his numerous followers argued that change is nothing more than the possession of a sequence of different properties at different times. This was called the *doctrine of changing form* (*forma fluens*). The opposition doctrine was (confusingly) called the *doctrine of change of form* (*fluxus formae*).[15] It is not easy to glean the difference between the doctrines from their names: what is the difference between changing form and change of form? The key lies in the shift between 'form' being the noun in the name of Ockham's doctrine, qualified as 'changing', whereas the name of the opposition doctrine uses 'change' itself as the main noun. The latter doctrine treats change itself as a subject, a characteristic of a thing. For instance, it is held by the change of form doctrine that an object which is darkening not only will have a given shade of, say, grey at a given time, but will also have a *changing* shade of grey. The property of being a darkening grey at a time is claimed to be a property of a thing at *that* time, and not just reducible to its various properties at *other* times.

The medievals discussed many kinds of change: darkening, cooling, getting heavier, and so forth. One which they found especially interesting was the change involved in a person getting better, be-

15 For an intriguing exposition and further references, see Dijksterhuis (1961). The Ockhamist doctrine has become orthodoxy among analytical philosophers in this century.

coming more perfect, improving morally. But for science, the crucial case was that of change of place – motion. The Ockhamist doctrine of changing form held that motion is no more than the occupation of successive places at successive times. The opposing doctrine of change of form insisted that a moving body possesses not only a position, at a given time, but also what amounts to an instantaneous velocity.

Newton's name for the differential calculus was the theory of "fluxions" – and like the medieval doctrine of *fluxus formae,* Newton's theory attributed instantaneous velocities to moving bodies. Let us call the anti-Ockhamist view the doctrine of flux.

The doctrine of flux, in essence, attributes a *vector* to a moving object at a time. The Ockhamist view, too, can be seen as employing vectors, but in a different, a less full-blooded sense. Ockhamists can use the sequence of positions of a body to define a vector of an abstract mathematical kind. Followers of Newton have done likewise in their interpretation of "instantaneous velocity". Using the sequence of positions of a body over time and defining a mathematicial limit, we can define the vector giving its so-called instantaneous velocity. But thus defined, the vector is a mathematical abstraction, reflecting a pattern displayed across time. It is not an intrinsic property of a body at a particular time.

In contrast, the doctrine of flux does attribute a vector to a moving body at a time, and this vector is construed not as merely a description of where the object has been or will be, but as an intrinsic property of an object at a particular instant. In addition to the Ockhamist second-order property of having various positions at various times, there is the first-order (perhaps varying) property of the object at each time. Consider an object which is in motion, and consider its properties at a given time. At a given time it has a position. It also has relational properties: it has a history and a destiny. It has the property of having been located in such-and-such places in the recent past and about to be located in other places. So far, the Ockhamist agrees. The question is then whether these properties are all that the instantaneous motion vector amounts to. The doctrine of flux denies that it is. There is a further property over and above an object's position, history, and destiny. Indeed, the vector is to a considerable extent independent of position, history, and destiny. The object could have had the same past positions and not have had the vector; and conversely, the object might not have had those past positions and yet still have had the vector. Similar comments apply to future positions of the object.

In typical medieval fashion, the flux theorists put their doctrine to theological use. Aristotle had argued that beyond the sphere of the stars there was nothing at all, not even empty space (indeed, especially not empty space, that being for Aristotle a logical impossibility). When combined with the Ockhamist theory of motion, Aristotle's cosmology entailed that the universe as a whole could not have any rectilinear motion. God, then, could not have set the universe into rectilinear motion. This sounded to some (including the Pope) as though it unduly restricted God's omnipotence. The doctrine of flux was used to provide a way out. God could have given the universe an instantaneous velocity, a vector – even though there was nowhere for the universe to go. This is entertainingly silly, but it is also instructive, because it highlights the independence of the "flux" vector and the mere list of "form–time pairs" (as we might characterize the Ockhamist view). The supporters of the flux doctrine held that a body could possess a motion vector even if it possessed no appropriate sequence of different positions at different times. The motion vector was conceived as logically independent of both prior and posterior positions of the body. Furthermore, flux theorists took the motion vector to *explain* the sequence of positions a body occupies, when indeed it does occupy such positions. The vectors explain the sequence of positions, not vice versa. We shall defend this core doctrine of the flux theory. The idea that an object could have an instantaneous velocity, without *any* variation whatever of position across time, may seem fanciful. But whatever links there may be between possession of a velocity vector at a time and positions at other times, we argue that such links arise because the velocity vector *explains* change of position, not because it is *defined* by change of position.

Although logically independent of one another, the velocity vector and the prior and posterior positions of a body will be significantly connected by physical laws. The presence of a vector at a time will contribute to an explanation of its subsequent positions. The flux doctrine reverses the order of explanation implicit in the Ockhamist view.

For Ockham, the object has a certain velocity just because it has been in different places and will be in still further different places.

Indeed, its having a velocity is not so much explained by its sequence of positions – rather, its having a velocity is nothing over and above its having a suitable sequence of positions. The having of a position is a first-order property (or predication, if you are a nominalist). Velocity is a second-order property (or predication), since it refers to or quantifies over a number of distinct first-order properties or predications. For Ockham, what *constitutes* velocity is a second-order pattern over first-order positions at times. Hence, velocity cannot explain an object's sequence of positions on Ockham's theory: we cannot say an object is now located to the right of where it was a moment ago *because* it was in motion a moment ago.

According to the flux theory, that gets the explanatory direction cart before horse. An object will be, say, a little higher a moment from now *because* it is now moving upwards. It will be a little higher because it is *now* moving upwards – and not just because it is here now and was lower a moment ago. To explain why it will move still higher, we must appeal to more than the fact that it has moved upwards to here. We must appeal to the fact that it is *still* moving upwards. The first-order properties of position are explained by another first-order property of instantaneous velocity. That is to say, the flux doctrine introduces a first-order property of instantaneous velocity *in addition to* the second-order Ockhamist property, and the flux doctrine uses the first-order property to explain the second-order one.

In order for first-order instantaneous velocity to "explain" second-order patterns of positions, there must be the right sort of relationship between first- and second-order properties. On the one hand, there must be some degree of logical independence between them if one is to "explain" the other – at least, if explanation is here to be construed as (roughly) scientific rather than purely mathematical. On the other hand, the logical independence cannot be so exaggerated that there is no relevant connection between explanans and explanandum. What the flux theory requires is the logical possibility of first-order instantaneous velocity without a second-order pattern of positions, and vice versa; but in addition, it requires some relevant connection between first- and second-order properties. Roughly speaking, the first-order property must entail some appropriate second-order property, but only when supplemented with extra information of some appropriate, non-question-begging sort.

Let us consider the logical independence of instantaneous velocity, on the one hand, and sequences of positions, on the other. In the first place, we should note that the possession of an instantaneous velocity, say east to west, is compatible with a variety of sequences of positions. The body may have been due east a moment before and may be due west a moment later. Yet it *may* never have been due east and *may* never be due west in the future. Suppose, for instance, that it is at the highest point of a parabolic trajectory. Then its instantaneous velocity is *horizontal,* yet all its prior positions and its future positions are *lower* than its present position. The Ockhamist takes this to show that velocity is to be defined not just in terms of a finite distance covered in a finite time, but rather as a mathematical limit of sequence of distance–time pairs.

The flux theory, however, argues that instantaneous velocity is in principle independent of the existence of distance–time pairs which tend to a mathematical limit. We should reflect on two cases. There may be instantaneous velocity at a time, without any sequence of *prior* positions generating an appropriate mathematical limit – since an object may be jolted into motion instantaneously. For similar reasons, there may be instantaneous velocity at a time without any sequence of *future* positions generating an appropriate mathematical limit – since an object in motion may be instantaneously brought to rest, say by colliding with another. This is at least a logical possibility (and that is all we require for present purposes). Consider a simplified Newtonian world with two perfectly rigid bodies colliding, one going from velocity V to rest at just the moment when the other goes from rest to velocity V.

Now consider three Newtonian rigid spheres a, b, and c. Suppose b and c are at rest and are in contact with one another. Sphere a moves with velocity V along the line joining the centres of b and c, and strikes b. Theory tells us that a will stop, b will not budge, and c will move off with velocity V. The velocity of a is transferred from a, through b, to c. There is a moment in time when b has velocity V – even though there is no appropriate series of past *or* future positions for b which will yield velocity V as a limit.

This demonstrates that it is possible to have a first-order property of instantaneous velocity *without* any appropriate Ockhamist second-order property. Instantaneous velocity does not entail any sequence of positions which generates a mathematical limit.

This demonstration in support of the flux theory can be challenged. We have formulated the argument within the framework of a Newtonian system, with absolute space and time. Things will be much more complicated in the theories of special or general relativity, let alone in quantum field theory. Nevertheless, we let our argument stand, in its simple Newtonian form. It shows that it is *not* absurd in *principle* to attribute a nonzero instantaneous velocity vector to an object which displays no variation whatever in its sequence of positions across time. This is enough to establish the logical point, that instantaneous flux vectors need not be construed as logically entailing the existence of a sequence of attributes which vary across time.

Conversely, it may be argued that the existence of an Ockhamist sequence of positions does not entail instantaneous velocity in the sense required by the doctrine of flux. There could be an Ockhamist sequence of positions which was explained *not* by the presence of instantaneous velocities, but by some other source. Consider, for instance, the movement of an image projected onto a screen from a movie projector. The image of Clint Eastwood's cigar, for instance, may occupy a sequence of positions which generates an Ockhamist pattern. Yet the image, at a given time, does not have any instantaneous velocity. According to Malebranche, the whole world is a bit like a movie, and God is a bit like the projectionist.[16] An object is created at one location at one time and is created at another location a moment later. At neither time is it really moving – it has no genuine instantaneous velocity in the sense required by the doctrine of flux. Its position at the later moment is not explained as causally dependent on its position and velocity at a previous moment. Rather, its position at the later moment is explained solely as causally dependent on the free choice of God.

In Malebranche's world, the motion of a body is not in fact a genuine causal process, in the sense referred to in Einstein's special theory of relativity. Einstein tells us that no object can be accelerated up to and beyond the speed of light. But an *image* can be. For instance, suppose a spot is projected by a laser onto the surface of one of Jupiter's moons, say Callisto, and then suppose the laser gun is jiggled here on earth. If the angles and distances are right, the spot on Callisto will whizz across the surface of that moon faster

16 Malebranche (1688/1923).

than the speed of light.[17] This is no threat to Einstein's theory, because the motion of the spot is not a causal process – and the spot is not an object. The spot at one time is not really the same thing as the spot at another time. When we say of the spot that "it" has moved, this is deceptive. There is really no single "it" which is in one place at one time and another place at another time. In the case of the spot, there is only the Ockhamist kind of velocity – a sequence of positions across time. The spot does *not* occupy the position it does because of its earlier position and its earlier instantaneous velocity. Rather, it occupies the position it does because of the position of the laser gun on earth. For the spot, Ockhamist velocity is all there is. Hence, the existence of a second-order pattern of positions does not entail the existence of a first-order instantaneous velocity.

Newton's three spheres show that instantaneous velocity does not entail an Ockhamist sequence of positions. The laser spot on Callisto shows that an Ockhamist sequence of positions does not entail the presence of instantaneous velocity. This establishes one of the things the flux theory requires: the logical independence of first-order from second-order velocity. Yet the flux theorist must beware. In order for the first-order velocity to explain the second-order velocity, there must be some fairly intimate link between them.

One plausible link is suggested by the spot on Callisto. The spot lacks instantaneous (first-order) velocity. It also lacks genuine numerical identity across time – it is not really a single thing which occupies different places at different times. Perhaps these issues – velocity and identity – are linked. It may be argued that numerical identity across time is dependent on the presence of causal links across time. And then it may be argued that an Ockhamist sequence of positions *plus* numerical identity of an object across time *does* entail instantaneous velocity. That is to say, it may be argued that when a sequence of positions is occupied by the *same object,* the object *must* have been moving if it was to get from one position to the next.

This means of linking first- and second-order velocities, however, is a dubious one. It attempts to draw a "metaphysical" link

17 Noncausal pseudomotion of this sort is discussed in various places, e.g., Reichenbach (1958).

between first- and second-order velocities – a link which is prior to the contingent laws of nature. Malebranche's world, however, stands as a counterexample, if you believe that world to be a logical possibility at all. In Malebranche's world, God creates the very same object over and over, but in different places at different times. For Malebranche, there are patterns of both positions and identities across time, yet there are no first-order velocities of the sort required by the doctrine of flux. Hence, there is no entailment between Ockhamist velocity and flux velocity, even if we add the assumption of numerical identity across time.

For this reason, we argue that the link between Ockhamist and flux velocities is provided not by a metaphysically essential link, but by the contingent laws of nature. Ockhamist sequences of positions are often explained by instantaneous velocities. The reason is simply that, given the laws of nature, there is (in most cases) no way of getting an object from one place to another except by imparting to it an instantaneous velocity.

Velocity is not the only vector which plays an explanatory role of the sort we have been discussing. Indeed, for purposes of theory, velocity is frequently given no separate mention, but is subsumed under momentum, which, of course, is also a vector. Furthermore, momentum is taken not as an intrinsic property (or "invariant"), but as relativized to a frame of reference. Force is yet another vector, and so is acceleration. All these vectors play significant roles in laws of nature, and it is this which gives them explanatory power. Each of these vectors constitutes an intrinsic characteristic of a thing at a time. Acceleration, for instance, is a characteristic of a thing at a time, and its possession helps to explain the pattern of past and future velocities of the thing. The connection between instantaneous acceleration and a pattern of velocities across time is mediated by their roles in laws of nature.

There are other quantities which can be defined in physical theory but which do not play any such explanatory role. For instance, consider the way an object accelerates as it falls toward the earth. The Galilean law of free fall assumes constant acceleration. However, this is not strictly correct. As the body moves closer to the earth, the gravitational force becomes stronger, and a stronger force entails a greater acceleration. Hence, the body's acceleration varies with time – and its acceleration varies at an accelerating rate. (Indeed, the relevant law is an exponential one, and it turns out that the rate of change of acceleration equals the rate of change of rate of change of acceleration.)

The rate of change of acceleration, however, does not play an explanatory role comparable to that of velocity, force, and so forth. There is thus no reason to posit an instantaneous vector for rate of change of acceleration which explains past and future accelerations. The Ockhamist pattern of accelerations over time is all there is to the "rate of change of acceleration" vector. The pattern of differing accelerations over time is not explained by a "flux" for rate of change of acceleration. The pattern is adequately explained by "flux" vectors of a lower order than that one. In the hierarchy of rates of change, only those of velocity and acceleration are instantaneous vectors which explain patterns across time. The hierarchy is merely Ockhamistic from there on up. There is no good reason for positing flux vectors, as well as Ockhamist ones, above the level of acceleration – at least not unless higher-level rates of change should come to figure in newly discovered laws of nature. This further illustrates the explanatory link between flux vectors and patterns across time. The link is not a tight logical or metaphysical one, but a looser, nomological one.

Explanation and the doctrine of flux

The doctrine of flux, it may be noted, rests comfortably with the Cartesian law of inertia: that an object continues with uniform motion in a straight line unless acted upon by a force. If we regard instantaneous velocity as a genuine, intrinsic property of an object at a time, there is an important sense in which continuing motion is *not* a change. One of the object's intrinsic properties, its velocity vector, is the same. Assuming that every change requires a cause, we can infer, therefore, that motion will change only if there is a cause, a force, which acts on the body. Thus, the flux doctrine implies a reversal of the Aristotelian view on motion. Aristotle held that motion was change, change of place, and so required continual force to maintain continual change of place. The flux doctrine takes change of place to be extrinsic change for a body, consequent upon lack of change of an *intrinsic* property, the velocity vector.

The flux doctrine has a number of explanatory advantages, at least in the case of motion. (For some cases of change, the Ockhamist view is very probably quite right. The process of darkening, for instance, is different from that of motion. Ockham may have been quite right about darkening, cooling, moral improvement, and many of the other medieval talking points.) Consider, for instance, a meteor striking Mars, and consider the problem of explaining why it creates a crater of precisely the size that it does. At the moment of impact, the meteor exerts a specific force on the surface of Mars. Why does it exert precisely that force? Because it is moving at a particular velocity relative to Mars. On the Ockhamist view, it exerts the force it does because it has occupied such-and-such positions at such-and-such times. In other words, the Ockhamist appeals to the positions the meteor has occupied in the past. But why should a body's *past* positions determine any force *now?* This requires the meteor to have a kind of "memory" – what it does to Mars depends not only on its current properties but also on where it has been. Given that the meteor is accelerating, the Ockhamist will need a very fancy story about how the action on Mars depends on previous positions. The meteor's action on Mars depends on the fact that the distances between successive positions have been proportional to the squares of the times between them.

For the Ockhamist, then, the impact on Mars will depend on the pattern of the sequence of positions before impact. The effect of a body at a time thus depends for the Ockhamist not only on the intrinsic character of the object at the time but also on its history. This cannot be ruled out as absurd, without argument. Nevertheless, it is an advantage of the flux doctrine that it does not require any pseudomemory in objects or any time lag in causation. The meteor exerts a given force, at a moment, because of the property it has at that very moment. This property is an instantaneous velocity, a vector, with both magnitude and direction. The recent history of the body may be epistemically important, furnishing the canonical method of determining what its instantaneous velocity must be. But it is the resulting velocity itself which is causally active, not the history which produced it and revealed it to us.

Consider another problem case for an Ockhamist: the spinning disc.[18] Suppose the disc is of perfectly uniform shape and made of a perfectly uniform substance. Each part of it is qualitatively indistinguishable from each other part. Consequently, as it spins, there is no

18 The problem of the spinning disc has been impressed upon us by the work of D. Robinson, some of which emerged in Robinson (1982).

change at all in the distribution of qualities over time. As it spins, each part is replaced by another part which is qualitatively identical. Yet the spinning disc is quite different from the stationary one. If you were to cut a slice out of the spinning disc, the gap would spin around; whereas if the disc were not spinning, the gap would not move. The two discs differ in their causal powers. What is the difference between the discs, which explains the contrast in their causal powers?

One difference between the discs is that, even though they exactly match in the distribution of qualities over time, they differ in the identities of the portions of matter which instantiate those qualities. In the spinning disc, even though the same quality is instantiated in a given place at two times, it may be instantiated by a different portion of matter. The time slice in a given place at one time and the time slice in that same place at a later time are not stages in the history of any one single, persisting individual. Tracking the identity of individual portions of matter on the spinning disc, we will trace out circles. Tracking identities on the stationary disc, we will not trace out circles – the portion of matter we start with will stay in the same place across time.

This account of the nature of the spinning disc entails a disputed doctrine about identity. It requires us to trace the identity of a portion of matter in a way that cannot be reduced to a tracing of some bundle of qualities. In short, it requires nonqualitative identities. Why should we say that a portion of the disc at one moment is identical with one portion rather than some other qualitatively identical portion? What distinguishes them is simply that one does, and the others do not, have a specific identity, or "thisness", or (using the medieval word) *haecceity*.[19]

We are not opposed to nonqualitative identities. We accept that the spinning disc does display a circular displacement of identities across time. Yet this is not the whole story. Even if one can accept nonqualitative identities, it is hard to believe that they generate the kinds of causal powers which the spinning disc possesses.

It is here that the doctrine of flux can play an explanatory role. The basis for the causal powers of the disc is the possession, by each portion, of an instantaneous velocity. It is because of the instantaneous velocities instantiated at each moment that the identities of

19 See, e.g., Adams (1979). The revival of this ugly medieval word owes much to Kaplan (1975).

portions of the disc track in circular paths. For these reasons, we argue that instantaneous velocity, a vector with both magnitude and direction, should be construed as a physical property, a universal, possessed by a body at a given moment of time. It is an intrinsic property, not a property relating the object to other times and places.

Yet in construing a vector thus, as an intrinsic property at a time, rather than a pattern displayed across time, we saddle ourselves with a problem. We have to explain what sort of property this is, and how it can be understood as having both a magnitude and a direction. It is not as though a property is like a pin in a pincushion, with the length of the pin constituting the magnitude and its orientation constituting its direction.

By supporting the theory of flux, in fact, we make it more difficult to account for the magnitude and direction of a vector. The Ockhamist theory does have an advantage on this score, at least on the face of it. If a velocity vector simply reflects a pattern of positions of a body across time, as Ockhamists suggest, then magnitude and direction emerge immediately from the relevant temporally extended pattern. The past and future positions of the body lie in determinate directions from the present position and at determinate distances from it. We can define instantaneous magnitude and direction for velocity by taking limits of sequences of magnitudes and directions for prior and posterior positions.

For physical reasons, we have argued, the flux theory should be accepted. On this theory, magnitude and direction for velocity cannot be defined in the Ockhamist manner. On the flux theory, we may attribute to a body some Ockhamist pattern of prior and posterior positions; and we may attribute an instantaneous velocity. But the latter attribution is not the same as the former; in attributing an instantaneous velocity we are not attributing any pattern of prior and posterior positions. So we cannot explain what constitutes the magnitude and direction of the velocity by appeal to that pattern.

How, then, can we explain what constitutes the magnitude and direction of a vector? The trick is to draw on a theory about the relations among properties.

Magnitude and direction

First, consider two-place relations between velocities. Consider, for instance, two points on the flux theorist's spinning disc. Suppose both of these points lie on the same radius, so that their instantaneous velocities have the same direction. Then these two velocities have something in common. Yet they differ from one another, since the point farther from the centre is moving faster than the one closer to the centre. All the points along a given radius, at a given moment, have something in common, yet they also differ. What they have in common, however, is not their velocity, for each has a different velocity. Rather, what they have in common is a second-order property: the property of having a velocity with such-and-such direction.

Consider, in contrast, the set of points on the disc which are located half-way between the rim and the axis of rotation. All these points have the same speed. That is to say, they all have a velocity of the same magnitude. But they are moving in different directions. Hence, what they have in common is not the possession of the same velocity. Rather, what they share is the second-order property of having a velocity of such-and-such magnitude.

Thus, vectors, construed on the flux theory as intrinsic properties, have second-degree properties, that is, properties of properties. When two vectors share one of these properties, we say that they have the same direction; when they share another of these properties, we say that they have the same magnitude. If they share both magnitude and direction, then "they" are really "it"; they are identical. This gives us an initial grip on the way that a flux theorist can understand the magnitude and direction of a vector.

We are arguing that, in the case of vectors, both second-degree and second-order properties are genuine universals, in the sense of Armstrong's *a posteriori* realism about universals. For instance, we argue that two points on the same radius of a spinning disc do have something in common. What they have in common is not merely that some linguistic item, a description, applies to both. There is something nonlinguistic present in both, a genuine universal – a second-order universal.

Two velocity vectors, then, may be the same in magnitude or may differ in magnitude, and the same holds for direction. But, of course, that is only part of the story. Not only can we say whether two velocities are the same or different in magnitude (or direction), but also we can say what manner of difference there is. When two velocities differ in magnitude, we must say whether the first is greater than the second, or vice versa, and indeed we must say *how much* they differ. Similarly, when two velocities differ in direction, we must say how much they differ: whether, for instance, one is directed only a few degrees clockwise (in a given plane) from the other, or whether it is directed say 90° clockwise from the other.

To obtain a framework for understanding such degrees of difference, we draw upon ideas which derive from Frege, Whitehead, Wiener, and Quine,[20] in fact upon the same framework we used in our discussion of quantities. We shall give a little more detail here, and shall further develop the details in Chapter 8. The central idea we borrow from them is that comparisons of "degrees" of difference derive from specific *relations between relations.*

Consider a relation, Q say. Suppose this relation holds between one individual x_0 and another x_1:

$$x_0 \; Q \; x_1.$$

Suppose it also holds between x_1 and some further individual x_2:

$$x_1 \; Q \; x_2.$$

Furthermore, suppose there is a sequence of individuals, each one standing in relation Q to the next:

$$x_0 \; Q \; x_1 \; Q \; x_2 \; Q \; . \; . \; . \; Q \; x_n.$$

When this chain of Q-relations holds, we may say that you can get from x_0 to x_n by n Q-steps. Whenever there exists some sequence of individuals enabling you to get from x_0 to x_n by n Q-steps, we may then abbreviate this assertion by the formula

$$x_0 \; Q^n \; x_n.$$

With this notation to call upon, let us compare two distinct relations, say Q and R. Infinitely many relationships can hold between two relations Q and R. Here are some examples: Q and R may be such that, for any x and y,

$$x \; Q^1 \; y \text{ if and only if } x \; R^2 \; y,$$

or

$$x \; Q^1 \; y \text{ if and only if } x \; R^3 \; y,$$

20 See Frege (1893); Whitehead and Russell (1910, esp. vol. 3, pt. 6: 'Quantity'); Wiener (1912); 'Whitehead and the Rise of Modern Logic' in Quine (1966), originally Quine (1941); and Bigelow (1988a).

or

$$x\ Q^2\ y \text{ if and only if } x\ R^3\ y,$$

and so on.

As Quine has noted, the first of these, $x\ Q^1\ y$ if and only if $x\ R^2\ y$, holds in the case where Q is the grandparent relation and R is the parent relation. If x and y are linked by the grandparent relation, they are linked in two steps by the parent relation. In contrast, the condition $x\ Q^1\ y$ if and only if $x\ R^3\ y$ is not satisfied where Q is the grandparent and R the parent relation. It is, however, satisfied where Q is the great-grandparent relation and R is the parent relation.

For present purposes, there are more important relations which satisfy the various conditions just illustrated. Let Q be the relation which holds between two individuals just when one is travelling 2 m/sec faster than another. Let R be the relation which holds when one is travelling 1 m/sec faster than the other. For this Q and R, the condition is satisfied that, for any x and y, $x\ Q^1\ y$ if and only if $x\ R^2\ y$. In a similar way, we can use "relative velocities" to instantiate any of the conditions of the form

$$x\ Q^n\ y \text{ if and only if } x\ R^m\ y.$$

These conditions require that *there be* enough relative velocities to serve as appropriate stepping-stones between one relative velocity and another. We trust that there are indeed enough relative velocities to serve. Whenever an object accelerates, it passes through continuum-many velocities, so all the velocities we need are in fact instantiated, we claim. Even if they are not instantiated in the actual world, that would not torpedo our theory. Parting company from one of the core doctrines of Armstrong's version of *a posteriori* realism about universals, we advocate a Platonist doctrine: that various physical properties and relations *exist* whether or not any individuals instantiate them in the actual world. (We face this issue again in our discussion of mathematics, in Chapter 8.) We thus assert that, whether or not anything actually has *had* all the required velocities, nevertheless *there are* enough relative velocities to make it the case that

$$x\ Q^n\ y \text{ if and only if } x\ R^m\ y$$

for appropriate relations of relative velocity, Q and R.

Non-Platonists may have more difficulty accepting the existence of all the stepping-stones required by conditions of this form. It is open to the faint-hearted to retreat here to a merely modal claim: that there *could be* sufficiently many things to yield the pattern required for the condition relating Q to R. However, we suggest that this modal fact does require some categorical basis for its truth. There has to be some genuine relation of proportion between Q and R in order for the modal condition to hold. Indeed, there has to be such a relation in order for Q and R to instantiate the required pattern *either* possibly *or* actually. We maintain that there is a genuine relation of proportion between universals such as those of velocity, and this relation is what *explains* the patterns which emerge of the form

$$x\ Q^n\ y \text{ if and only if } x\ R^m\ y.$$

Thus, we maintain, for any such condition there are relations of relative velocity – whether instantiated or not, actual or merely possible – which satisfy that condition.

Techniques such as these can be used to describe infinitely many relations among relations. These can appropriately be called *proportions* among relations. The definitions of all of these proportions employ straightforward generalizations of the standard definitions of such relations as coextensiveness, inverse, transitive closure, and so forth.

However, we are left with a problem when we try to bring this framework to bear on physical vectors like velocity. The definition of proportions applies to relations, not properties. Yet we have argued for the flux theory, according to which a vector, like velocity, is a property rather than a relation. To explain the nature of a vector, we need somehow to bridge the gap between *relations* which stand in definable proportions and the *properties* which we have argued vectors must be.

The bridge we need is generated by an important correlation between properties and relations. Whenever two individuals have the same property, this constitutes the holding of a relation between the individuals – the individuals match in a certain respect. Schematically, when Fx and Fy, then there is a relation, R_F say, such that $x\ R_F\ y$.

More generally, even when two individuals have distinct properties, this (at least sometimes) constitutes another sort of relation between them. When Fx and Gy, there exists a relation, R_{FG} say, such that $x\ R_{FG}\ y$. So we claim. For present purposes we do not need to rely on the fully general claim that any properties whatever, F and G, generate a relation R_{FG}. We need only rely on the claim that, when F and G are vectors of the same sort, they generate a relation R_{FG}.

Let us return to the spinning disc. Let us first restrict our attention to points whose velocity is in the same direction – points lying along the same radius of the disc. Each of these points has a different velocity. But because each has the velocity it does, there will be relations among them. The ones farther from the centre will be moving *faster* than those closer in. More specifically, ones farther out will be moving a determinate amount faster. One may be 1 m/sec faster than another. Yet another may be π m/sec faster. And so on.

In particular, we may compare each point on the radius with the point at the centre which is not moving at all. A point x with such-and-such velocity will be moving a specific amount faster than the central point y. Because the point x has the property it does, it will stand in a specific relation to y – it will be moving 'so-much faster than' y.

Thus, the properties of points on the radius, their velocities, generate relations of relative velocity. These relations stand in proportions to one another, defined according to the Frege–Whitehead–Wiener–Quine theory. Then the *properties,* velocities, stand in a proportion to one another if and only if the corresponding *relations,* relative velocities, stand in that proportion. For instance, the velocity we call 2 m/sec is twice that which we call 1 m/sec because the corresponding relation, '2 m/sec faster than', is twice the relation '1 m/sec faster than'. Hence, we can bring the Frege-to-Quine theory to bear on velocities in virtue of the natural correspondence between velocities and relative velocities.

Velocities have both magnitude and direction. We have focussed so far on differences in magnitude, keeping direction constant. We have made comparisons of this sort by drawing our subject-matter from the points along one given radius of a spinning disc.

Now let us consider differences in direction, keeping magnitude constant. Consider a collection of points all of which are the same distance from the centre of the spinning disc. These points have the same speed, but in different directions. In virtue of their different directions, they will stand in determinate relations. In particular, one will be 90° clockwise from another, others will be 45° anticlockwise from yet others, and so on. A relation of 'so many degrees clockwise from' will hold among velocities; but in consequence of this, a similar relation will hold among the things which have those velocities. For instance, points equally distant from the centre of the disc will be related to one another in distinctive ways: one will have a motion so many degrees clockwise from that of another. These relations stand in proportions to one another, by straightforward application of the Frege-to-Quine theory.

Let us arbitrarily select one of the points on the circle we are considering: its velocity will have some determinate direction, say east. Then each other velocity around the circle will be oriented some specific number of degrees clockwise from the designated velocity. Hence, each of the velocities around the circle can be paired with a specific relation. We can then say that two velocities stand in a determinate proportion with respect to direction. Such proportion holds between two velocities just in case that proportion holds between the two correlated relations. We can thus apply the Frege-to-Quine theory of proportions by associating each property with a relation and then defining proportions among such relations.

Let us take stock, then, of the relations we have described among velocities. Take two points x and y along the same radius of the spinning disc. Their velocities will stand in a specific proportion with regard to magnitude. In virtue of this, we may say that x and y are related in a distinctive way to one another – say, that the velocity of x has r times the magnitude of the velocity of y. Abbreviate this as

$$x \; P_r \; y.$$

Now compare y with a point z which is the same distance as y from the centre of the disc. There will be a distinctive relation between y and z in virtue of y's possession of a velocity so many degrees clockwise from that of z. Abbreviate this as

$$y \; P_\theta \; z.$$

Put together the two relations

$$x \; P_r \; y,$$
$$y \; P_\theta \; z,$$

and we obtain a derivative relation between x and z. Call this derivative relation P^*. We may define P^* by saying

$$x \; P^* \; z \text{ iff, for some } y, \; x \; P_r \; y \text{ and } y \; P_\theta \; z,$$

where P_r and P_θ are the specific magnitude and direction proportions already described.

On the spinning disc, any two points will be related by a "two-step" proportion of the same form as P^*. This is so in virtue of the intrinsic properties, the instantaneous velocities, of each of the points on the disc. These properties of instantaneous velocity count as *vectors* because they stand to one another in a family of two-step proportions of the form of P^*.

The two-step proportions displayed among velocities can be generalized to "n-step" proportions. This permits a physical grounding, along the lines of the flux theory, for vectors of a more general sort. Mathematically, a vector like the velocity of a point on a spinning disc can be represented by an ordered pair of real numbers. Generalizing, *all* ordered n-tuples of real numbers are said to represent vectors. These more general vectors can be physically grounded by generalizing from two-step relations of proportion to n-step relations of proportion. That is to say, *some* of these mathematical "vectors" may correspond to vectors in the sense required by the doctrine of flux. Not all mathematical n-tuples will correspond to vectors of the flux sort. Some may correspond only to the Ockhamist sort of second-order properties of things. Mathematically, it is useful to amalgamate several different flux vectors which characterize an individual and represent them as a single n-tuple which represents the individual's *state*. Usually these n-tuples will not correspond to any single flux property over and above the aggregate of separate flux properties which generated them. This will be useful for various purposes. However, in some cases physical theory might benefit from a postulation that there is a first-order flux property as well, which will then play a distinctive explanatory role in the theory.

Vectors and vector spaces, as abstract mathematical machinery, are exceedingly useful in physical science. The reason they are so useful is that physical objects have various intrinsic properties which are aptly called vectors – in a different sense than in abstract pure mathematics. Some mathematical vectors correspond to such intrinsic physical properties. Why is it appropriate to call these properties vectors? The answer is: because of the rich network of relations of proportion into which they are embedded. These properties are related to one another in these ways, and furthermore, it is essential to the natures of these properties that they enter into just those relationships. If two of these properties did not stand in the proportion in which they do stand to one another, they would not be the properties they are.

An explanatory science requires the attribution of such properties as instantaneous velocities, which are vectors. These should be recognized as genuine physical properties, not just as useful jargon or abstract mathematical structures. They are as independent of language and mind as are the moving bodies which instantiate them. Furthermore, these properties are related to one another in a very tight pattern. In virtue of this pattern of interrelations, they stand in *proportions* to one another. And that is why real numbers, and complex numbers too, are so useful in science.[21]

2.7 STRUCTURAL UNIVERSALS

We turn now, in a manner of speaking, from physics to chemistry. In physics, we deal with vectors; and vectors are universals which stand in a complex pattern of proportions to one another – compound proportions, as we have called them. In chemistry, too, we deal with universals, and these universals stand in complex patterns of relations to one another. Chemistry deals with the elements that make up Mendeleyev's Periodic Table and with the compounds, ions, and so forth which can be formed from the elements. Chemical compounds are structures which are formed from the elements. The property of *having* such a structure is a universal which is related in quite distinctive ways to the universals which determine the elements. It is a *structural* universal.

Any theory of universals worth its salt must be able to offer an account of structural universals. Structural universals are referred to by predicates such as 'being methane' or 'methane'. Methane molecules consist of a carbon atom and four hydrogen atoms bonded in a particular configuration. We assume that this is not just a contingent property of methane molecules, but an essential property. Something would not be a methane molecule if it did not contain carbon and hydrogen atoms in that configuration.

Methane molecules instantiate the universal methane. This universal is intrinsically related to three other universals: being hydrogen, being carbon, and being bonded (or simply hydrogen, carbon, and bonded). Necessarily, something instantiates methane if and

21 This section is based on Bigelow and Pargetter (1989b).

only if (roughly) it is divisible into five spatial parts c, h_1, h_2, h_3, h_4 such that c instantiates carbon, each of the h's instantiates hydrogen, and each of the c–h pairs instantiates bonded, and none of the h–h pairs instantiates bonded.

David Lewis provides an excellent survey of the reasons a realist about universals needs an account of structural universals.[22] We will not restate the case here. Suffice it to say that it is widely believed by those sympathetic to a realist theory of universals that an account of structural universals is required. We share both the sympathy and the belief. But, as we shall see in Chapter 4, we would like a theory of universals to provide a reductionist account of possible worlds and other possibilia,[23] an account which will provide an analysis of modal notions. For this reason, we aim for an account of structural universals free of such modal notions. Hence, we seek a theory which not only accounts for the relationships among the universals methane, hydrogen, carbon, and bonded, and which can, of course, be extended to other structural universals used in science and in our ordinary macroscopic account of the world, but at least in principle could be used in theories in which possible worlds are considered to be highly complex, structural universals. We will not further consider such world-property theories here, but only seek an account of structural universals which we will use in Chapter 4.

If any such theory is to work, it will have to be possible to give an account of structural universals and, indeed, to give an account which does not itself presuppose the very modalities which it aims to explain. Hence, it is crucial to our project that we find a satisfactory theory of structural universals. That is why we shall now focus, for an extended period, upon humble compounds like methane.

Given the intrinsic nature of methane, there is a complex pattern of entailments that requires explanation. If a molecule instantiates methane, it necessarily follows that there are parts of the molecule which instantiate the other universals hydrogen, carbon, bonded; also that there are four times the number of instantiations of hydrogen than of carbon; also that the bonded relation is instantiated four times by carbon–hydrogen pairs, never by hydrogen–hydrogen pairs, and that no hydrogen atom is bonded more than once; and so

22 Lewis (1986b).

23 So we are sympathetic to the approaches of Forrest (1986a), Bigelow (1988b), and Armstrong (1989).

on. With such a rich pattern of entailments to explain, the threat of modal circularity is great, and the chance of explaining them all using no more than the 'is part of' relation is slim.

There are two lessons we should learn from Lewis if we are to find an adequate theory of structural universals. We have imposed on ourselves the requirement that modal notions not be used. Lewis teaches us that mereology (the theory of the relations of parts to wholes) will be insufficient to the task of developing a structural universal from its constituent universals, and he also teaches us that, if we are to have an adequate theory, adequate for the purposes we have set ourselves, we must not resort to magic.

The first lesson: Suppose a structural universal is simply a mereological composite of its constituent universals. Methane, then, is nothing more than a mereological sum of hydrogen, carbon, and bonding. This at once gives us some of our required pattern of entailments. If a molecule instantiates methane, the universal methane is present; and if the universal is present, its mereological constituents will be present too. Hence, it would seem, whenever methane is present, hydrogen is also present. And that is, it seems, one of the entailments we wish to explain, one of the entailments which is constitutive of methane being a *structural* universal.

But the problem is that each methane molecule has four hydrogen atoms, and thus the universal methane must have the universal hydrogen as a part, in some sense, four times over. How can a universal have another universal as its part four times over? There cannot be four of the universal hydrogen. There is only one such universal. Either methane, as a universal, contains the universal hydrogen, or it does not. Likewise for the universal carbon and the relations of chemical bonding. Either these are parts of the universal methane, or they are not. Being universals, there is only one of each of them – each may have many *instances,* but there is only one *universal.* Clearly the universal methane does not contain the many instances of the universal hydrogen. As a universal, it is not a mereological sum of individuals. If it is to be construed as a mereological sum at all, it must be a mereological sum of the three universals hydrogen, carbon, and bonding. But there are only three of these; there are not four separate hydrogen *properties* or bonding *relations.* So as a mereological sum, methane can only be the sum of the three universals hydrogen, carbon, and bonding. And yet if we say that the universal methane has only three parts (unlike the molecules which instantiate it), then we cannot explain, for instance, the difference between methane and butane, for butane molecules are also made up of hydrogen (ten) and carbon (four) atoms bonded (thirteen times) in a particular configuration. Hence, at best the mereological picture of structural universals is incomplete, and it cannot provide an adequate basis for a theory of structural universals.

The second lesson: Two kinds of magic spring to mind in response to the failure of mereological composition. One is to postulate a kind of nonmereological composition, a sui generis mode whereby many distinct things can be made from exactly the same parts.[24] If the parts once combined are no longer present, we have something that cannot with any propriety be called composition, while if they are still there, it is incomprehensible how the one universal can be present many times over. (Peter Forrest's response to Lewis has generally been expressed in terms of some nonmereological mode of "composition"; and we wish to avoid that element in Forrest's theory. We also wish to be more scrupulous than Forrest in avoiding modalities. There is nevertheless a formal similarity between Forrest's theory and our own, as we shall see later.)

A second kind of magic would be invoked if we were to insist that structural universals, like methane, are not mereologically composite, but are simple, unstructured properties of individuals, having no parts at all. Methane is not composed of hydrogen, carbon, and bonding. At most, what we mean when we talk loosely of methane having constituent parts is that an instance of the universal methane necessarily contains instances of the other universals carbon, hydrogen, and bonding. (Things are, in fact, messier than this, for the instances must occur in a *certain way.* For instance, butane and isobutane contain the same number of instances of the same simple universals, but differ in structure because of the manner in which these universals are combined.)

But why should these entailments hold? Why should there be a necessary connection between methane being instantiated and carbon being instantiated if the two universals are wholly distinct (unstructured) individuals? This is magic. Lewis rightly insists that it is unacceptable (magic) to insist that entailments hold among

24 This has been the approach of Forrest (1986b) and Armstrong (1986).

universals, even though there is nothing in the natures of these universals themselves to explain why these entailments should hold.

If we allow either kind of magic, we could arrive at our complex pattern of entailments. Instances of methane will have instances of carbon and the right number of instances of hydrogen and bonding. Butane and isobutane will need a little more work, since these have the same number of instances of these universals, but with some instances of some extra universals added, undoubtedly the job can be done.

What is wrong with magic? Magic is used as an alternative to an explanation of the complex pattern of entailments. We seek a theory of structural universals that explains this pattern, but a theory involving magic *does nothing more* than say that the entailments are there. To list a set of entailments is not to explain them. Explanation must, of course, stop somewhere. As we shall acknowledge later, everyone needs a little magic somewhere. There is a sense in which even mereology is magical. The trouble with the acquiescence in unexplained entailments is that the magic comes too soon and is not sufficiently general. When the magic is invoked in one special case and does no work anywhere else, it is worth trying to eliminate it.

Let us compare our problem of structural universals with the problem of quantities. With quantities, too, there will be a complex pattern of entailments. Necessarily, an object with a specified determinate mass also has the determinable mass, while anything with the determinable mass must have some specific determinate mass. And the full pattern of relationships and entailments is far more complex than this.

We have attempted to make explanatory sense of this complex pattern of entailments by what we have called the three-level theory of quantities. At level (1) we have our population of individuals. At level (2) we have the specific determinate relationships (universals) among individuals. At level (3) we have relations, called proportions, which hold among these determinate relationships.

This theory of quantities can be a useful blueprint for dealing with the problems associated with structural universals. We have defended this as a theory of quantities, largely because it will account for the complex of entailments and relationships that are noted with quantities like mass and relative velocity. If we are right that it will do the job, it is to be hoped that an analogous account might do a similar job for structural universals.

Let us consider again the three metaphysical levels within the theory of universals. At level (1) we have material individuals which *have* the property of being butane or being methane or being isobutane, and so on. These are then methane molecules or butane molecules, and so on. These individuals, of course, have parts with different properties and relations.

At level (2) we have the properties and relations of these material individuals. Being hydrogen, being bonded, being carbon, being bonded to a carbon atom, and being methane are clearly properties of individuals. Some of these may be so-called first-order properties and relations which are taken to be possessed *simpliciter,* that is, not in virtue of the possession of any other property or relation. But some of them may be *second-order* properties or relations of individuals – ones that an individual has in virtue of having some other properties or relations.

At level (3) we have the relations (or proportions) which hold among the properties or relations of individuals, be they first- or second-order properties of those individuals. An example is furnished by the relation 'having the same number of instances as', which Frege employed in defining cardinal numbers.[25] This relation among properties is one of a family, including 'having twice as many instances as', 'having four times as many instances as', and so on.

These "numerical" proportions will also hold among more complex level (2) properties and relations. Consider the conjunctive properties 'being hydrogen and being part of this molecule' and 'being carbon and being part of this molecule'. If the molecule in question is methane, these two properties stand in a relationship which is characterized by the proportion 4:1.

Within such a framework we can characterize the structural universal methane as a relational property of an object, or more particularly of a molecule. It relates the molecule to various *properties*. These properties are being carbon, being hydrogen, being bonded. Being methane, then, is to be identified with a highly conjunctive *second-order* relational property of an individual (molecule): the property of having a part which has the property of being hydrogen,

25 Frege (1893).

and having a part which is distinct from the first part which has the property of being hydrogen, and . . .

The property of being methane clearly stands in a pattern of internal relations of proportion to other properties: being hydrogen, being bonded, being carbon, and so on. These properties are not "constitutive" in the mereological sense. Nevertheless, it is of the nature of methane that it stands in these relations to these properties. Standing in these relations is an *essential* characteristic of methane. These relations are then what we call *internal* relations. The related terms could not exist without standing in those relations.[26]

The entailment relationships constitutive of structural universals, like methane, should, we suggest, be explained not by mereological composition, but by reference to these essential or internal relations among properties and relations. Thus, being methane is so related to being carbon that being methane cannot exist without standing in that relation to being carbon.

This theory does not make use of either of the kinds of magic we rejected earlier, and yet the theory explains the complex pattern of relationships and entailments characteristic of structural universals. Hence, the theory answers the challenge presented by Lewis to structural universals.

Our theory rests on the reality of internal relations of proportion holding among properties and relations. Some find such relations essentially mysterious. It may also legitimately be questioned whether modal magic has been smuggled into our theory by the use of these essential, level (3) relations.

But we feel that the charge of magic is inappropriate. The internal relations of proportion are universals, and in no way more mysterious than universals in general. Universals can share properties and can stand in relations to one another. Proportions are just such relations. They do useful work of many sorts and are not wheeled in simply to save structural universals.

It might be urged that the objection is not just that they are second-degree relations, but that these relations of proportion are "internal", that is, essential to the identities of the terms they relate.

26 We use the term 'internal relation' to characterize a relation which is essential to the relata. There is another use of this term, meaning 'supervenient on the intrinsic properties of the relata'. On this interpretation, similarity is an internal relation, whereas on our usage it is not.

The relation between the relata is one of the things which make up the essences of those relata. The universals which stand in such relations could not exist without standing in those relations; if they did not stand in those relations, they would not be those things.

The case for internal relations (in our sense) among individuals is strong. The case is strong both for essential relations among universals and for essential relations among individuals. Spatial and temporal relations between regions of space and intervals of time, for instance, can plausibly be construed only as essential or internal relations. These regions or intervals would not be the regions or intervals they are if they were not related to one another in the ways they are. And once we have internal relations among individuals, should we hesitate to accept them among properties of individuals and relations among individuals? Now we can add an additional argument. If we are to have a viable theory of universals, we need structural ones. We offer a plausible account of structural universals, an account requiring internal relations of proportion. Hence, if we are to have a viable theory of universals, we should accept such internal relations.

Forrest's theory of structural universals has strong formal analogies with our appeal to higher-order relations. Forrest speaks of a quasi-mereological n-place *operation* which takes n quasi-parts and "combines" them into a quasi-whole:

$$\text{Operation } (a_1, \ldots, a_n) = a_{n+1}.$$

However, this is equivalent to there being an $(n + 1)$-place *relation:*

$$R(a_1, \ldots, a_n, a_{n+1}).$$

Where Forrest speaks of an n-place operation, we speak of an $(n + 1)$-place relation among universals. Our disagreement with Forrest lies principally with his construal of this operation/relation as quasi-mereological.

A further objection may be advanced. We have undertaken to explain entailments involving properties at the first-order level, and we have done so by appealing to higher-order properties and relations. Yet it may be objected that we have, in fact, presupposed entailments among these higher-order properties and relations. Hence, our theory embodies a vicious circularity, or regress, explaining one class of entailments by another class of entailments which are equally problematic, in the same way, and for the same reasons.

For instance, we argue that there is an essential relation between being methane and being carbon:

Necessarily, (being methane) R (being carbon).

It is because they stand in that relation, that there is an entailment between something's being methane and that thing's containing a part which is carbon:

(I) Necessarily, for any F and G, if (F) R (G), then any instance of F has a part which is an instance of G.

Principle (I) involves necessity. This necessity is to be grounded in the essences of the universals, thereby (it is hoped) avoiding unexplained modal primitives. And yet isn't any appeal to essences an appeal to a kind of modal magic?

Our answer is yes and no. It is "white magic", but not the "black magic" that we have sought to keep out of an adequate theory of structural universals. There is a sense in which we all need a little magic in our lives. What varies is only our judgements about which is black and which is white.

Consider the familiar part–whole relation: the tail is part of the donkey. Most of us feel no pressing need to offer an analysis of what this claim amounts to. If the part–whole relation were to be used to explain structural universals, entailments, and so on, most of us (including Lewis) would not object to the use of that relation as "magical". Yet in a sense it is. Suppose that entailment were to be explained in terms of the part–whole relation. Suppose F's entailing G were to be explained in terms of the universal F's containing the universal G as a part.[27] There may be various objections to such a theory. But the objection that it appeals to something magical is misplaced. Yet a construal of entailment in terms of a part–whole relation *will* entail a principle of the same form as the allegedly magical principle (I), namely,

(II) Necessarily, for any universals F and G, if being G is part of what it is to be F, then any instances of F will be instances of G.

This principle has exactly the same form as the one enunciated earlier, (I), concerning the relations between being methane and

27 See Bigelow (1988b) for a defence of a mereological theory of entailment.

being carbon. If principle (I) is magical, then surely so is (II). It is perhaps a whiter magic, but it is magic all the same. We take it to be white magic, though there are some who would not. What you take to be white magic will be whatever you are comfortable to take as a primitive.

We cannot escape magic altogether. Even nominalists require magic of their own. The cost of many versions of nominalism includes the acceptance of unexplained, primitive modalities. That is magic; and we take it to be black magic. Even if a nominalist avoids primitive modalities, as Lewis succeeds in doing through his concrete modal realism, there will nevertheless be a residual need for distinctions between so-called natural classes and heterogeneous classes of individuals. That distinction is magic and, to us, a black magic that a good theory of universals avoids.

Our theory explains structural universals by appeal to a network of higher-order universals. The relationships among higher-order universals have not yet been presented in a way which avoids all (implicit) appeal to modalities. But we believe that the modalities involved are similar to those which rest directly on the part–whole relation. Ultimately they may be reducible to the part–whole relation, a kind of special mereology for universals which, at least for most of us, would make them as white as magic can be. Lewis has shown that structural universals are not built out of simpler universals in a straightforward way; but he has not shown that they cannot be explained using a part–whole relation on higher-order universals. Yet even if reduction to the part–whole relation fails, we maintain that the residual magic in our theory is "pretty bloody white". The unexplained elements in our theory are not mysterious in the very same way as the things they purport to explain.

Our hope is that the residual modality might be resolved once we take account of the two levels of a quantification essentially involved in an adequate theory of universals. There is first-order quantification over "somethings", where those things include both individuals and universals, and there is second-order quantification over "somehows" that these things stand to one another. It seems likely that, once we sort out the ramifications of these two levels of quantification, we will have a powerful enough theory to explain the modalities remaining in a theory of structural universals. But this is a matter we shall return to later, in Section 3.7.

The hierarchy of universals, then, can explain a great deal: quantities, vectors, and structural universals. In Chapter 4 it will be our task to bring the hierarchy of universals to bear on the problem of explaining modality. But first we must sort out the rich modal language that we need to express the modal notions that are essential for science. Before facing modal facts, we shall marshal a modal language. With the help of a regimented team of modal truth-bearers, we will then be well equipped to explore the realms of modal truthmakers.[28]

[28] This section is based on Bigelow and Pargetter (1989c), which benefited from comments by David Armstrong, Peter Forrest, and particularly David Lewis.

3

Modal language and reality

3.1 RULES AND REFERENCE

What is the network of properties and relations in virtue of which a word refers to a thing? What turns one object in the world into a representation of other objects? We shall say very little about this matter: it would take us too far into cognitive science and away from the ontological concerns which are the topic of this book. Our sympathies are with a broadly causal, or functionalist, account of the nature of representation and reference. We shall simply assume that there is such a thing as reference, to be elucidated sooner or later, but not here.

Taking reference for granted, we then ask: what things in the world *are* the referents of words in language? More specifically, what are the referents of the symbols used in science? Having defended scientific, metaphysical, and modal realism, the answer a realist must give is determined. Clearly, the referents of scientific language include individuals, universals, and possibilia. Included among universals are higher-level properties and relations, like quantities and vectors and structural universals, relations of proportion (real numbers), and – let us not forget – *sets*. Sets are especially important in semantics, for reasons which we shall explain.

A crucial question for semantics is: what, if anything, do predicates correspond to? Given a true sentence, with a name in it which refers to an individual, say, we may divide the sentence into two parts: the name and the rest. We may call the rest a predicate. Given that the name refers to an individual, what does the rest refer to?

Our answer is complex. In the first place, we say that there need not be *anything* corresponding to the predicate. That is, there need not be any *thing,* there need not be anything which can be named or quantified over in a first-order language. There will be *somehow* that the predicate represents the individual as being. But this *somehow* need not correspond to *something* which is related to the individual. This *somehow* requires a higher-order quantifier; it cannot be expressed in a first-order language. We shall address this at greater length later, when we outline the logic of higher-order quantification.

Thus, corresponding to a predicate there will be somehow that an individual must be, if the predicate applies to it, but there *need not* be something which is related to that individual. At least sometimes, a somehow will correspond to no something whatever.[1]

Nevertheless, in the case of *many* predicates there *is* something, or several things, which are referred to, which must exist if the predicate is to apply to a thing. Very often, a predicate will apply to a thing only if *there is* something which it has in common with other things. Predicates do *very often* correspond to universals. And the better our science becomes, the more often will our predicates correspond to the universals – the physical quantities, relations, and so forth – existing in nature.

For the purposes of semantics, however, it is a nuisance if we cannot be sure that a given predicate corresponds to a universal. It would be frustrating if we could not begin doing semantics until we had worked out an *a posteriori* realist theory of universals. A great many predicates apply to a range of things for which it is very doubtful whether there is any *one* property common to all and only those things. Think how varied dogs are, for instance, and how much some dogs resemble wolves, others resemble jackals or coyotes, and others resemble foxes. Is there really some one property common to all and only the things we call dogs? It is plausible that a

1 We noted the difference between somehow and something in Chapter 2, where we noted that instantiation is not a universal of the same sort as others. Instantiation is somehow that particulars stand to universals, not something which stands somehow to various pairs of particular and universal. We noted a precedent in Frege's distinction between object and concept.

We now add further precedents: Strawson (1959) spoke of a "nonrelational tie" between particular and universal; Mackie (1973) granted predicates some existential force, without reification of what they signify; and Armstrong (1978) defended what he calls nonrelational immanent realism and asserted that the qualification 'nonrelational' signifies that instantiation is not a universal in the same sense as those which "have instances". Strawson, Mackie, and Armstrong are all saying that some predicates, at least, correspond to no "thing". We think they were on the right track, and higher-order quantification provides the best articulation of what they are apprehending.

predicate, like 'is a dog', does not correspond to any one universal: at least, not if we restrict our attention to universals low in the type-theoretical hierarchy.

If, however, we look higher in the type-theoretical hierarchy of universals, and if we find sets there, it is much easier to be confident that most predicates will refer to, or correspond to, *something*. Even if, for instance, dogginess is not a simple, low-level universal, there is still the set of dogs. We can presume that *most* predicates correspond *at least* to a set.

Even if we believe in sets, we still cannot assume that every predicate corresponds to a set. The predicates 'is a member of' and 'instantiates', for instance, do not refer to any set. There is no set X containing all and only the pairs $\langle x, y\rangle$ for which x is a member of y. Set-theoretical paradoxes show that there can be no such set. Similarly for the predicate 'is a set': set-theoretical paradoxes show that there can be no set containing all and only sets. The predicate 'is a set' does not correspond to a set containing all and only the things it applies to. There is *somehow* that all sets are, but that does not entail that there is *something* to which they all belong.

There is no guarantee that a predicate corresponds to a set, or universal, which exhausts its extension. Nevertheless, it streamlines semantics if we can assign a referent to all but a very few predicates. For this reason, set theory is a windfall for semantics. Set theory emerges from mathematics, not from semantics; yet it turns out to have important applications in semantics. The usefulness of set theory in semantics, in fact, provides reinforcing reasons for accepting set theory.

Set theory streamlines semantics because it permits us to rest a heavier explanatory burden on reference in formulating a correspondence theory of truth. Reference plays a central role, but there is room for considerable flexibility about exactly what role it should play.

Truths can be expressed by sentences. In a sentence, some of the terms refer to *things* in the world. In addition, there is *somehow* that those terms are related to one another, syntactically, to constitute a sentence. For instance, the referential terms may occur one right after the other in time (if spoken), or one right above the other (as in Chinese writing), or one on either side of such-and-such an inscription; or any one of an indefinite number of other relationships may hold among the referential terms in a sentence. A sentence will be true just in case (1) the referential terms are related to one another somehow; and (2) the things referred to by those terms are related to one another somehow; and (3) there is a general rule (of correspondence) relating how the terms stand to one another and how their referents stand to one another.[2]

It follows from this that one of the fundamental distinctions in semantics will be that between

A. an assignment of referents to basic terms of the language and
B. the specification of componential rules for determining the referent of a compound expression, from the referents of its constituents and their syntactic relations together with the corresponding relations among the referents of those constituents.

The line between assignment of referents and specification of componential rules can be drawn in different places. It is possible to include more things as referents for expressions and thereby to simplify the componential rules; and it is also possible to minimize the number of things we admit as referents, provided that we compensate by increasing the complexity of the componential rules.

Tarski had strongly nominalistic leanings; and the dominant tendency in semantics has from the start been one of minimizing the number of referents assigned to expressions.[3] Referents are assigned to *logically proper names* – to individual constants and individual variables. Reluctantly, reference has been extended to predicates – and predicates have been semantically paired with sets. More generally, any *open sentence* is paired with at least one set.

Individual constants, predicates, and open sentences, then, are to be assigned referents. But other symbols, standardly, are not. The negation sign, the disjunction and conjunction signs, and the quantifiers $(\forall x)$ and $(\exists x)$, for instance, are not assigned any referents. Their semantic input is fed into the componential rules. For instance, consider the negation sign $\sim$. We do not assign any referent to $\sim$. Rather, we lay down a componential rule which specifies that when $\sim$ is placed in front of a sentence, the truth value of the compound sentence is the opposite of the truth value of the component sentence:

2 Here we echo Wittgenstein (1922) and his picture theory of meaning, but we are not harking back to a simplistic and superseded idea. The core of the so-called picture theory persists through mainstream formal semantics, from Tarski (1956) through recent theories as diverse as those of Cresswell (1973) and Field (1972).
3 See especially Tarski (1956).

If α is any sentence, then $\sim\alpha$ is true if α is false, and $\sim\alpha$ is false if α is true.

Disjunction, conjunction, quantifiers, and so forth are treated similarly. They are not assigned a referent, but rather a separate componential rule is stated for each.

When philosophers consider various extensions of the predicate calculus, they usually leave the minimal referential apparatus unaltered and add extra componential rules. The paradigm case of this is provided by the introduction of the modal operators 'necessarily' and 'possibly', as we shall see. These are not taken to refer to anything; rather, they are accompanied by a clause which specifies the way of determining the referent (truth value) of 'necessarily α' from the referent of α, whatever sentence α may be.

It is worth asking whether there are alternative ways of dividing the semantic labour. Consider first the attempt to minimize the assignment of referents. Standard Tarskian semantics is already very grudging in the range of symbols which are treated referentially. Could it be more grudging still? In particular, could we avoid assigning referents to predicates and open sentences, provided that we made compensating adjustments to the componential rules? The answer is that we could. In fact, we could go further and take away reference altogether, packing it all, in disguised form, into the componential rules.[4]

However, one wonders what point there would be in doing so. The things in the world are there, whether or not we manage to avoid explicitly mentioning them when doing semantics. If your only reason for believing in certain things were that you could not do semantics without explicitly mentioning them, reference-free semantics would free you from the need to believe in anything at all. But if you already believe in things such as sets for other reasons, there is no advantage in eliminating them, at the cost of complicating the componential rules.

Furthermore, it may be suspected that the nominalist, in attempting to remove sets from semantics, is really just disguising them rather than removing them. Sets may no longer be assigned as the referents of predicates. Yet new componential rules are introduced which have the same effect in the long run. It may be suspected that the referents of predicates, sets, are really hidden in the nominalistic componential rules. Instead of blatantly referring to sets, in subject

4 For one extreme version of such a semantic theory, see Bigelow (1981).

position, when doing semantics, the nominalist takes care to construct a predicate which wraps up reference to a set with an extra, predicating function. Instead of a compound expression, like

refers to the set of rabbits,

for instance, they may take care to say,

applies to all rabbits, and only rabbits.

The latter expression does not explicitly mention sets; so nominalists feel happier. Yet the existence of their purportedly set-free paraphrase does not prove that semantics can be done without sets; for it remains to be shown whether their paraphrase is genuinely set-free, as they think, or merely disguised talk of sets. For instance, the nominalistic reformulation would obviously be no real ontological economy if

applies to all and only rabbits

were merely an abbreviation of

refers to a set containing all and only rabbits.

The meaning of 'applies to' must be independently debated before the mere existence of a paraphrase can prove anything ontological.

The decision whether there are sets, therefore, does not rest on the choice of a preferred way of doing semantics. If, for nonsemantic reasons, you do not believe in sets, you must find a set-free way of doing semantics. If you believe in sets, for nonsemantic reasons, you may suspect that even the nominalist's allegedly set-free semantics merely disguises reference to sets.

Yet the attempt to minimize referents is not the only project one could pursue. It is also possible to attempt to simplify the componential rules, even at the cost of assigning referents to a greater number of symbols. This project is taken up by the development of λ-categorical languages, along the lines initiated by Montague grammar and clearly developed in Cresswell's *Logics and Languages* and many later publications.[5]

To illustrate the overall strategy, let us consider again the semantics of negation. It is possible to eliminate any special componential rule for negation and to institute instead a single, extremely general

5 See especially Montague (1974) and Cresswell (1973, 1985a,b).

componential rule. The cost of this simplification of the rules is an addition to the referential layer of the theory. Counterintuitive as this may seem at first, the negation symbol can be assigned a referent. Let us compare the standard componential rule for negation, given above, with an alternative semantics which assigns a referent to the negation symbol.

In an alternative semantic scheme, we may assign to the negation symbol a specific set-theoretical function, a function which maps the truth value 'true' onto the truth value 'false' and which maps the truth value 'false' onto the truth value 'true'. For any symbol α, let $\mathcal{V}(\alpha)$ be the *semantic value,* or referent, of α. If $\alpha = \sim$, we can explain the semantic value of α, $\mathcal{V}(\alpha)$, as follows. Let 1 be the truth value 'true', and 0 be the truth value 'false'. Then we may define the semantic value of negation, $\mathcal{V}(\alpha)$, to be that function $\omega_\sim$ such that

$$\omega_\sim(1) = 0, \qquad \omega_\sim(0) = 1.$$

Now consider a compound expression, $\sim\beta$, where β is some sentence. We assign to $\sim$ the value $\mathcal{V}(\sim) = \omega_\sim$, and we assign to β a truth value $\mathcal{V}(\beta)$, where we have either that $\mathcal{V}(\beta) = 1$ or that $\mathcal{V}(\beta) = 0$. We can then obtain the truth value of the compound expression $\sim\beta$ from the referents of its components, by letting the function which is the referent of $\sim$ take as an argument the truth value which is the referent of β. We can say that

$$\mathcal{V}(\sim\beta) = \mathcal{V}(\sim)(\mathcal{V}(\beta)).$$

This componential rule is a special instance of a single, very general rule. To see this, consider next the case of conjunction. Take $\wedge$ to be the symbol for conjunction. The referent for this symbol can be defined to be a two-place function,

$$\mathcal{V}(\wedge) = \omega_\wedge,$$

such that

$$\begin{aligned} \omega_\wedge(1, 1) &= 1, \\ \omega_\wedge(1, 0) &= 0, \\ \omega_\wedge(0, 1) &= 0, \\ \omega_\wedge(0, 0) &= 0. \end{aligned}$$

Then we can determine the truth value of the conjunction

$$\beta \wedge \gamma$$

from the references of β, γ, and $\wedge$ by the rule

$$\mathcal{V}(\beta \wedge \gamma) = \mathcal{V}(\wedge)(\mathcal{V}(\beta), \mathcal{V}(\gamma)).$$

This rule for $\mathcal{V}(\beta \wedge \gamma)$ has the same form as that for $\mathcal{V}(\sim\beta)$. The referent of one of the symbols is taken to be a function, and the referents of the other symbols are taken to be its arguments. The general form of such rules is this: Any compound expression α can be taken to be a sequence of component expressions:

$$\alpha = \langle \delta, \beta_1, \beta_2, \ldots, \beta_n \rangle.$$

One of these components, δ, is singled out as the *operator;* the others are its *arguments*. The general, componential rule for any such expression is

$$\mathcal{V}(\alpha) = \mathcal{V}(\delta)(\mathcal{V}(\beta_1), \mathcal{V}(\beta_2), \ldots, \mathcal{V}(\beta_n)).$$

That is, the referent (or semantic value) of the compound α is obtained by applying the referent of the operator δ to the referents of the arguments β_1, β_2, . . . , β_n.

This one general rule is not quite the only componential rule required. In λ-categorical languages, there is one other very general rule, a rule of λ-abstraction. We will not give the details here. Yet the fact remains that in λ-categorical semantics there are very few componential rules. The minimization of rules has been achieved by expanding the range of symbols to which we assign referents.

For a realist, this sort of semantics can be defended as the most ontologically honest. Its ontological commitments are laid bare, and not disguised as mere componential rules which are presumed without argument to be ontologically innocent.

Yet a realist need not insist on doing semantics one way rather than another. Suppose we do adopt the orthodox format in which, for instance, negation is assigned its own componential rule rather than a referent. No harm is done, from a realist perspective. Neglecting to mention explicitly a truth function when doing the semantics of negation is not to be construed as denying its existence. And indeed, the realist can expect to find sentences which need semantic analysis and which explicitly mention truth functions in subject position. And here, the realist will wheel in truth functions as referents, even if the semantic framework avoids assigning these as the referents of the negation symbol.

When we turn to the semantics of modal claims, the same choice of semantic frameworks must be faced. At one extreme, we can follow the lead of Montague and assign referents to modal operators like 'necessarily' and 'possibly'. We could tailor our semantic theory so that it assigns to each such operator a specific function. Conversely, we could tailor our theory in such a way that we bury all explicit references under a tangle of componential rules which allegedly allow us to remain neutral on ontological issues. Our strategy will be an intermediate, moderate, and orthodox one. We shall describe the semantics for modal claims in a way which explicitly assigns referents only to names (individual constants and variables) and open sentences.

On this strategy, the semantics for modality will augment orthodox Tarskian semantics in two ways. The initial Tarskian presumption has been that we should begin with a restricted domain of entities which we can refer to by names (or, in logic, individual constants) and pronouns (or, in logic, individual variables) – a domain consisting only of individuals which can be located in our space and time and, furthermore, which occupy only one region of space at any one time. Predicates, then, refer to sets of these individuals, or to sets of sets.

We shall argue that the domain for nouns or pronouns should be taken to include more than this. It should include not only individuals with a single location in our space and time, but also *possibilia:* things which are not actually located anywhere, ever, but *could have been*. Furthermore, the domain should include what we have called *universals*. (In the end, we shall argue that possibilia are universals.) And then, once nouns and pronouns are interpreted as ranging over universals and possibilia, we may interpret predicates as referring to sets constructed out of these.

Having expanded the range of referents for nouns, pronouns, and predicates, we turn our attention to other symbols, ones which introduce explicit *modal* qualifications into sentences: symbols meaning 'necessarily', 'possibly', 'probably', and so forth. We will not introduce a new range of referents for these terms. Rather, we will introduce a new componential rule for each such symbol. These componential rules will determine the referents of complex expressions containing the symbol, given the referents of the expressions with which it has been combined.

The resulting structure is the orthodox semantics for modal logic. It is generated by expanding Tarski's classical semantics in two ways. First, the range of referents for individual constants and predicates is expanded to include possibilia. Second, a new componential rule is added for each modal operator.

Our justification for formulating the semantics in this way will not be based on the idea that this is the only way of doing modal semantics. So we will not be arguing for the existence of possibilia entirely on semantic grounds. Rather, as we have already argued, there are good nonsemantic grounds for believing in possibilia. Given the existence of possibilia, we are free to formulate semantics in a way which explicitly draws on them and puts them to work. Having put them to work in semantics, however, we can then turn the tables and claim that the usefulness of possibilia in semantics provides another good reason for believing in their existence.

3.2 MODAL AXIOMATICS

We shall sketch a fairly orthodox axiomatic system. To fix notation and delimit scope, we list the basic symbols of the language:

Individual constants, variables, and predicates
Logical constants, including truth-functional operators:

$\sim$	('not')
$\wedge$	('and')
$\vee$	('or')
$\supset$	('if–then', or material implication)
$\equiv$	('if and only if')

Quantifiers:

$\exists$	('There is a . . .' or 'For some . . .')
$\forall$	('For every . . .' or 'For all . . .')

Modal operators:

$\Box$	('It is necessary that . . .')
$\Diamond$	('It is possible that . . .')

Subjunctive conditional operators:

$\Box\!\!\rightarrow$	('If it were the case that . . . then it would be the case that . . .')
$\Diamond\!\!\rightarrow$	('If it were the case that . . . then it might be the case that . . .')

Using these basic symbols, we can build sentences in the usual manner, by the usual formation rules. We shall also employ standard conventions for dropping some of the parentheses in a formula, when this facilitates comprehension.

Having generated the sentences of this language, we may ask which of these sentences are true. One project we can pursue is the study of *axiomatic theories* formulated in this language. We may ask which sentences could be taken as *axioms,* which *rules of inference* could be allowed, and which *theorems* could be proved from those axioms and rules of inference.

Given ingenuity, a sense of adventure, the need to discover something new in order to establish a reputation, and time, logicians are sure to breed an endless stream of new axiomatic systems and deviant logics, and to discover unexpected uses for many of them. It is not to be expected that there will be any consensus about which sentences of our language are true or about which are the appropriate rules of inference. Nevertheless, it is worth singling out one set of axioms as historically the most salient – the axioms of classical logic, descending to us through Russell and Whitehead's *Principia Mathematica.*[6]

First, we note that some of the things we could state as *axioms* are normally stated, rather, as *definitions.* Given the symbols of negation, $\sim$, and disjunction, $\vee$, as undefined or *primitive,* we define all the other truth-functional connectives. For any sentences α and β,

D1. $(\alpha \wedge \beta) = \text{df} \sim(\sim\alpha \vee \sim\beta)$,
D2. $(\alpha \supset \beta) = \text{df} (\sim\alpha \vee \beta)$,
D3. $(\alpha \equiv \beta) = \text{df} ((\alpha \supset \beta) \wedge (\beta \supset \alpha))$.

Similarly, given $\exists$ as primitive, we define $\forall$. For any sentence α and any variable x,

D4. $\forall x\alpha = \text{df} \sim\exists x\sim\alpha$.

Given $\Diamond$ we define $\Box$:

D5. $\Box\alpha = \text{df} \sim \Diamond \sim\alpha$.

We define $\Diamond\!\!\rightarrow$ in terms of $\Box\!\!\rightarrow$, in the manner Lewis suggested:

D6. $(\alpha \Diamond\!\!\rightarrow \beta) = \text{df} \sim(\alpha \Box\!\!\rightarrow \sim\beta)$.

6 Whitehead and Russell (1910).

After these definitions come four axioms which relate solely to the truth-functional operators. Let *Fa, Fb,* and *Fc* be three atomic sentences; then the following sentences can be set down as axioms:

A1. $(Fa \vee Fa) \supset Fa$,
A2. $Fb \supset (Fa \vee Fb)$,
A3. $(Fa \vee Fb) \supset (Fb \vee Fa)$,
A4. $(Fb \supset Fc) \supset ((Fa \vee Fb) \supset (Fa \vee Fc))$.

(These are descendants of the original axioms of Russell and White-head.)

To these axioms we add two transformation rules, or rules of inference:

R1. Rule of substitution: The result of uniformly replacing any atomic sentence in a theorem by any given sentence will also be a theorem.
R2. Modus ponens, or rule of detachment: If $(\alpha \supset \beta)$ and α are theorems, then β is a theorem.

(By "uniformly" replacing an atomic sentence, we mean replacing *every* occurrence of that atomic sentence, within a given sentence, by the *same* sentence.)

Next, we add two axioms for the quantifiers:

A5. If β is a sentence with an individual constant or free variable, and $\forall x\alpha$ generalizes β by uniformly replacing that constant or variable in β by x and then binding it with $\forall x$, then

$$\forall x(\alpha \supset \beta)$$

is an axiom.

A6. Any sentence of the form

$$\forall x(\alpha \supset \beta) \supset (\alpha \supset \forall x\beta)$$

is an axiom, provided that α contains no occurrences of x.

(To A5 and A6 we add that uniformly replacing the variable x by any other variable will also yield an axiom.)

To these we add a further rule of inference:

R3. Universal generalization: If α is a theorem, so is $\forall x\alpha$; and so is the result of replacing x by any other variable.

This completes the axioms and inferential rules relevant to the truth-functional operators and the quantifiers. Given these axioms and rules, we can derive a class of *theorems*. The set of all sentences in our language partitions into two subsets: the set of theorems and the set of nontheorems.

For many purposes, the complications introduced by quantifiers are irrelevant. Hence, it is also useful to distinguish another subset of the set of all sentences. We can define the set of all theorems which can be derived by axioms A1–A4 and rules R1–R2 alone. This will be called the *propositional fragment* of the full quantificational theory. This fragment is more fully understood than the full quantificational theory. Some central questions which have been answered for the propositional fragment remain open for the full quantificational system.[7] For this reason it is easier to restrict many of our claims to the propositional fragment. Some of these claims can be extended to the full system, but others have been proved, as yet, only for the propositional fragment.

Quantificational theory (or just its propositional fragment) can now be enlarged by adding axioms and rules of inference which relate to the modal symbols $\Box$ and $\Diamond$. The axioms and rules given so far have been the classical ones; and although they have variants and rivals, they do have an outstanding salience over others. When we consider axioms governing $\Box$ and $\Diamond$, there is no longer any single list of axioms that stand out as the most salient.

This is due partly to the fact that we lack full knowledge about the nature of necessity. But there is another reason too. There is a large variety of quite distinct senses of necessity. Axioms which hold of one sort of necessity may fail to hold of another. Imagine that you are in Melbourne. Consider the question of whether you could get from Melbourne to London in twenty-five hours, starting right *now.* How do you interpret it? You could reach for the telephone to find out when the next flight leaves Tullamarine. You could check your bank balance to see if you could afford it. You could reflect on whether you could bring yourself to do such a thing. And so forth. You might describe these various responses to the question by saying that they are considerations of whether it is *practically* or *financially* or *psychologically* possible for you to get to London in twenty-five hours, starting now.

If asked whether you could get to London in ten *minutes,* however, it would be unnecessary to worry about whether it is psychologically or financially or practically possible. It is clearly *technologically* impossible at the present time. And could you get to London in *thirtieth of a second?* Given the distance from Melbourne to Lon-

7 See Hughes and Cresswell (1984, esp. chap. 6).

don and the speed of light, it is *physically impossible* to get to London that fast. The laws of physics rule it out, without need for extra facts about available technology and so forth. If we ask whether it is possible to leave Melbourne now and get to London *ten years ago,* it may then be debatable whether this is not just physically, but *logically,* impossible. Some people think that it is ruled out not just by the laws of physics, but by the law of noncontradiction alone, without help from anything beyond logic.

The permissible senses of necessity and possibility can be extended further afield to include even such things as *legal* possibility, *moral* necessity, and so on. And there are other concepts which are not comfortably expressed by 'necessarily' or 'possibly' but which are nevertheless worth considering as alternative interpretations of $\Box$ and $\Diamond$. For instance, we may construe $\Box$ as 'always' and $\Diamond$ as 'sometimes'.

With this warning, we may approach the axioms for $\Box$ and $\Diamond$ with an expectation that there may be many sets of axioms, each of which is true, or arguably true, for some interpretation of $\Box$ and $\Diamond$. It is wise, in fact, to cease to worry too much about which senses of $\Box$ and $\Diamond$ make each set of axioms true. The axiomatic systems are interesting objects of study in their own right, and often applications turn up for a given system which could never have been foreseen at the time the system was first studied.

The first modal axiom which we consider is so basic that it is shared by all "normal" modal systems:

A7. Axiom K: $\Box(\alpha \supset \beta) \supset (\Box\alpha \supset \Box\beta)$,

where α and β are any sentences. (The tag 'K' is for 'Kripke', who pioneered the sort of semantics which make this axiom absolutely central to modal logic, as we shall see.)

To this axiom, we add one inferential rule:

R4. Necessitation: If α is a theorem, so is $\Box\alpha$.

This reflects the idea that, if something is a theorem of logic, it is logically necessary; and if it is logically necessary, it is necessary in any "normal" sense of the term.

When axiom K and the rule of necessitation are added to the propositional fragment of our system (A1–A4 plus R1–R2), the result is called system K. This embraces all the theorems of the propositional fragment and more.

The next salient axiom is as follows:

A8. Axiom D: $(\Box\alpha \supset \Diamond\alpha)$.

This axiom is especially suited to the notion of moral necessity, or *obligation,* and its companion notion of *permissibility.* On this interpretation of $\Box$ and $\Diamond$, A8 says that what is obligatory is permissible. The resulting system thus makes a start in characterizing the logic of obligation, or *deontic* logic – hence, the label D. System D contains all the theorems of K and more.

After axiom D comes the following:

A9. Axiom T: $(\Box\alpha \supset \alpha)$.

This axiom requires that whatever is necessary is true. Note that, although this is a very plausible axiom for most notions of necessity, it is not plausible under the interpretation of $\Box$ as 'obligatory'. Although what is obligatory is permissible, we cannot assume that what is obligatory is true. The theorems of T, the system obtained by adding axiom T to system K, include all the theorems of D and more.

After T come two systems: B (for 'Brouwer', the intuitionist mathematician, though the system has little to do with intuitionism), and S4 (the fourth of the systems which Lewis and Langford found most salient in the early, presemantic, pre-Kripke days of axiomatic modal logic). System S4 contains all the axioms and rules of T, plus the following:

A10. Axiom S4: $(\Box\alpha \supset \Box\Box\alpha)$.

('If something must be true, then it must be true that it must be true'.) System B contains all the axioms and rules of T, plus the following:

A11. Axiom B: $(\alpha \supset \Box\Diamond\alpha)$.

('If something is true, then it must be true that it is possible'.)

Some of the theorems of S4 are not theorems of B, and some of the theorems of B are not theorems of S4.

If we add to T the axiom

A12. Axiom S5: $(\Diamond\alpha \supset \Box\Diamond\alpha)$,

we can prove both S4 and B as theorems. The resulting system, S5, contains all the theorems of S4 and all the theorems of B *and nothing more*. Given axiom T plus axiom S5, we can prove both S4 and B; but conversely, too, given T plus both S4 and B, we can prove S5. Yet T plus S4 without B does not enable us to prove S5; and T plus B without S4 does not enable us to prove S5. (Axiom S5 is intuitively 'If something is possibly true, then it must be true that it is possibly true'.)

The axiomatic systems surveyed so far can be displayed as follows:

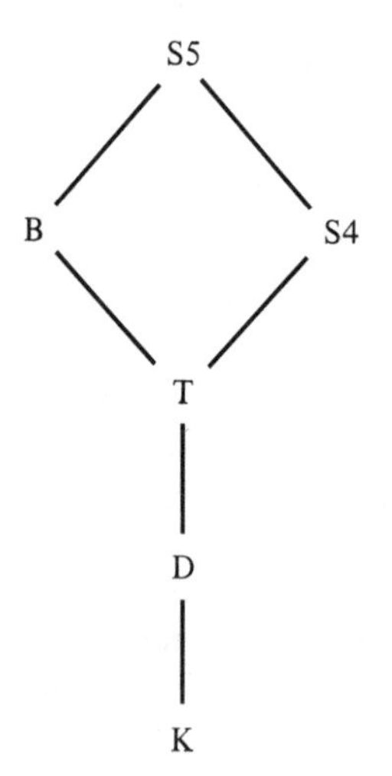

(A downward line connecting two systems indicates that the upper system contains all the theorems of the lower one.) The system S5 is plausible as a characterization of the notion of logical necessity. If we take □ to mean 'can be proved by logic alone', and ◇ to mean 'cannot be disproved by logic alone', then axiom S4 becomes

If you can prove something by logic alone, then you can prove by logic alone that you can prove it by logic alone,

which is plausible. And B becomes

If something is true, then you can show by logic alone that it cannot be disproved by logic alone,

which again is plausible. It is surprising that both these can be proved from the S5 axiom, which becomes

If something cannot be disproved by logic alone, then it can be proved by logic alone that it cannot be disproved by logic alone.

However, the reason for the importance of the systems K, D, T, B, S4, and S5 is not entirely that their axioms sound plausible under natural interpretations. The real reason for their importance lies in their relationship to semantics. Before Kripke's semantics, there were systems S1, S2, and S3, which at that time seemed the salient steps up to S4 and S5, rather than K, D, T, and B. But since Kripke, our interest in modal systems has been coloured by keeping one eye on their accompanying semantics.

The axioms stated so far can be divided into those concerned with truth functions, those concerned with quantifiers, and those concerned with modal operators. There are also, however, some axioms which cut across this classification, "mixed" principles which concern the interaction between modal operators and quantifiers. The key axiom of this sort is

A13. The Barcan formula: $\forall x \Box \alpha \supset \Box \forall x \alpha$.

It could be argued, with initial plausibility, that the intended interpretations of $\Box$ falsify the Barcan formula. Take $\Box$ to mean 'It is logically necessary that', for instance. Now consider some versions of atheism. On these versions, it is maintained that everything there is, is located in space and time – there are no angels, spirits, or gods (or abstract entities for that matter, though this is not crucial). So we have

$$\forall x(x \text{ is spatial}).$$

But, it may be added, for any given spatial thing – a wrench, say – it is logically impossible for that spatial thing to have been anything other than a spatial thing. To state it paradoxically, if that wrench had not been spatial, it would not have been that wrench. And so

$$\Box(\text{If that wrench exists at all, then it is spatial}).$$

This is true of each spatial thing that exists, according to our atheist. In the jargon, being spatial is an *essential property* of each of the things that actually exists. So we may generalize:

$$\forall x \Box(\text{If } x \text{ exists, then it is spatial}).$$

This has the form

$$\forall x \Box \alpha,$$

and the Barcan formula states that

$$(\forall x \Box \alpha \supset \Box \forall x \alpha).$$

Hence, the Barcan formula, if the atheist accepts it (along with modus ponens), would commit her to

$$\Box \forall x \alpha,$$

that is,

□(Everything is such that, if it exists, then it is spatial).

Yet many atheists would reject this claim. This claim would commit them to the belief that the existence of nonspatial things, like God, is a *logical* impossibility rather than a contingent falsehood.

Despite this plausible line of attack, we recommend that the Barcan formula be accepted, especially by modal realists, for the most fundamental interpretations of □. The reason for believing the Barcan formula emerges clearly only when we turn to semantics. When we try to spell out the semantics for a denial of the Barcan formula, we find ourselves *assuming* the Barcan formula in the language we are speaking while doing the semantics. Consequently, if a modal realist undertakes to give a semantics for his modal realism, the Barcan formula will hold in the language in which he states his modal realism. An honest, introspective modal realist, therefore, has to believe the Barcan formula when the □ is interpreted as the strictest of all kinds of necessity.

This means, in effect, that the modal realist should deny that *it is contingent what things there are*. It may be contingent what things there are in the actual world, because it is in some sense contingent which world is actual. But remember that the modal realist claims that 'there are' nonactual possibilia. Consequently, for the modal realist, 'There are such-and-suches' is logically equivalent to 'There are or could have been such-and-suches'. If modal realists think that the existence of God is logically possible, they must accept 'There is *or* could have been a God' – and hence they *must* accept 'There is a God', under their own interpretation of 'There is'. On this understanding of the Barcan formula, we commend it to the modal realist.

There are other axioms which interrelate modal and quantificational symbols. We will not recommend any of these; but just to illustrate the kinds of issues that arise, we present a formula which Hughes and Cresswell call the principle of predication:[8]

$$\forall x(\Box\alpha \vee \Box{\sim}\alpha) \vee \forall x(\Diamond\alpha \wedge \Diamond{\sim}\alpha).$$

This principle in effect divides all properties (or all conditions α that an individual x might meet) into two kinds: *essences* and *accidents*. We discuss this principle, even though we reject it, because it furnishes a useful illustration for future reference, when we discuss set theory more fully in Chapter 8.

An essential property is one which an object either has essentially or lacks essentially. Things that have such a property could not exist without it: things that lack such a property could not have it without ceasing to be the things they are. To give an Aristotelian example, a fish could not be the thing it is without being fishy; but a fowl (or any other nonfish) could not possibly be a fish without ceasing to be the thing it is.

Properties which are not essential are accidental: any given thing might have such a property or lack it. The property of 'having been thought of by Aristotle' might seem initially to be such a property – though if you think too hard about it, counterexamples begin to emerge.

What the principle of predication rules out are properties which are essential properties *for* one object, but merely accidental *for* some other object. Yet it is a mistake to rule such properties out of court. Consider, for instance, the property of 'being awake for the first hour of the year 1600'. It would seem that this is an accidental property for Descartes – he might have had it, and he might have lacked it. (He was three years old at the time.) Yet it is arguably what we might call a *counteressential* property for, say, Descartes's toy soldier – that toy soldier could not have been awake without ceasing to be what it is. If this is correct, there is something which is not necessarily awake nor necessarily not awake (namely, Descartes), so the first disjunct of the principle of predication is false. Yet there is also something (a toy soldier) for which we cannot say both that it could be awake and that it could be not awake. So the second disjunct of the principle is false. Hence, we do not recommend adding it as an axiom, unless your interest in it is purely

8 See Hughes and Cresswell (1968, pp. 184–8).

formal or you can find a different interpretation of $\Box$ and $\Diamond$ which makes it true.

Systems for modal conditionals

Having set down axioms for $\Box$ and $\Diamond$, we may consider what axioms should be set down for $\Box\!\!\rightarrow$ and $\Diamond\!\!\rightarrow$. Before doing that, however, we should ask whether there is in fact any need for new axioms for $\Box\!\!\rightarrow$ and $\Diamond\!\!\rightarrow$, or whether we might be able to save ourselves trouble by simply using the materials already available.

The first thing to note is that *counterfactual* or *subjunctive* conditionals cannot be identified with the material conditional, i.e., $\supset$, which is subject to the axioms of the propositional fragment A1–A4.

To obtain an initial grip on the difference between counterfactuals and other conditionals, compare the counterfactual

If van Gogh had not painted *Sunflowers*, then someone else would have

with the corresponding (*indicative*) conditional:

If van Gogh did not paint *Sunflowers*, then someone else did.

The first of these sentences is false, and the second is true, so obviously they must differ in meaning.[9]

The distinction we are interested in seems to correlate with the so-called subjunctive mood, often marked in English by auxiliary verbs like 'were' and 'would'. There is evidence, however, that sometimes the important distinction is marked not by mood, but by tense. Dudman has argued, more specifically, that when there is a forward shift in tense as we move from the 'if' clause to the 'then' clause, then the conditional is a counterfactual in the intended

9 The example in Lewis (1973a) referred to the assassination of President Kennedy. It has received extensive attention in the literature, too much to review thoroughly here. Harper, Stalnaker, and Pearce (1980) collect some of the debates over conditionals. See also Jackson (1987). We find Lewis's central argument, the one we have recast with reference to van Gogh, conclusive in establishing the difference in meaning between the two sentences given. Objections raised against Lewis may cast doubt on some further consequences he draws from the example, but they do not shake the conviction that the sentences given do differ in truth value and hence in meaning.

sense.[10] But these issues need not delay us here: we shall return to the problems of natural languages later in this section.

Counterfactuals often invite us to consider possibilia even when we know them not to be actual. They thus draw us into modal realism even more fully than does decision theory, which surveys possibilities in order to select which one to make actual. In decision making, the possibilities surveyed are not all known not to be actual; each one may turn out to be actual – although, of course, in general we know that all but one will turn out to have been counterfactual. The more places we look for counterfactuals, the more we find them, and the more important they turn out to be.

In everyday life, we take great interest in characteristics of people, which, when spelled out, contain counterfactuals. Consider courage or honesty, for instance. Even if you have never been offered a bribe, you may wonder what you would do if you were. In a play by Oscar Wilde, a character named Ernest, who thinks his name is Jack, says to someone he fancies, 'But you don't really mean to say that you couldn't love me if my name wasn't Ernest?' Gwendolen is multiply mistaken when she dismisses his counterfactual: 'Ah! That is clearly a metaphysical speculation, and like most metaphysical speculations has very little reference at all to the actual facts of real life, as we know them'. Metaphysical they may be, but such counterfactuals are both knowable and worth knowing.

Counterfactuals are also crucial for a language rich enough for us to express full-fledged science. Many physical properties are what we call *dispositional* properties, which really means that they involve counterfactuals. Consider the chemical property of *solubility,* for example: something is soluble in turpentine, say, when it would dissolve if put in turpentine. Even properties which are, arguably, not "dispositional" *in themselves* (i.e., even "intrinsic" properties) are often nevertheless closely bound up with counterfactuals. Consider the physical property of inertial mass and its close relationship to counterfactual questions about how much an object *would* accelerate if subjected to such-and-such a force.

To round off this plea for counterfactuals, consider also the way that physical laws are often framed in terms of "ideal systems". The law of inertia, for instance, tells us how an object would move if it

10 See Dudman (1983, 1984).

were experiencing no forces at all – if it were an ideal "inertial system".

Finally, it is worth noting that the history of science is studded with so-called *thought experiments,* which have been every bit as important as real experiments. And thought experiments are riddled with counterfactuals. Consider one of Aristotle's arguments against the existence of any vacuum in nature. Experience supports the rule that speed of motion for an object is inversely proportional to the resistance of the medium through which it moves. Aristotle argued that it follows that, if there were a vacuum, motion through it would meet no resistance and therefore would have to proceed at infinite speed. This is absurd, so Aristotle concluded that there could not possibly be a vacuum. Reasoning of this sort has recurred throughout the history of science. Lucretius, for instance, argued against Aristotle's view of the finitude of the universe. He reasoned that, if one were to stand at the edge of the universe and throw a javelin outwards, either it would fly farther outwards or it would stop. If it were to fly outwards, that would show that there was at least empty *space* beyond the supposed end of the universe. If it were to stop, something must have resisted its motion – again implying the existence of something beyond the supposed end of the universe. Note that the reasoning here, as with Aristotle, concerns what *would* result if certain events *were* to occur: we are dealing not with actual experiments, but with thought experiments. Galileo used thought experiments to great effect, as did Newton. Harvey used thought experiments in his investigations of the circulatory system. Darwin used thought experiments. And contemporary physics furnishes a large stock of intriguing thought experiments. There is the imaginary voyage at near the speed of light, for instance, and the return to meet a greatly aged twin brother; there is Schrödinger's allegedly half-dead cat that jumps into the state of being a fully dead and decomposing cat only when someone observes it; and the list goes on. In all such cases, important theoretical issues are explored by way of counterfactuals, by considering what *would* ensue if certain conditions *were* met.

Thus, there is a wide range of important jobs for subjunctive conditionals, and they all call for a conditional which is essentially modal. They require us to focus on possibilities which may not be actualized. For this reason we cannot identify subjunctive with material conditionals.

But perhaps a definition could be constructed using a material conditional together with a modal operator. Given the sort of interpretations we have in mind, there is one initially tempting way we could try to define $\Box\!\!\rightarrow$. We could define $(\alpha \Box\!\!\rightarrow \beta)$ as

$$\Box(\alpha \supset \beta)$$

for some yet to be specified kind of necessity, $\Box$. Whereas $(\alpha \supset \beta)$ is called a material implication, we call $\Box(\alpha \supset \beta)$ a *strict* implication, or *entailment*. We might be tempted by such a definition, because the syntax of a counterfactual in English seems to combine an implication ('if–then') with some sort of modal claim ('. . . then it *would* be or *must* be that β'). However, there are good reasons for resisting such a definition.

One good reason rests on the fact that the axioms for $\Box$ and $\Diamond$ and so forth ensure that strict implication is *transitive*. That is, from the premisses

$$\Box(\alpha \supset \beta),$$
$$\Box(\beta \supset \gamma),$$

we may always infer

$$\Box(\alpha \supset \gamma).$$

Counterfactuals, however, are not transitive. Consider an example. Suppose that, if you were to change your occupation, you would become a lumberjack:

(You change occupation $\Box\!\!\rightarrow$ you become a lumberjack),
$$(\beta \Box\!\!\rightarrow \gamma).$$

This may be true; nevertheless, you might also agree that, if you were offered a job in Hollywood costarring with Paul Newman, you would change your current occupation, pull up stakes, and go to Hollywood:

(You are offered a part $\Box\!\!\rightarrow$ you change occupation),
$$(\alpha \Box\!\!\rightarrow \beta).$$

So we may agree to the two premisses

$$(\alpha \Box\!\!\rightarrow \beta),$$
$$(\beta \Box\!\!\rightarrow \gamma);$$

nevertheless, we may refuse to conclude

$$(\alpha \mathrel{\Box\!\!\rightarrow} \gamma),$$
(You are offered a part □→ you become a lumberjack).

So we must deny transitivity for counterfactuals. Hence, we should not define counterfactuals as strict conditionals.

Analogous arguments can be mounted against other attempts to define counterfactuals in terms of material implication bolstered by 'necessarily' and 'possibly'. The conclusion to draw is that axioms A1–A13 are not enough to delimit the set of counterfactuals which are necessarily true. For this reason, we must add further axioms involving □→ and ◇→.

Among the intimidating profusion of axioms we could explore, the salient core of axioms are the following, which we adapt from Lewis.[11] The Lewis axioms are preceded by an additional rule of inference:

R5. Deduction within conditionals: If it is a theorem that

$(\beta_1 \wedge \beta_2 \wedge \cdots \wedge \beta_n) \supset \gamma,$

then it is a theorem that

$((\alpha \mathrel{\Box\!\!\rightarrow} \beta_1) \wedge (\alpha \mathrel{\Box\!\!\rightarrow} \beta_2) \wedge \cdots \wedge (\alpha \mathrel{\Box\!\!\rightarrow} \beta_n)) \supset (\alpha \mathrel{\Box\!\!\rightarrow} \gamma).$

The axioms we take from Lewis are then the following:

A14. $(\alpha \mathrel{\Box\!\!\rightarrow} \alpha)$,
A15. $(\sim\alpha \mathrel{\Box\!\!\rightarrow} \alpha) \supset (\beta \mathrel{\Box\!\!\rightarrow} \alpha)$,
A16. $(\alpha \mathrel{\Box\!\!\rightarrow} \beta) \supset (\alpha \supset \beta)$,
A17. $(\alpha \mathrel{\Diamond\!\!\rightarrow} (\beta \wedge \gamma)) \supset ((\alpha \wedge \beta) \mathrel{\Diamond\!\!\rightarrow} \gamma)$,
A18. $((\alpha \mathrel{\Diamond\!\!\rightarrow} \beta) \wedge ((\alpha \wedge \beta) \mathrel{\Diamond\!\!\rightarrow} \gamma)) \supset (\alpha \mathrel{\Diamond\!\!\rightarrow} (\beta \wedge \gamma))$.

The last two axioms replace a single axiom,

$$((\alpha \mathrel{\Box\!\!\rightarrow} \sim\beta) \vee ((\alpha \wedge \beta) \mathrel{\Box\!\!\rightarrow} \gamma)) \equiv (((\alpha \wedge \beta) \mathrel{\Box\!\!\rightarrow} \gamma) \supset (\alpha \mathrel{\Box\!\!\rightarrow} (\beta \supset \gamma))),$$

for which Lewis apologizes, since it is 'long and obscure'.[12] Lewis's axiom has the formal merit of being framed entirely in terms of □→, which he takes as primitive relative to ◇→. Axioms A17 and A18, however, have the merit that they sound true if we read ◇→ as 'If it were that . . . then it might be that . . .'. A17 says, 'Supposing that β and γ might be true together, if α were true, it follows

11 Lewis (1973a, esp. the app.).
12 See Lewis (1973a, p. 132).

that γ might be true if α and β were'. A18 says, 'Supposing β might be true if α were true, and that γ might be true if α and β were true together, then it follows that β and γ might be true together if α were true'. This is complex, but when you grasp it, it does have the ring of truth.

Lewis includes a further axiom of *strong centring:*

$$(\alpha \wedge \beta) \supset (\alpha \mathrel{\Box\!\!\rightarrow} \beta).$$

We do not endorse this axiom. Our primary reason is a formal one: strong centring is inconsistent with principles we take to be true concerning probability, discussed in Section 3.6. We also note, however, that intuition speaks against strong centring, and protests against it have been heard from time to time.

Consider Anthea, an Australian who, like most Australians, thinks that Auckland is the capital of New Zealand. She is also under the delusion that governments of all nations will see her as a threat to their national security if she writes to anyone in their capital city. She has written to someone in Wellington. Then she thinks to herself, 'If Wellington were the capital of New Zealand, then counterespionage agents would read my letter'. As it happens, Wellington is the capital of New Zealand, and counterespionage agents do in fact read her letter – not, however, because they know her from a bar of soap, but merely because they have selected her letter at random for a spot check. The strong centring axiom then entails that Anthea's counterfactual is true. Let

$$\alpha = \text{Wellington is the capital of New Zealand}$$

and

$$\beta = \text{Agents will read Anthea's letter.}$$

Then $(\alpha \wedge \beta)$ is true, so by the strong centring axiom, $(\alpha \mathrel{\Box\!\!\rightarrow} \beta)$ is true. Intuitively, however, it is not true that, if Wellington were the capital of New Zealand, agents would read Anthea's letter.

Another sentence, called *conditional excluded middle,* is worth considering as a possible axiom:

$$(\alpha \mathrel{\Box\!\!\rightarrow} \beta) \vee (\alpha \mathrel{\Box\!\!\rightarrow} \sim\beta).$$

Lewis urges that sentences of this form are not always true. But the other pioneer of counterfactual logics, Stalnaker, has defended it against all comers.[13] The kinds of cases we must consider include ones like the following. There is some temptation to say that it may be *false* to say,

If I had gone to a movie last night, then I would have seen *The Fly*.

Yet it may also be *false* to say,

If I had gone to a movie last night, then I would not have seen *The Fly*.

We may wish to say,

If I had gone to a movie, then I might have seen *The Fly;*
and if I had gone to a movie, I might not have seen *The Fly*.
$(\alpha \Diamond\!\!\rightarrow \beta) \wedge (\alpha \Diamond\!\!\rightarrow \sim\beta)$.

If so, then we deny something of the form

$$(\alpha \Box\!\!\rightarrow \sim\beta)$$

and also something of the form

$$(\alpha \Box\!\!\rightarrow \beta).$$

So we deny both disjuncts of $(\alpha \Box\!\!\rightarrow \beta) \vee (\alpha \Box\!\!\rightarrow \sim\beta)$; hence, it seems, we must deny conditional excluded middle.

In addition to noting the presence of intuitive arguments against conditional excluded middle, it is worth noting one of its formal consequences. If it were accepted, the difference between 'would' and 'might', between the $\Box\!\!\rightarrow$ and $\Diamond\!\!\rightarrow$ counterfactuals, would collapse. It is clearly true that $(\alpha \Box\!\!\rightarrow \beta)$ entails $(\alpha \Diamond\!\!\rightarrow \beta)$. (There are complications in the case in which α is contradictory, but these do not cancel the entailment.) But the converse entailment can also be derived, if we accept conditional excluded middle. Suppose we have $(\alpha \Diamond\!\!\rightarrow \beta)$. By the definition of $\Diamond\!\!\rightarrow$, this means that $\sim(\alpha \Box\!\!\rightarrow \sim\beta)$. This is the denial of one of the two disjuncts in conditional excluded middle. If we affirm that disjunction and deny one disjunct, we must affirm the other disjunct. That is, if we deny $(\alpha \Box\!\!\rightarrow \sim\beta)$, we must affirm $(\alpha \Box\!\!\rightarrow \beta)$ – hence the supposition that $(\alpha \Diamond\!\!\rightarrow \beta)$ has entailed that $(\alpha \Box\!\!\rightarrow \beta)$ under the assumption of conditional excluded middle. So not only does 'would' entail 'might', but 'might' also entails 'would'. Hence, conditional excluded middle

13 Stalnaker's system, including conditional excluded middle, appeared first in Stalnaker (1968); he defends it against Lewis's critique in Stalnaker (1981).

makes $(\alpha \mathrel{\Box\!\!\rightarrow} \beta)$ and $(\alpha \mathrel{\Diamond\!\!\rightarrow} \beta)$ logically equivalent. This, we urge, is undesirable.

Intuition, however, is an unreliable guide here. With regard to a large number of principles, like conditional excluded middle, our initial modal opinions must be treated as fluid and revisable. Occasionally, some of our clearest intuitions turn out even to contradict one another. Consider, for example, the intuitive confidence many people have had in the principle of *distribution of disjunction:*

$$((\alpha \vee \beta) \mathrel{\Box\!\!\rightarrow} \gamma) \supset ((\alpha \mathrel{\Box\!\!\rightarrow} \gamma) \wedge (\beta \mathrel{\Box\!\!\rightarrow} \gamma)).$$

Every example you try seems obviously true. 'If you were to eat or drink, you would be my prisoner; so if you were to eat, you would be my prisoner, and if you were to drink, you would be my prisoner'. Yet the formula for distribution of disjunction cannot be added consistently to our axiom system, or to any plausible revision of it.[14]

Intuitive counterintuitions of this sort should not, however, be taken as constituting, on their own, refutations of an axiom system. There are often ways of explaining such intuitions without revising our axioms. Here is an example of one way that we might try to explain the intuitions supporting the distribution of disjunction. Suppose that some conditional statements, 'If A, then B', in a natural language, are equivalent to quantifications over situations: 'In all situations where A holds, B holds'. Under this supposition, it is possible to read natural-language renderings of distribution of disjunction as having the logical form

$$\forall x((Ax \vee Bx) \mathrel{\Box\!\!\rightarrow} Cx) \supset (\forall x(Ax \mathrel{\Box\!\!\rightarrow} Cx) \wedge \forall x(Bx \mathrel{\Box\!\!\rightarrow} Cx)).$$

This is uncontroversially a logical truth. If natural-language assertions of distribution of disjunction seem obviously true, that is some evidence that their logical form is captured by this latter, quantificational formula, rather than by the earlier unquantified formula. 'In any situations, if you were to eat or drink, you would be my prisoner; so in any situation, if you were to eat, you would be my prisoner, and in any situation, if you were to drink, you would be my prisoner'. This is a logical truth; and it is also plausible as an English paraphrase of the invalid, unquantified formula for

14 See Ellis, Jackson, and Pargetter (1977).

distribution of disjunction. This, therefore, furnishes one possible explanation of why an invalid formula seems intuitively to be valid.

There may be some better explanation of the plausibility of distribution of disjunction; but whatever our explanation, we cannot accept the intuition supporting distribution of disjunction at face value. Extroverted judgements, expressed by the unquantified formula for distribution of disjunction, must be overruled by introverted reflections on the axiomatic theories involving that formula. This is another instance of the interplay between language and ontology that was highlighted in Chapter 1.

Because there is some slack between our initial opinions and our considered theory, there is room for a large input from semantics. Given a promising semantic theory we can measure this theory against our axiomatic systems. At least sometimes, we may be justified in overruling intuitions in the interests of a tidy semantics.

It turns out that a robustly realist correspondence theory of truth backs up our classical axioms, or something very close to them. And that is one of the most powerful arguments for a conservative, classical logic.

3.3 MODAL SEMANTICS

We have now outlined a language, together with informal indications of some intended interpretations. We have demarcated a set of basic symbols which may be combined according to rules to form more complex expressions. We have set down a string of sentences, axioms, which we claim to be true. And we have given a set of rules which allow us to deduce theorems from our axioms. These rules are intended to guarantee that the theorems be true, provided that the axioms are true.

A realist semantics for the axiomatics above will begin with a set $\mathfrak{D}$ of individuals, the *domain*. Some of these individuals will be actual. It is not to be merely assumed, however, that all are actual. Some members of $\mathfrak{D}$ may be merely possible individuals. A restriction of $\mathfrak{D}$ to actualia would have to be separately argued, but will be anticipated as a special case of the more general theory which permits both actualia and mere possibilia. Set $\mathfrak{D}$ is to contain anything we might want to talk about – all the things there are (in the broad sense of 'there are', which is logically equivalent to 'there are, were, will be, or could have been').

Among the members of $\mathfrak{D}$ are especially important individuals called possible worlds. We make minimal assumptions about the nature of possible worlds, about whether they are abstract or concrete, primitive or reducible, and so forth. We call the set of possible worlds $\mathcal{W}$.

Sets $\mathcal{D}$ and $\mathcal{W}$ play different roles in the semantics for the formal language described earlier. Names (individual constants) and individual variables will be interpreted as referring to members of $\mathcal{D}$. When these members of $\mathcal{D}$ are described in a sentence, however, the resulting description may be true in one possible world and false in another. For this reason, the interpretation of both predicates and sentences will involve members of $\mathcal{W}$.

A sentence may be thought of as describing this world, the world we live in, the actual world – and it describes this world truly or falsely. It ascribes, in a broad sense, a *property* to the world – a property which the world may or may not actually have. There is a *set* of things which have this property. This is the set of possible worlds in which the sentence is true. Corresponding to each sentence, then, is a set of possible worlds. It is convenient (and traditional in one subculture) to call a set of possible worlds a *proposition*. The set of propositions may be called $\mathcal{E}$. We could leave it open whether $\mathcal{E}$ contains all possible subsets of $\mathcal{W}$ or only some; but for our purposes we will allow it to include any subset of $\mathcal{W}$. (In classical axiomatic presentations of probability theory, $\mathcal{W}$ is called the *sample space*, its members are called *elementary events*, and $\mathcal{W}$ is often written as Ω. Members of $\mathcal{E}$ are subsets of $\mathcal{W}$ and are called *events* – hence the symbol $\mathcal{E}$.)

Consider, then, an atomic sentence α. Let $\alpha = Fb$. There is an individual referred to by b. And there is to be a set of possible worlds in which α is true. The set of worlds in which α is true should be determined jointly by the interpretation of the name b and the interpretation of the predicate F. The interpretation of F must somehow enable us to determine the interpretation of α, given the interpretation of b.

What, therefore, should the interpretation F be taken to be? The most straightforward answer is

A predicate F will be interpreted by being assigned a *function* which maps a given member of $\mathcal{D}$ onto a corresponding member of $\mathcal{E}$.

That is, we will pair a predicate with a propositional function – a function from individuals to propositions. There are various alternative ways of interpreting predicates, but all turn out, if adequate, to be equivalent to an assignment of propositional functions to predicates. They must be, since the interpretation of a predicate must, in one way or another, determine an interpretation of a sentence given an interpretation for the name on which it is predicated. Given a domain $\mathcal{D}$ and a set of worlds $\mathcal{W}$, we can thus generate interpretations for names, predicates, and atomic sentences.

To interpret modal sentences involving $\Box$, $\Diamond$, $\Box\!\!\rightarrow$, and $\Diamond\!\!\rightarrow$, we will need something in addition to $\mathcal{W}$ and $\mathcal{D}$. We will have to appeal to some sort of relationship among possible worlds. We will call the required sort of relationship *accessibility.* For the moment we make minimal assumptions about the nature of this relationship. Unravelling the nature of accessibility is the key problem we address later in the book, particularly in Chapters 5 and 6.

For maximal generality, we will allow accessibility to come in *degrees*. Think of one world as more, or less, accessible from another, depending on how *far-fetched* it would be from the standpoint of that other world. The more far-fetched a world, the less "accessible" it is. At least much of the time (and perhaps always – we will discuss this later), the more *similar* a world is to a given world, the more accessible it is from the standpoint of that world. The fewer, and the more minor, the changes that would be required to transform one world into another, the more accessible that other world is.[15]

By allowing accessibility to come in degrees, we are in effect introducing a whole family of accessibility relations – one for each degree. And we are also imposing some sort of ordering on these various distinct accessibility relations. Accessibility, thus construed, is a *quantity.* The nature of quantities was discussed in detail in Chapter 2. For convenience (as we find with most quantities) it pays to assign *numbers* which represent the various degrees of acces-

15 Similarity between worlds is what Lewis (1973a) takes as the crux of what we call *degrees of accessibility.* We grant that similarity coincides with degrees of accessibility often enough; and so we exploit similarity in an initial indication of the kind of things which degrees of accessibility amount to. We are, however, impressed by a simple but powerful argument by Tichý (1975) which shows that degrees of accessibility do not always march hand in hand with degrees of similarity. Pollock (1976) speaks of how great the changes are that would have resulted in one world rather than another. This is a different way of getting a grip on degrees of accessibility. We find it roughly as problematic as Lewis's appeal to similarities. We return to the issue in Chapter 4.

sibility. It does not matter which numbers we choose for which degrees of accessibility, as long as the required ordering of degrees of accessibility is mirrored by the order of the corresponding numbers. To any pair of worlds, u and v say, we may therefore assign a number, which we will call

$$d(u, v)$$

and which will represent the degree of accessibility of v from the point of view of u. So d will be a function which maps pairs of worlds in $\mathcal{W}$ onto real numbers.

It is convenient to think of d as giving a kind of *distance* between worlds. Together, $\mathcal{W}$ and d constitute a *logical space*. Provided that d meets certain minimal assumptions appropriate to any "distance" measure, $<\mathcal{W}, d>$ counts technically as a metric space. Hence, we should remember that the *lower* the number assigned to a pair of worlds, the *closer* they are – and thus the *more* accessible one is from the other.

We leave open the possibility that there may be a pair of worlds, u and v say, for which v is quite simply *not accessible* from u. Its accessibility is "zero"; so the distance between u and v must be very large. Perhaps we could set $d(u, v)$ as infinity. But it is in some ways easier to say simply that d assigns *no* number to the pair $<u, v>$ – that $d(u, v)$ is *undefined*. Hence, we take the distance function d to be a partial function. If the distance between worlds is infinite, there will be a *minimal* degree of accessibility between them, but this is to be distinguished from there being *no* accessibility between them. One world is *not accessible* from another only when there is no number at all which represents a distance between them.

Note that any relation on $\mathcal{W}$ – that is, any "on–off" relation which does not come in degrees – can be represented by a numerical function of the form of d. Given any relation R, which holds between various worlds in $\mathcal{W}$, we may define a corresponding function d such that

$$d(u, v) = 0$$

just when u and v stand in relation R, and

$$d(u, v) = 1$$

just when u and v do not stand in relation R. Traditionally, in modal logic, accessibility has been taken to be an on–off relation. So it is important to note that this case is encompassed within a more general account of degrees of accessibility, as given by a numerical function d.

Given any numerical function d, we may define a large number of on–off accessibility relations. One of these will be especially important for our theory of laws of nature. It is the one Lewis calls an "outer modality" in *Counterfactuals*.[16] We will call it R_0:

$$u \; R_0 \; v \text{ if and only if } d(u, v) \text{ is defined.}$$

By such techniques as these, we can use a numerical function d to define a rich family of accessibility relations. This promises to be useful when we come to analyse the many distinct kinds of necessity described earlier – practical, moral, legal, physical, logical necessity, and so forth. But the most urgent need for degrees of accessibility derives from the counterfactual conditional $\Box\!\!\rightarrow$. When we consider what would be true if some condition held, we consider possibilities which are just far-fetched enough to permit that condition to hold, but no more far-fetched than that. Degrees of accessibility are precisely what we need for the semantics of counterfactuals.

To sum up, then, our semantics rests, so far, on a domain $\mathcal{D}$, a set of possible worlds $\mathcal{W}$, and a numerical function d which measures degrees of accessibility.

In standard modal logic, the pair of a set of worlds $\mathcal{W}$ and an accessibility relation R is called a *frame*. Adapting that terminology to our purposes, we shall call the pair

$$\langle \mathcal{W}, d \rangle$$

a *frame*. A frame furnishes us with the logical space within which we construct an interpretation of our language.

When we are trying to determine which sentences are *logically necessary,* we will try to determine which sentences remain true no matter how we reinterpret names and predicates. The reinterpretation of names and predicates may involve altering the domain $\mathcal{D}$. But the *frame* remains constant under all such reinterpretations. The most apt of several plausible accounts of logical necessity is one which equates it not just with truth in all possible worlds under one given interpretation, but with truth in all possible worlds under any interpretation which leaves the frame of interpretation intact. That

16 Lewis (1973a, p. 23) and elsewhere. We put outer modalities to work in Chapter 5, in the analysis of laws of nature.

is why the frame is singled out as a salient entity in its own right, even though it is only one contributor to an overall interpretation of our language.

A full interpretation of our language will be provided by what is called a *model*. A model is constituted by three things:

$$\langle \mathcal{F}, \mathcal{D}, \mathcal{V} \rangle,$$

where $\mathcal{F}$ is a frame, $\mathcal{D}$ a domain, and $\mathcal{V}$ a function which draws on $\mathcal{F}$ and $\mathcal{D}$ to provide an interpretation for each expression of the language. The definition of the *valuation function* $\mathcal{V}$ is complex. It has to be complex, because it must be a recursive definition which assigns an interpretation, or *semantic value,* to each of the infinitely many complex expressions that can be constructed in the language.

The definition of $\mathcal{V}$ begins by assigning a semantic value to each of the "nonlogical" basic symbols of the language – to individual constants, individual variables, and predicates. Then rules are given which determine the semantic value of any complex expression from the semantic values of its components. The interpretations of the "logical" symbols of the language are given by these recursive rules.

A formal definition of a valuation function may be given as follows. First, we assign a semantic value to the basic, nonlogical symbols, as sketched earlier in this section:

V1. If α is an individual constant or individual variable, then $\mathcal{V}(\alpha) \in \mathcal{D}$.
V2. If α is a one-place predicate, then $\mathcal{V}(\alpha)$ is a function such that, for any individual $b \in \mathcal{D}$, $\mathcal{V}(\alpha)$ maps this individual onto a subset of $\mathcal{W}$:

$\mathcal{V}(\alpha)\ (b) \subseteq \mathcal{W}$,

and if α is an n-place predicate, then $\mathcal{V}(\alpha)$ is a function which maps any n-tuple of individuals in $\mathcal{D}$ onto a subset of $\mathcal{W}$.

These rules can be subjected to further restrictions of various sorts in order to facilitate completeness proofs for different subsets of the classical axioms A1–A18. For instance, without restrictions to V1 and V2, the semantics is best matched with a logic which *includes* the Barcan formula. If, for reasons we sketched earlier, you doubt the truth of the Barcan formula, you must add restrictions to V1 and V2.

One important way of generating such restrictions is as follows. The domain $\mathcal{D}$ of possibilia may be divided into subdomains. For each world w there will be a domain $\mathcal{D}_w$, consisting of all and only the possibilia existing *in that world*. Different restrictions may then be set on the nature of this family of subsets. It might be required that this family form a partition, that is, that each of the subdomains $\mathcal{D}_w$ be disjoint from the others, with no individual existing in more than one possible world and every individual existing in some possible world. Such is David Lewis's counterpart theory, which we shall meet again later. Another important proposed restriction requires a specific sort of interaction between the subdomains and the accessibility relation among worlds. It can be required that, if one world w is accessible from another u, then w cannot contain any individuals which u lacks. Rather, w must contain only rearrangements of the constituents of u. Hence, when w is accessible from u, we must have $\mathcal{D}_w \subseteq \mathcal{D}_u$. (Notice that this proposal flatly contradicts Lewis's counterpart theory, which requires $\mathcal{D}_w$ and $\mathcal{D}_u$ to be disjoint.)

Given some specific proposal concerning the ways in which the domain $\mathcal{D}$ divides into subdomains, this information can be fed into the details of the valuation rules V1 and V2. It is sometimes suggested, for instance, that in order for an object to have a property in a world, it must exist in that world. (This precludes, for instance, there being any such property as that of being nonexistent, being fictional, etc.) If we were to adopt such a view, we would require that a predicate map an individual in world w onto a set of worlds each of which contains that very individual. In keeping with that approach, we might require that a many-place predicate should never yield a true proposition unless all the individuals it relates come from the same world. Such restrictions are worth exploring, even though our modal realism will, arguably, militate against any restrictions when we are dealing with the most fundamental modalities.

Clauses V1 and V2 cover the interpretation of the "referential" component of our basic vocabulary. Next, we state rules which determine the interpretation of compound expressions given the interpretation of their components. There will thus be one clause in the definition of $\mathcal{V}$ corresponding to each of the familiar formation rules.

First, there is a rule corresponding to the formation of simple, atomic subject–predicate sentences:

V3. If α is an atomic sentence, that is, if

$\alpha = Fb$
(or $\alpha = Fx$),

where F is a predicate and b (or x) is an individual constant (or variable), then

$\mathcal{V}(\alpha) = \mathcal{V}(F)(\mathcal{V}(b))$
(or $\mathcal{V}(\alpha) = \mathcal{V}(F)(\mathcal{V}(x))$).

Next, there will be clauses in the definition of $\mathcal{V}$ which fix the interpretation of the truth-functional connectives – negation, conjunction, and their kin:

V4. If $\alpha = {\sim}\beta$, then $\mathcal{V}(\alpha) = \mathcal{W} - \mathcal{V}(\beta)$.
V5. If $\alpha = (\beta \vee \gamma)$, then $\mathcal{V}(\alpha) = \mathcal{V}(\beta) \cup \mathcal{V}(\gamma)$.

The next clause defining $\mathcal{V}$ reflects the intended interpretation of the quantifiers:

V6. If $\alpha = \exists x\beta$, then

$$\mathcal{V}(\alpha) = \bigcup_{b \in \mathcal{D}} \mathcal{V}_{[b/x]}(\beta),$$

where $\mathcal{V}_{[b/x]}(x) = b$ and $\mathcal{V}_{[b/x]}(\gamma) = \mathcal{V}(\gamma)$ for any $\gamma \neq x$.

The symbolism here requires some explanation. For each member b of the domain $\mathcal{D}$, we determine the set of worlds in which β would be true if x referred to b. This set of worlds is denoted by

$$\mathcal{V}_{[b/x]}(\beta).$$

In each of these worlds, β is obviously true for *some* value of x, namely, b, so $\exists x\beta$ is true. This applies whichever member b of $\mathcal{D}$ we choose. So we take the set of worlds in which β is true of b and form its union with all sets of worlds in which β is true of each other member of $\mathcal{D}$. These will all be worlds in which β is true of something. Hence, the set of worlds in which β is true of something – the set of worlds where $\exists x\beta$ is true – includes all worlds in the union of all the sets $\mathcal{V}_{[b/x]}(\beta)$, where b is any member of $\mathcal{D}$. We summarize this by saying

$$\mathcal{V}(\exists x\beta) = \bigcup_{b \in \mathcal{D}} \mathcal{V}_{[b/x]}(\beta).$$

Note that this rule is formulated in a way which treats all possibilia, all members of $\mathcal{D}$, equally. It is not required that the individual which makes β true exist in the very *same* world in which $\exists x\beta$ is true. For instance, it may be true in the actual world that 'There is something referred to by Andersen' in virtue of there being something in some nonactual world which satisfies the description 'referred to by Andersen'. If you wish to interpret the quantifier in a different way, you may do so by exploiting a structure of subdomains on $\mathscr{D}$. You may then ensure that $\exists x\beta$, 'There is something such that β is true of it', is true in a world w only when there is something satisfying β *and* that something *exists in world w.*

We will not restrict the rule for quantifiers in that way. There is no need to build such restrictions into the syntax. We may acknowledge that in some contexts 'There are . . .' does not mean unrestrictedly that there are so-and-sos in the domain consisting of all possibilia. Often 'There are . . .' means 'There are in the actual world . . .'. Certainly this is so. But often 'There are . . .' is more restricted than that in its meaning. Sometimes it means 'There are in the actual world, at the present time, within easy reach of us here in Collingwood . . .'. There are standard methods of restricting the scope of quantification without altering its semantics. We just need to add further conjuncts to the expression within the scope of the quantifier. We do not have to build these into the semantic rule for the quantifier itself. And as modal realists, not only do we not *have* to do so, but we *should* not do so.

The next formation rules we must accommodate concern the modal operators, $\Box$ and $\Diamond$. Here, we need two different interpretations. The simplest interpretation is what we call the Leibnizian one. Leibnizian necessity is truth in *all worlds*. Its interpretation will be given by the following clause in the definition of the valuation function $\mathscr{V}$:

V7. If $\alpha = \Diamond\beta$, then $\mathscr{V}(\alpha) = \mathscr{W}$ if β is true in some possible world, and otherwise it is the empty set, that is, $\mathscr{V}(\alpha) = \phi$.

Given this interpretation of $\Diamond$, we fix the interpretation of $\Box$ by defining it in terms of $\Diamond$:

$$\Box\beta =_{\text{df}} \sim\Diamond\sim\beta.$$

In addition to Leibnizian necessity, we must define necessity of another kind. When something is true in *all accessible worlds,* it may be said to be necessary in one important sense, even if it is not true in all worlds. When $\Box$ and $\Diamond$ are to be interpreted as expressing the idea of truth in all accessible worlds, rather than truth in all worlds, we will generally alter the notation as a warning of the interpretation which is in the offing. We will generally write $\boxed{\mathrm{N}}$ and $\Diamond_{\mathrm{N}}$ in place of $\Box$ and $\Diamond$. (The 'N' is for 'nomic' or 'natural'; we will argue that this is the kind of necessity which is possessed by laws of nature.) The interpretation will then be fixed by the clause

V8. If $\alpha = \Diamond_{\mathrm{N}}\beta$, then $\mathcal{V}(\alpha)$ is the set of worlds $w \in \mathcal{W}$ such that β is true in some world u which is accessible from w – some world u such that $d(w, u)$ is defined.

Given this interpretation for $\Diamond_{\mathrm{N}}$, the interpretation of $\boxed{\mathrm{N}}$ is then fixed by being defined in terms of $\Diamond_{\mathrm{N}}$ (just as $\Box$ was defined in terms of $\Diamond$):

$$\boxed{\mathrm{N}}\beta =_{\mathrm{df}} \sim \Diamond_{\mathrm{N}} \sim\beta.$$

The final clause we require in the definition of $\mathcal{V}$ is one which provides an interpretation for the counterfactual conditional $\Box\!\!\rightarrow$:

V9. If $\alpha = (\beta \Box\!\!\rightarrow \gamma)$, then $\mathcal{V}(\alpha)$ is the set of all worlds $w \in \mathcal{W}$ such that there is some world u in which β is true and γ is true,

$u \in \mathcal{V}(\beta)$ and $u \in \mathcal{V}(\gamma)$;

and every world v in which β is true and γ is false is less accessible from w than u is, that is,

If $v \in \mathcal{V}(\beta)$ and $v \notin \mathcal{V}(\gamma)$, then $d(w, v) > d(w, u)$.

This rule is complex, but the idea behind it is simple. A counterfactual $(\beta \Box\!\!\rightarrow \gamma)$ is true in a world when the nearest β-worlds are all γ-worlds. For instance, the counterfactual 'If I were a blackbird, I'd whistle and sing' is true in the actual world (when uttered by, say, Violet) because the nearest worlds in which Violet is a blackbird are worlds in which she is whistling and singing. We speak here of the "nearest worlds" in the plural, because we do not wish to presuppose that there is any unique "nearest" world in which Violet is a blackbird. The world in which Violet is a blackbird on the letter-box and the world in which she is a blackbird on the pear tree may be equally near the actual world. The question whether such "ties" can occur relates to the debate we have mentioned over Stalnaker's axiom of conditional excluded middle. The complexity of rule V9 is due partly to our desire for generality. If we cannot assume that there is a unique nearest world meeting a condition, we must speak of what holds in worlds where the condition holds, which are *nearer* than others where the condition holds. We require that a counterfactual ($\beta \;\Box\!\!\rightarrow \gamma$) will be true in a world when there is a collection of worlds in which β is true and γ is true, which are *nearer* than any worlds in which β is true and γ is false. And that is the import of V9.

Given this semantics for the 'would' conditional $\Box\!\!\rightarrow$, the interpretation of the 'might' conditional $\Diamond\!\!\rightarrow$ is fixed by the definition

$$(\alpha \;\Diamond\!\!\rightarrow \beta) = \mathrm{df} \sim(\alpha \;\Box\!\!\rightarrow \sim\beta).$$

This completes the semantics for a basic language containing the resources for describing the core of the modal structure of the world. Let us take stock. We have argued for realism: for scientific, metaphysical, and modal realism. We have argued that these realisms support a correspondence theory of truth, or semantic realism. And a correspondence theory of truth takes the form of a specific kind of semantic theory, a semantic theory built around the word–world relationship of reference. Then, after realism has instigated such a semantic theory, a feedback loop emerges. Theoretical developments in semantic theory may justify revisions and extensions of the realist theories with which we began.

Interplay between ontology and semantics, which we predicted on theoretical grounds in Chapter 1, is illustrated in practice in this chapter. The language we have described so far has been given a referential semantics. Yet while describing the semantic framework we favour, we have found it appropriate to make fairly frequent allusions to ways in which the semantics could be modified. For instance, we argued that a straightforward semantics for quantifiers would support the Barcan formula. Then if we assert (an instance of) the Barcan formula, standing on its own, we will not be making a semantic assertion at all. We will no longer be *mentioning* the symbols it contains; we will be *using* them. And we will then be making an ontological assertion about what things there are in the world. Such an ontological assertion is supported, therefore, by semantic considerations. Semantics does not have uncontestable authority, of course. If independent grounds can be marshalled to contest the claim made by the Barcan formula, we may be justified in revising our semantics for quantifiers. We have attended to ways in which that may be done. Thus, the argument from semantics to ontology is a defeasible one. Nevertheless, it does have significant weight in most cases. This interplay between semantics and ontology will be further explored in the following section.

3.4 COMPLETENESS, CORRESPONDENCE, AND REALISM

The sentences of our language may be partitioned into two classes: the theorems and the nontheorems. All the theorems are, we claim, true. Some of the nontheorems may be true, some may be false, depending on the interpretation given to the names and predicates those sentences contain. Indeed, some of the nontheorems may be necessarily true, true in all possible worlds. If α is a nontheorem, it may yet be true, and indeed $\Box\alpha$ may be true. Of course, $\Box\alpha$ will be a theorem if α is a theorem; but $\Box\alpha$ may be true even if it is not a theorem. In this case, however, its truth will depend on the interpretation given to the names and predicates α contains.

Consider an example, to orient our intuitions, though we do not pretend to prove anything here from such intuitions. It is true that electrons have a negative charge: let α be an atomic sentence which asserts this. Then α will be true, though not a theorem. Arguably, α is *necessarily* true – there could not possibly be an electron without a negative charge. If it lacked a negative charge, it would not be an electron. So $\Box\alpha$ will be true on at least one salient interpretation of one kind of necessity. We may say, perhaps, that α is "metaphysically" necessary. Yet we also say that it is not "logically" necessary – since, being a nontheorem, its truth is not guaranteed by the axioms and inference rules of "logic" alone.

We are hoping you will agree that an electron could not possibly lack a negative charge. This will be denied by some. We might vary the example to overcome some such objections. But there are those who will resist all such examples, saying, 'It ain't necessarily so' no matter *what* nontheorem 'it' might be.[17] Leave these doubts, however, for the time being. The point remains that you must not simply assume that *only* theorems are necessary truths. Such a claim would have to be separately argued.

Thus, a distinction should be drawn between necessary truths which happen also to be theorems and necessary truths which are not theorems. It is convenient to call the former *logically* necessary;

17 See Putnam (1962).

the latter may be said to have some other sort of necessity – perhaps the term 'metaphysical necessity' sets the right tone.

The boundary between logical and nonlogical truths is not easy to draw without circularity. Consider, for instance, a proposal that we simply add α to our list of axioms. Then $\Box\alpha$ *will* be a theorem after all (by the rule of necessitation). Yet this would be something of a cheat: there would remain a significant difference between α and the theorems we derive *without* allowing α as an axiom. The difference is that the truth of α does depend on the interpretation that we give to specific predicates – to 'electron' and 'negatively charged' (or the regimented counterparts of these English expressions). If we reinterpret the symbol 'electron' so that it applies to the set of protons, then α ceases to be true. It is a distinctive feature of the theorems we derive *without* permitting α as an axiom that they remain true no matter how we reinterpret the names and predicates they contain. (It will be explained shortly just how it can be established that this is indeed so.) Thus, the theorems of the system of genuine axioms (i.e., excluding α) have their truth guaranteed *solely* by the interpretation that we give to symbols *other than* names and predicates. The symbols other than names and predicates are conveniently classed as "logical" symbols. Thus, the genuine theorems have their truth guaranteed by the interpretation of the *logical* symbols alone. Hence, they may be said to be *logically* necessary.

There are two distinct ways of characterizing logical necessity: an axiomatic way and a semantic way. These two ways illustrate the two stages of realism described in Chapter 1: introverted and extroverted realism. On the one hand, we may select a list of axioms and principles of inference; and we may then say that something is logically necessary when it follows by those principles from those axioms. In this case, something is logically necessary just in case it is a *theorem*. This is an axiomatic characterization of logical necessity. On the other hand, we may identify logical necessities by a semantic route: something is logically necessary when it is true in virtue of the interpretations given to purely logical symbols alone. The traditional notion of a *tautology* is a semantic notion of this kind; and tautologies are prime examples of logically necessary truths.

Something has gone dreadfully wrong somewhere if the axiomatic and semantic characterizations of logical necessity fail to agree over whether all theorems are logically necessary. It is essential that the truth of all theorems be guaranteed by the interpretation of the logical symbols alone. Otherwise, we must admit either that the interpretation offered is a *mis*interpretation or that what we took to be a purely logical axiom or rule of inference is not purely logical after all. The most urgent thing to establish, then, is that all theorems derived from the axioms *are necessary* according to the semantics. To establish this is to establish what is called a *soundness* result.

But there is another thing we may seek to establish as well. Consider all the things whose truth (in all worlds) is guaranteed by the interpretation of logical symbols alone. These will all be logically necessary in the semantic sense. Will they also be logically necessary in the axiomatic sense? Will they be provable as theorems? If in addition to soundness we can establish that all semantically logically necessary truths are theorems, this is called a *completeness* result. A completeness result is good news for an axiomatic system. It assures us that the axiomatic system generates all the necessary truths which intuitively count as purely "logical" necessities.

The laying down of an axiomatic system may be regarded as a way of exhibiting an interpretation of the logical symbols without ascending to a metalanguage in which we explicitly discuss the semantics of the object language. The axiomatic approach provides what may be called *implicit definitions* of the logical symbols. Anyone who fails to agree to the axioms thereby demonstrates (in all probability) that he is misinterpreting the logical symbols. Anyone who understands the logical symbols will (other things being equal) see that the axioms are necessarily true and that the rules of inference necessarily preserve truth. A person can demonstrate understanding by *affirming* axioms, even if that person never made any semantic remarks *about* those axioms.

The axiomatic approach can be, and often is, purely formal, involving nothing but manipulation of symbols without regard to their meanings. But it is also possible to formulate a theory axiomatically using a language which you speak and in which you do mean what you say. You can then state axioms and use rules of inference without ever referring to your own symbols. You can *use* your language without *mentioning* it. Furthermore, you can mean something by your symbols, without having first of all to state explicitly an interpretation for them. Indeed, it cannot be a general requirement that you explicitly state an interpretation for a language before you use the language. That would lead to a vicious infinite regress, since you would have to state the interpretation of your symbols using some language, and that language, too, would stand in need of interpretation. Sooner or later we must come to a language we simply speak and mean something by without any prior affirmation of semantic rules. It is possible to develop a theory axiomatically in that language. This we may call *extroverted* axiomatics.

In extroverted axiomatics, there is no formalistic concern about "metatheorems", no concern about the mathematical properties of the system of symbols we are using. Nor is there any explicit concern with what our symbols mean. We know what they mean, in the sense that we understand them and we mean something by them. But though we have meant things by using them, we have never said anything about them, never said what they mean, never even thought about what they mean. We just state axioms and derive theorems, without (yet) reflecting on what we are doing. In purely extroverted axiomatics, *acceptance* of the axioms and *understanding* of the symbols are inseparable. This sort of implicit definition must precede explicit definitions – since we must already understand and mean something by some language before we can begin giving any explicit definitions of anything at all. What we are calling extroverted axiomatics here is, of course, just a revisitation of the extroverted realism we discussed in Chapter 1.

There is a sense, then, in which the axiomatic characterization of logical truths is fundamental. Parasitic on extroverted axiomatics will be introverted reflection on what we have been doing and on what we have meant by what we have been saying. This yields the semantic characterization of logical necessity. If our explicit semantic theory then deems things to be necessarily true which do *not* follow from our extroverted "implicit definitions", then this should worry us. It may mean that our explicit semantic theory is *mis*interpreting our language, is failing to describe accurately what we have been meaning by our symbols when we were using them rather than mentioning them. Conversely, if our explicit semantics guarantees the truth of all and *only* the theorems we have asserted extrovertedly – that is, if we can establish a completeness result – then this reassures us that our semantics is not reading anything into the language which was not already there.

Thus, a completeness result is good news for an explicit semantic theory. But of course it is also, to a degree, good news for an axiomatic system. Even if we do understand a language, we will still be fallible in our judgements about which statements are true – even in the case of axioms. Self-evidence and *a priori* certainty are an unattainable ideal. Prima facie, failure to assent to an axiom demonstrates failure to understand it; but this presumption can be overridden. There may be indirect reasons for granting that someone does understand logical symbols, even if she is wrong in her judgements about which axioms are true. We are not infallible, even over the axioms of logic. Our fallibility over axioms gives another reason for interest in completeness results. A completeness result can bolster our confidence in an axiom system which did seem to capture all the truths of logic when we formulated our logical theory purely extrovertedly.

The existence of completeness proofs for our proposed modal axioms furnishes a kind of *correspondence theory of truth,* or *semantic realism,* which backs up the (extroverted) modal realism which we express when we state our modal axioms and derive our modal theorems.

It can be shown that *all* the sentences which our semantics guarantees to be true in all worlds, without any restrictions on $\mathcal{W}$ or d, are theorems which can be derived using only the truth-functional axioms A1–A4; the quantificational axioms A5–A6; the modal axiom K,

A7. $\Box(\beta \supset \gamma) \supset (\Box\beta \supset \Box\gamma)$;

the Barcan formula,

A13. $(\forall x\Box\alpha \supset \Box\forall x\alpha)$;

and the counterfactual axioms

A14. $(\alpha \mathrel{\Box\!\!\rightarrow} \alpha)$,
A17. $(\alpha \mathrel{\Diamond\!\!\rightarrow} (\beta \wedge \gamma)) \supset ((\alpha \wedge \beta) \mathrel{\Diamond\!\!\rightarrow} \gamma)$,
A18. $((\alpha \mathrel{\Diamond\!\!\rightarrow} \beta) \wedge ((\alpha \wedge \beta) \mathrel{\Diamond\!\!\rightarrow} \gamma)) \supset (\alpha \mathrel{\Diamond\!\!\rightarrow} (\beta \wedge \gamma))$.

This is our fundamental completeness result.[18]

This can be followed up with a sequence of further completeness results, corresponding to the various restrictions we may wish to

18 Lewis (1973a) gives the proof for the counterfactual extension of modal logic (the second printing gives a correction to overcome a problem raised by Krabbe, 1978), and Hughes and Cresswell (1968) give the proofs for the precounterfactual components. This applies not only to the fundamental result above, but also to the further completeness results which follow.

place on $\mathcal{W}$ and d. We will say that degrees of accessibility are *weakly centred* (Lewis's term) when no world is more accessible from a given world than that world is from itself. We may meet this requirement most conveniently by

$$d(w, w) = 0.$$

The weak centring restriction on accessibility guarantees that some extra sentences will be true, in all worlds, in addition to those guaranteed by the axioms above. These extra necessary truths will be derivable as theorems if we add the following axioms:

A9. $(\Box\beta \supset \beta)$,
A16. $(\beta \;\Box\!\!\rightarrow \gamma) \supset (\beta \supset \gamma)$.

The first of these axioms corresponds to the requirement that accessibility be *reflexive* – that every world be accessible to some degree from itself. The second of these axioms corresponds to the requirement that no world be *more* accessible from a world than that world is from itself. This leaves open the possibility that several worlds might be "zero distance" from a world w.

Lewis imposes a further condition on the degrees of accessibility in his semantics for counterfactuals: the condition he calls *strong centring*. This requires that no world be as accessible from a given world as that world is from itself. We may meet this requirement by saying not only that $d(w, w) = 0$ for each world w, but also that

if $w \neq u$, then either $d(w, u)$ is undefined or else $d(w, u) > 0$.

The addition of this semantic condition permits the proof of a completeness theorem for the axiomatic system which is obtained by adding to all the above axioms the axiom of strong centring:

$$(\alpha \wedge \beta) \supset (\alpha \;\Box\!\!\rightarrow \beta).$$

This yields Lewis's favoured axiomatization of counterfactual logic, VC. We do not endorse strong centring. The system we endorse is the one Lewis calls VW ('V' is for 'variably strict' and 'W' for 'weakly centred').

The axioms we endorse, by the end of Chapter 5, are ones for which a completeness result can be proved. This is a desirable outcome for a realist. We will now describe a general technique for probing completeness results – the method of canonical models. This method is philosophically significant, and not only in virtue of the importance of the completeness result which the method establishes. Canonical models are important for a number of reasons, which will emerge not only in this chapter but also in the next.

Canonical models

There are a variety of techniques for proving completeness theorems, but one of these is especially noteworthy. It is worth explaining in some detail, since it casts much light on the relationship between semantics and ontology and on the nature of modal realism. In fact, one of the plausible reductionist accounts of possible worlds is the one we discuss in Chapter 4 under the label 'book theories'. Such accounts identify possible worlds with precisely the sorts of entities which are employed in canonical models: maximal consistent sets of sentences. However, whatever ontological status you assign to them, canonical models can certainly be used to establish completeness.[19]

A completeness theorem is one which proves that, if some sentence is guaranteed to be true under the specified semantics, then that sentence can be proved as a theorem. How can this be proved? How can we prove that *every* such sentence is a theorem? One way to prove this is by proving its contrapositive. Instead of proving

If α is guaranteed to be true by the semantics, then α is a theorem,

we can prove that

If α is *not* a theorem, then α is not guaranteed to be true by the semantics.

We can then prove that a nontheorem is not guaranteed to be true by the semantics alone, by finding an interpretation, compatible with those semantics, which makes it false.

A canonical model provides an interpretation which is guaranteed to make each nontheorem false in at least one world. In a canonical model, we construct so-called possible worlds out of sentences of our language. For each axiomatic system, we construct a corresponding set of possible worlds.

We begin with the set of all theorems of the axiomatic system in

19 The technique of canonical models was discovered after the publication of the classic work of Hughes and Cresswell (1968). It is described, however, in their companion work (Hughes and Cresswell, 1984).

question. Characteristically, there will be some sentences α for which neither α nor $\sim\alpha$ is a theorem. Such a sentence can be added to the set of axioms to give another consistent set of sentences. We may then add all further sentences which can be derived from this set using the logical rules of inference. This yields a *consistent extension* of the set of theorems. To this consistent extension, we may generally add a further sentence, β say, and further derivable consequences to yield another consistent extension.

It can be proved that, for any of the axiomatic theories we are concerned with, there is a *maximal consistent extension*. (This is called Lindenbaum's lemma.) A maximal consistent extension of a set of theorems is a consistent set of sentences, a set which includes all theorems and is such that, for *any* sentence γ, either γ is in the set or $\sim\gamma$ is in the set. Hence, no further sentence can be added to the set without creating an inconsistency.

We then take $\mathcal{W}$ to be the set of all maximal consistent extensions of the axiom system we started with. Each maximal consistent set of sentences is a possible world. A possible world in which a sentence α is true, then, is defined to be a maximal consistent set which contains α as a member.

The important feature of this construction is that every nontheorem is guaranteed to be false in some world. If α is a nontheorem, then either it is the negation of a theorem or it is not. If it is the negation of a theorem, it is false in all worlds. If it is not the negation of a theorem, $\sim\alpha$ can be added to the set of theorems to generate a maximal consistent extension. And in that maximal consistent extension, α will be false. So there will be at least one maximal consistent extension, that is, a world, in which α is false. Hence, whether or not α is the negation of a theorem, there is sure to be some maximal consistent extension in which α is false, some world in which it is false.

Hence, any nontheorem will be false in some possible world if we take possible worlds to be maximal consistent extensions of the set of theorems. This, therefore, establishes that every sentence which is true in all worlds is a theorem. And this, we may recall, is what a completeness theorem is designed to establish.

This account of completeness proofs leaves some gaps to be filled. We have presumed that a sentence is true in a world just when that world contains that sentence as a member. Yet this is to pre-empt the semantic clauses V1–V9, which specify interpretations for the logical symbols. We must show that, if we take a sentence to be true in just those worlds which contain it, we will nevertheless find that all the designated semantic clauses are satisfied. We must also define accessibility relations among our constructed worlds. And (here is the hard part) we must show that the accessibility relation, thus defined, does meet the restrictions we are interested in. If we are constructing a completeness proof for S5, for instance, we must show that the defined accessibility relation is reflexive, transitive, and symmetric.

The canonical model for S5 is of special interest, since as we noted earlier many have seen it as providing a reductionist account of possible worlds. We are tempted by some sort of reductionism, but not by a reduction of worlds to maximal consistent sets of sentences. Consider the canonical model for S5. You might wonder whether this model captures the Leibnizian idea that necessary truths are true in all possible worlds. In fact, it does not. In the canonical model, a sentence is necessarily true in a world just when it is true in all worlds which are accessible from that world. But *not* all worlds are accessible from any given world. Possible worlds, construed as maximal consistent extensions of S5, fall into a number of distinct equivalence classes. A world in one such class is accessible from all others in that class; but a world in one class is never accessible from any worlds in a distinct equivalence class.

We will sketch a brief proof to show that in the canonical model not every world is accessible from every other world. Consider an atomic sentence *Fa*. Either it or its negation may be consistently added to the theorems of S5 and then augmented to yield a maximal consistent set, or possible world. Thus, we can construct a world in which *Fa* is true. If this world were accessible from all other worlds, then $\Diamond Fa$ would be true in all worlds. Yet any sentence which is true in all worlds in the canonical model must be a theorem. Hence, it would follow that $\Diamond Fa$ would have to be a theorem of S5. Yet we know it is not. If it were a theorem, rule R2 of universal substitution would ensure that $\Diamond \alpha$ was true for any α, even if α were, say, $(\beta \wedge \sim\beta)$.

If the intended interpretation of S5 is the Leibnizian one, as we think it is, it follows that the canonical model does *not* capture the intended interpretation of S5. This helps to support the view we defend, that possible worlds are not sets of sentences. It also supports our assertion that accessibility, the kind of accessibility relevant to alethic modalities, is *not* an equivalence relation. Logical necessity is truth in all worlds, not merely truth in all accessible worlds.

3.5 EXTENSIONS

Within the language we have described so far, we may distinguish three constituents:

(a) a vocabulary of names and predicates whose interpretations are all, so far, completely unconstrained,
(b) atomic subject–predicate sentences,
(c) recursive rules for generating compound sentences from atomic sentences.

There are three ways in which this language can be extended:

(a) We can leave the syntax of the language entirely unchanged, and we can designate specific names and predicates and give them fixed, specific interpretations. As with the logical terms, we can fix an interpretation for a given name or predicate in two ways: either we can lay down axioms involving that name or predicate, or we can add a semantic rule for that symbol. Ideally, these two strategies should yield the same outcome.
(b) We can introduce more complexity into the structure of atomic sentences. In particular, we could introduce functional terms, or operators.
(c) We could extend the modes of generating compound sentences. Two such extensions require special mention: the extension of modal operators to yield probability theory and the extension of quantification from first-order to higher-order quantification.

Extensions of type (c) will be discussed later, in Sections 3.6 and 3.7. For the present, we shall introduce several salient extensions of types (a) and (b).

Identity

Without altering the language we have described, it is useful to single out one of the two-place predicates and designate it as an *identity* predicate. A two-place predicate is written before its two arguments, as in

$R(x, y)$, or
$R(a, b)$, etc.

The identity predicate is written as =. Strictly, then, identity claims should be written as

$$=(x,\ y),\ \text{or}$$
$$=(a,\ b),\ \text{etc.},$$

but it is conventional to write them as

$$(x = y),\ \text{or}$$
$$(a = b),\ \text{etc.}$$

And we abbreviate

$$\sim=(x,\ y)$$

as

$$(x \neq y).$$

The identity predicate is obviously crucial to mathematics (think of the importance of equations) and thereby to science more generally.

There are two very natural axioms to lay down governing the identity predicate. The first is that everything is self-identical:

A19. $\forall x(x = x)$.

The second requires that, if something is true of something and not true of something, then the thing that it is true of cannot be identical with the thing that it is not true of. We can summarize this by the axiom schema

A20. $(\alpha \wedge \sim\alpha_{[\sigma/\lambda]}) \supset (\sigma \neq \lambda)$.

In this schema, λ is any name or variable occurring in α (so that α asserts something to be true of the referent of λ). Similarly, σ is any name or variable and $\alpha_{[\sigma/\lambda]}$ is the formula which results from replacing every occurrence of λ in α by σ. Hence, $\alpha_{[\sigma/\lambda]}$ asserts something to be true of the referent of σ. What it asserts to be true of the referent of σ is exactly what α asserts to be true of the referent of λ. Hence,

$$(\alpha \wedge \sim\alpha_{[\sigma/\lambda]})$$

asserts that what is true of the referent of λ is not true of the referent of σ. If this is so, then the referents of λ and σ cannot be identical – hence, the axiom.

These axioms have a surprising consequence. They entail that there are no such things, strictly speaking, as contingent identities. If it is true that $(x = y)$, it is necessarily true that $(x = y)$; and if $(x \neq y)$, it is necessarily true that $(x \neq y)$. That is, the following can be proved as theorems:

NI. $(x = y) \supset \Box(x = y)$,
NNI. $(x \neq y) \supset \Box(x \neq y)$.

('NI' stands for 'necessity of identity', and 'NNI' for 'necessity of nonidentity'.) Furthermore, the semantic rule which matches these axioms is one which makes any identity or nonidentity claim true in all possible worlds, or in none, but never anything in between. The semantic value of = will be a function, $\mathscr{V}(=)$, such that for any member c of domain $\mathscr{D}$, we will have

$$\mathscr{V}(=)\ (c,\ c) = \mathscr{W},$$

where $\mathscr{W}$ is the set of all possible worlds. Hence, if $\mathscr{V}(a) = c$ and $\mathscr{V}(b) = c$, then we have

$$\mathscr{V}(a = b) = \mathscr{W},$$

and so the sentence $(a = b)$ will be true in all possible worlds. If, however, there is no member c of $\mathscr{D}$ such that $\mathscr{V}(a) = c$ and $\mathscr{V}(b) = c$, then we have

$$\mathscr{V}(a = b) = \phi$$

where ϕ is the empty set, and so the sentence $(a = b)$ will be true in no possible worlds.

It follows from the suggested axioms for identity and from the associated semantics that all identity claims are either necessary or impossible. This is surprising. It furnishes another illustration of the interplay between semantics and ontology. A streamlined and plausible semantics commits us to an ontological view, a view on the necessity of identities. And this ontological view may seem implausible, on independent grounds.

There have been many occasions in the history of science in which it has seemed that we have discovered that things we thought to be distinct were in fact identical. It was discovered that the sun we see rising on one day is in fact the same body as the one we saw the day before. It was discovered that heat is molecular motion; that lightning is electric discharge; that genes are strands of DNA; and so on. If these identities were discovered through scientific research, it would seem that they must be contingent truths. Hence, it is plausibly argued, not all identities are necessary.

If you are persuaded by such arguments, you can use this as grounds for seeking a revision of the semantics. And, indeed, such a revision is not too hard to find. Our problems flowed from the presumption that things are distinct when something is true of one but not true of the other. We can revise this by requiring that things are distinct when something *nonmodal* is true of one but not true of the other.

By means of such restrictions as this, we can generate a range of systems of contingent identity. It is interesting that some of these systems verify NNI while continuing to falsify NI. Evidently it is harder to identify two distinct things than to suppose that a single thing might be two distinct things. It is harder to allow that New York might be the same city as Miami than to allow that Miami might be two distinct cities.

Our own recommendation, however, is that on this issue we should let semantics call the tune, and we should simply accept the conclusion that there are no contingent identities. Instead of finding ways of adjusting the semantics to avoid that conclusion, we should seek ways of explaining why identities seem to be contingent even though they are not.

Russell's theory of descriptions provides a very powerful means by which the apparent contingency of some claims involving identity can be reconciled with the necessity of genuine identities.[20] Claims of roughly the form

$$\text{the } F = \text{the } G$$

can be analysed as contingent claims to the effect that the properties corresponding to F and G are coinstantiated by a single thing, and this is, in general, a contingent matter. This is compatible with its being necessarily true, of whatever this thing may be, that it is identical with itself and is distinct from all other things. The contingency of claims involving both identity and descriptions is compatible with the necessity of identity itself. By appealing to this fact, most appearances of contingent identities can be explained away. Consequently, we defend the straightforward axioms and semantics

20 Russell (1903).

for identity. This again illustrates the way in which, as we argued in Chapter 1, introverted realism may reinforce the extroverted realism from which it sprang.

Set membership

Identity has been central to mathematics and science for thousands of years. It is less plausible to suggest that sets, too, have been active in mathematics and science since ancient times. Their presence has become explicit and pervasive only since the second half of the nineteenth century. But very soon after their first appearance under their modern name, sets came to permeate every last corner of mathematics. In the theory of identity we saw an interplay between semantics and metaphysics. An interplay of this sort has been even more prominent in the theory of sets.

Sets, we have claimed, are universals of a special sort. When a set contains entities, all those entities have a property in common, the property of 'being one of these things'. And that common property is just what the set consists in. Being a *member* of that set simply amounts to *instantiating* that property. On this view, set theory is a fragment of the theory of universals. It is only a fragment of the theory of universals because sets are only some of the universals that there are. The set-membership predicate differs from the instantiation predicate only in being more restricted in its application, so that it covers only those cases of instantiation in which the universal being instantiated is one of the special universals we call a set.

The way that set theory emerges from the language we described earlier is by the addition of a new two-place predicate, $\in$, along with some axioms. The most popular list of axioms constitutes ZF, or Zermelo–Fraenkel set theory.

The set-membership predicate, like the identity predicate, is a two-place predicate, and so should strictly be written in front of its two arguments:

$$\in(x, y).$$

But it is conventional to rewrite this as

$$(x \in y),$$

which asserts that x is a member of y and which we interpret as a special case of the assertion that x instantiates y.

The special two-place predicate ϵ is distinguished from others not by its syntax but by adding new axioms and by restrictions on its interpretation. Some of the new axioms tell us what sets and set membership would have to be like if there were to be such things. Other axioms (the axiom of infinity, in particular) tell us that there are some such things (the von Neumann natural numbers, in particular).

The result is a very rich theory which provides the basic framework within which mathematics is discussed. We argue that mathematics is also grounded in other ways – for instance, in the theory of quantities. But it cannot be denied that set theory, too, provides an important foundation for mathematics, and hence for science. We shall explore set theory more deeply in Chapter 8.

Operators

In mathematics we sometimes find such sentences as

$$(3 + 5) = 8.$$

Such sentences cannot occur in the basic language we have outlined so far. This is because the symbol + in the above sentence plays a role which no symbol in our language plays. The symbol + is called an *operator.* When concatenated with two names of numbers, '3' and '5', for instance, it yields a term '(3 + 5)', which refers to another number. That is, + makes a new name of a number out of two names of numbers. None of the symbols described so far does that. We have names and symbols (called predicates) which make a sentence out of one or more names, and symbols which make a sentence out of one or more sentences (e.g., negation, disjunction, and the quantifiers). But we have nothing which makes a name out of one or more names.

In a sense, it is not really necessary to expand the language to make room for operators like +. We can say what we need to say without operators. Corresponding to each operator there is a matching predicate. For instance, corresponding to + is a predicate which we may call R_+. When an operator applies to two terms to yield a third, there will be a relationship holding among these three things. For instance, when the operator + applies to two numbers to yield a third, there will be a relationship holding among these three numbers. When '+' combines with '3' and '5' to yield '(3 + 5)', which refers to the number 8, then there will be a relationship among the numbers 3, 5, and 8. We will have

$$R_+(3, 5, 8).$$

This is a three-place relation, which can be read as

3 augments 5 to give 8.

In general, for an n-place operator O, there will be an $(n + 1)$-place predicate R_o. Then instead of saying,

$$O(x_1, \ldots, x_n) = x_{(n+1)},$$

we may say,

$$R_o(x_1, \ldots, x_n, x_{(n+1)}).$$

So there is no *need* to add operators to our language, since anything we say with operator O we can say with the corresponding predicate R_o. Yet there is another way of taking the correspondence between O and R_o. We may use it to clarify the semantics of operators, so that we need feel no unease about adding them to the language. Operators correspond semantically to relations. But the relations are employed in a different way in the semantics. Instead of using the relation to determine the truth value of a sentence, we use the relation to determine the referent of a compound referential term. For instance, O and R_o correspond to the same relation (universal). But the syntactic rule for R_o is that

$$R_o(x_1, \ldots, x_{(n+1)})$$

is *true* iff $x_1, \ldots, x_{(n+1)}$ stand in the required relation. In contrast, the syntactic rule for O is that

$$O(x_1, \ldots, x_n)$$

refers to $x_{(n+1)}$ iff $x_1, \ldots, x_{(n+1)}$ stand in the required relation.

This notion of an operator can be generalized further. In probability theory we meet assertions like the following:

$$P(p \vee q) = P(p) + P(q) - P(p \wedge q).$$

In this, we have the operator +. But we also have terms like '$P(p \vee q)$' which refer to numbers. Hence, the role of 'P' is to attach itself to '$p \vee q$', or 'p', or whatever, to result in a compound referring phrase. The term to which P attaches here, however, is not a name, but a sentence. Thus, P converts a sentence into a referring phrase. It is therefore different from $+$. Whereas $+$ is a *two names to a name* operator, P is a *sentence to a name* operator.

The equivalence between an operator, like $+$, and a relation is one which has an analogue for the operator P. Modal language introduces sentential operators like $\Box$ and $\Box\!\!\rightarrow$. These can be adapted to yield probabilistic sentential operators of the form 'It is probable to degree n that . . .'. Such operators attach to sentences to yield sentences. A recasting of their semantics can easily yield a *name* instead of a sentence. The operator P will be such that, for any sentence p,

$$P(p)$$

will refer to the number n just in case

It is probable to degree n that p

is true.

The problem which remains is that of giving the semantics for

It is probable to degree n that . . . .

That is the topic of the following section.

3.6 PROBABILITY

It is possible to extend our modal language further, to accommodate probability theory. It is desirable for us to do so for many reasons. And when we do so, completeness results emerge which are closely analogous to those we have met in modal logic. We shall find that the normal method of expressing probability claims, however, involves a radical expansion of the language. Probability theory usually employs expressions like

$$P(\alpha),$$

read as 'the probability that α'. This expression refers to a number. Hence, the symbol 'P' is one which attaches to a sentence to make a referring expression – an expression of the same sort as a name (and capable of being bound by a quantifier). As we have indicated, it would be possible to augment our language to make room for probability theory, as normally presented. But this would obscure the modal issues we want to put on centre stage. Hence, we employ a way of introducing a fragment of probability theory which inflicts minimal distortions on simple predicate logic and dramatically highlights certain neglected continuities between the modalities (possibility, necessity, and counterfactuals) and probability theory.

Instead of writing,

$$P(\alpha) = r,$$

where r is a real number, we write,

$$\textcircled{r}\alpha,$$

read as 'It is probable to degree r that α'.[21]

When something is necessary, that is, true in all *accessible* worlds, it is probable to degree 1, and when something is not possible, that is, not true in any accessible world, it is probable to degree 0. Otherwise, its probability is between 0 and 1. For technical reasons, we must allow that some things have zero probability, even though they are not impossible. Or, more accurately, we must allow this if we are unwilling to believe in *infinitesimal* probabilities, where an infinitesimal is a number x so small that $(nx) < 1$ no matter how large n is. If we were to accept infinitesimals, we could define the modals $\boxed{N}$ and $\Diamond_N$ in terms of probabilities:

$$\boxed{N}\alpha =_{df} \textcircled{1}\alpha,$$
$$\sim \Diamond_N\alpha =_{df} \textcircled{0}\alpha.$$

But not wanting to buy into realism about infinitesimals in this context (we have enough realism to defend as it is), we keep $\boxed{N}$ and $\Diamond_N$ as distinct vocabulary from the probability operators.

Hence, we add to the vocabulary for our language a list of symbols of the form

$$\textcircled{r}$$

– one such symbol for each real number r between 0 and 1. There are uncountably many such symbols since there are uncountably many real numbers between 0 and 1. That makes the vocabulary of our language much larger than we expect a vocabulary to be. We could artificially constrain the language to allow only finitely many symbols $\textcircled{r}$, but this would be inconvenient, and for our purposes it will do no harm to imagine there to be uncountably many of them. Ideally, we should treat $\textcircled{r}$ as a compound expression in which a

21 The symbols $\textcircled{r}$, and $\textcircled{r}\!\!\rightarrow$ introduced later, are borrowed from unpublished work by Lewis; they are anticipated in print in Sobel (1985).

variable or other expression referring to a number r is combined with another symbol to yield a sentential connective $\textcircled{r}$. For our purposes, however, the internal complexity of $\textcircled{r}$ is only a distraction. So we shall pretend that there are simply infinitely many such distinct, primitive symbols.

In addition to these probabilistic counterparts of $\boxed{N}$ and $\Diamond_{N}$, there are probabilistic counterparts of the counterfactual conditional $\Box\!\!\rightarrow$. We write these conditionals in the form

$$\textcircled{r}\!\!\rightarrow,$$

where r is any real number from 0 to 1. We read

$$(\alpha\ \textcircled{r}\!\!\rightarrow \beta)$$

as 'If α were true, β would be probable to degree r'. When $0.5 < r < 1$, say, then

$$(\alpha\ \textcircled{r}\!\!\rightarrow \beta)$$

entails that, if α were true, then β *might or might not* have been true:

$$(\alpha\ \Diamond\!\!\rightarrow \beta)$$

and

$$(\alpha\ \Diamond\!\!\rightarrow \sim\beta);$$

but it would more likely have been true than false.

The appropriate axioms for the probability operators $\textcircled{r}$ can be gleaned from the classic Kolmogorov axioms, which opened the way for the dramatic mathematical reduction of probability theory to a branch of measure theory.[22] This enabled science to use the full force of the theory of the calculus within probability theory – a world-shaking revolution whose full significance has seldom been recognized. The Kolmogorov axioms (simplified to allow for only finite disjunctions) translate into our modal language as follows:

A21. $(\boxed{N}\alpha \supset \textcircled{1}\alpha)$;
A22. $(\sim \Diamond_{N}\, \alpha \supset \textcircled{0}\alpha)$;
A23. $(\sim \Diamond_{N}(\alpha \wedge \beta) \wedge \textcircled{p}\alpha \wedge \textcircled{q}\beta)) \supset \textcircled{r}(\alpha \vee \beta)$,
where $r = p + q$.

These assert that, if something is necessary – true in all accessible worlds – its probability is 1; if something is impossible, its probabil-

22 For a presentation of probability theory see Kolmogorov (1956).

ity is 0; and the probability of a disjunction of incompatible possibilities is the sum of their probabilities taken separately.

For probabilistic counterfactuals, the corresponding axioms are as follows:

A24. $(\alpha \boxright \beta) \supset (\alpha \; \textcircled{1}\!\!\rightarrow \beta)$;
A25. $(\alpha \boxright \sim\beta) \supset (\alpha \; \textcircled{0}\!\!\rightarrow \beta)$;
A26. $((\alpha \boxright \sim \Diamond_{N}(\beta \wedge \gamma)) \wedge (\alpha \; \textcircled{p}\!\!\rightarrow \beta) \wedge (\alpha \; \textcircled{q}\!\!\rightarrow \gamma)) \supset (\alpha \; \textcircled{r}\!\!\rightarrow (\beta \vee \gamma))$, where $r = p + q$.

Earlier, we endorsed weak centring in the interpretation of counterfactuals, but we resisted strong centring. One of our motives was prompted by an observation brought to our attention by Frank Jackson. If we are to hold the above principles for probabilistic counterfactuals, we cannot simultaneously endorse the strong centring axiom,

$$(\alpha \wedge \beta) \supset (\alpha \boxright \beta),$$

and also the plausible assumption that, when $0 < r < 1$, then $(\alpha \; \textcircled{r}\!\!\rightarrow \beta)$ entails both that $(\alpha \diamondright \beta)$ and that $(\alpha \diamondright \sim\beta)$. If $(\alpha \; \textcircled{r}\!\!\rightarrow \beta)$ entails $(\alpha \diamondright \sim\beta)$, then (by the definition of $\diamondright$) this entails $\sim(\alpha \boxright \beta)$. But by strong centring, $\sim(\alpha \boxright \beta)$ entails $\sim(\alpha \wedge \beta)$. The upshot is that 'If α, then probably β' entails that α and β cannot both be true. This is obviously mistaken; and so we must resist either strong centring or the assumption that 'If α, then probably β' entails that, if α then it might be that β and it might be that not-β. Jackson urged that we reject the latter assumption when the 'might' conditional is interpreted as $\diamondright$. We choose, rather, to resist strong centring.

The semantics for the probability axioms requires that we introduce what is technically called a *measure* over the collection of all propositions. A measure is a function P which assigns a number $P(A)$ to each set in a family of sets and meets certain further requirements. These further requirements are as follows.

The measure must be defined over a family of sets, a real number being assigned to each set in the family. This family of sets must satisfy certain requirements. First, all the sets in the family must be subsets of some given set $\mathcal{W}$, which is also in the family. Second, the family must include the empty set. Third, the family must be closed under complements: if A is in the family, so is $(\mathcal{W} - A)$. Fourth, the family is closed under unions – in fact, under countable unions: if A and B are in the family, so is the union of A and B, $(A \cup B)$, and if there is a countable sequence of sets $A_1, A_2, \ldots$, each of which is in the family, then their union

$$\bigcup_{i=1}^{\infty} A_i$$

is also in the family.

Not only must a measure be defined over a family of sets meeting the above requirements, but it must also assign numbers to these sets in such a way as to meet the following conditions. First, the measure must assign the number 1 to the set $\mathcal{W}$, and 0 to the empty set ϕ:

$$P(\mathcal{W}) = 1 \text{ and } P(\phi) = 0.$$

Second, when two sets in the family are disjoint, the number assigned to their union must be the sum of the numbers assigned to each of them separately:

$$\begin{aligned} \text{If } (A \cap B) &= \phi, \text{ then} \\ P(A \cup B) &= P(A) + P(B). \end{aligned}$$

This condition is called *additivity.* In fact, we should require *countable* additivity – we should require that, if a sequence of sets $A_1, A_2, \ldots$ in the family are pairwise disjoint, that is,

$$(A_i \cap A_j) = \phi, \text{ for } i \neq j,$$

then the number assigned to their countable union equals the sum of the numbers assigned to each of them separately:

$$P\left(\bigcup_{i=1}^{\infty} A_i\right) = \sum_{i=1}^{\infty} P(A_i).$$

These defining characteristics of a *measure* very neatly match our axioms for the probability operators in our modal language. If we take $\mathcal{W}$ to be the set of possible worlds, then all its subsets are what we call propositions (or "events", if you wish to echo the terminology of statistics); and these do form a family which satisfies all the above requirements.

To give semantics for $\textcircled{r}\alpha$, we must draw on the measure P:

V10. $\mathcal{V}(\textcircled{r}\alpha)$ is the set of worlds w such that, if A is the set of worlds accessible from w in which α is true, then $P(A) = r$.

Intuitively, the idea is that, in a world, the probability of a proposition is a measure of the proportion of accessible worlds in which it is true. For the sake of thoroughness, we shall also give the semantics for ⓡ→. When you get the hang of it, it is quite intuitively graspable, but it is not worth taking the space here to explain the details. Hence, we simply state the rule:

V11. $\mathcal{V}(\alpha$ ⓡ→ $\beta)$ is the set of worlds w such that

$$\lim_{i \to \infty} \frac{P(B_i)}{P(A_i)} = r,$$

where each A_i is a set of worlds where α is true:

$A_i \subseteq \mathcal{V}(\alpha)$,

and each B_i is the set of worlds in A_i for which β is true:

$B_i = A_i \cap \mathcal{V}(\beta)$,

and the sets of worlds A_i monotonically converge on w:

$A_i \subseteq A_{(i-1)}$;

and if $u \in A_i$ and $v \in (A_{i-1} - A_i)$, then

$d(w, u) < d(w, v)$.

Intuitively, the idea is that (α ⓡ→ β), 'If α were so, then β would be probable to degree r', is true in a world just when r is the proportion of the nearest α-worlds which are β-worlds – or r is the limit to which the proportions converge as we consider the proportions of α-worlds which are β-worlds for nearer and nearer sets of α-worlds.

In our semantics for our modal language, *degrees of accessibility* are required for counterfactuals. We also require an "on–off" accessibility relation R for the interpretation of [N] and ⟨N⟩. We now find that we also require a *measure*, P, if we are to interpret the probability operators ⓡ.

We could interpret our modal language, then, by calling upon a sequence of separate elements:

$$\langle \mathcal{W}, \mathcal{D}, d, R, P \rangle,$$

where $\mathcal{W}$ is the set of worlds, $\mathcal{D}$ is the set of individuals, d gives degrees of accessibility between worlds, R is an on–off accessibility relation between worlds, and P is a measure over subsets of $\mathcal{W}$. But it is worth noting that, if we begin with degrees of accessibility d we get an on–off accessibility relation R for free. Just as the postulation of the set $\mathcal{W}$ of worlds need not be accompanied by a separate postulation of propositions, so too the postulation of d need not be accompanied by a separate postulation of R. Given d, we can generate R; and given any R we want to obtain, we can tailor a d which will generate it.

How does the probability measure P fit into the scheme? Can we get it for free, too? Yes, we can. Given a distance measure d on the members of $\mathcal{W}$, we can obtain a measure on the subsets of $\mathcal{W}$. (And given any measure P we want to obtain for subsets of $\mathcal{W}$, we can tailor a d which will generate P.) The technique for generating P from d involves sophisticated mathematical proofs, due to Hausdorff and others.[23] But the intuitive idea is a bit to grasp.

Take any world w and any given degree of accessibility n, and you can form a *sphere* around w: the set S_n of all worlds whose distance from w is *less* than the given degree:

$$S_n = \{u: d(w, u) < n\}.$$

Take any subset A of $\mathcal{W}$ and you can obtain an estimate of the *proportion* of $\mathcal{W}$ occupied by A by comparing the number of spheres of a given degree that it takes to cover A with the number it takes to cover $\mathcal{W}$. (A collection of spheres is said to *cover* a set A just when the set formed by the union of all these spheres contains A as a subset.) By performing the comparisons with smaller and smaller spheres, we obtain progressively better and better estimates of the proportion of $\mathcal{W}$ that is occupied by A. The best estimate is given by the limit of the sequence of better and better approximations.

This procedure is formally analogous to one of estimating the volume of a vat by counting the number of basketballs required to fill it, or the number of marbles, and so on. This is precisely the same sort of method used to define, in a nonarbitrary way, a measure of the area or volume of a region, given the *distances* between the various points the region contains. There are very nasty hitches to be unravelled along the way if we are to obtain a measure on propositions from distance between worlds. But mathematicians have already done all that unravelling for us.

23 For further work on the derivation of a probability measure from an accessibility metric, see Bigelow (1976, 1977, and 1979). See also Munroe (1953) and Friedman (1970).

Hence, given degrees of accessibility between worlds, we get a probability measure for free. Conversely, given any probability measure we wish to use to interpret some domain of probabilistic discourse, we can tailor degrees of accessibility in such a way as to generate just the measure we need.

Hence, the semantics we offer, based on a "logical space",

$$\langle \mathcal{W}, \mathcal{D}, d \rangle,$$

can be used to interpret probability discourse, whichever sort of probability we chance to meet. And there are different sorts of probability, just as there are different sorts of possibility.

In particular, there are epistemic and nonepistemic sorts of probability – subjective and objective probabilities – degrees of rational belief, and physical propensities. It is reasonable to expect that these distinct sorts of probability would rest on quite distinct probability measures. And these distinct probability measures would presumably have to be generated by quite distinct measures of degrees of accessibility.

The question then arises whether the degrees of accessibility we must draw on to interpret a given sort of probability discourse are the *same* as the degrees of accessibility we employ in interpreting counterfactuals. We conjecture that objective probabilities, in the sense of physical propensities, will rest on the same degrees of accessibility as counterfactuals.

The relationship between counterfactual-related accessibilities and *epistemic* probabilities is more problematic. There are reasons for thinking, however, that the relationship should be a fairly close one. There is a principle which David Lewis calls the *Principal Principle,* which draws subjective and objective probabilities together.[24] We shall explain in some detail how this principle relates the two kinds of probability, because the existence of such an essential tie between the two kinds of probability provides a further powerful argument in support of the modal realism we have been defending.

The Principal Principle begins from the recognition that probability appears under two guises. There are, on the one hand, the degrees of rational belief for a person to hold at a time. Lewis calls this probability concept *credence*. And there are, on the other hand,

24 See Lewis (1980).

the *chances* of various possible outcomes of various chance set-ups. The chance of an outcome is to be distinguished from the degree of rational belief for a person in that outcome. If a coin is biased, the chances of one outcome are not the same as the chances of the other outcome; but for a given person at a time, it may not be reasonable to hold unequal degrees of belief in the two outcomes. Even when a person knows that the coin is biased, if that person does not know which way it is biased, then it might be unreasonable for her to hold degrees of belief which exactly match the chances.

Although chance is to be distinguished from credence, there are links between them. Note, for instance, that although chance need not equal credence, there may be special cases in which the degree of rational belief for a person does match the chance of an outcome. This may be so as a matter of sheer luck. For instance, a person might believe a coin to be biased but not know which way, and so it may be that the reasonable degrees of belief for that person are equal for either possible outcome; yet the coin may, in fact, be a fair one, so that the chances of either outcome are exactly equal. In that case, by luck, the person's credences will match the chances.

And yet sometimes credence may match chances not by luck but by good management. If a person is sufficiently well informed, if, for instance, she knows exactly how much and which way a coin is biased, then her rational degrees of belief may exactly match the chances. Such a person must be well informed but not too well informed. In the extreme case, if such a person knew everything, she would already know the outcome of the chancy occurrence, and so her rational degree of belief in that outcome would be certainty. Yet the chances of that outcome may, in fact, be less than certainty. In order for credences to match chances (by good management), we have to suppose a person to be fully informed *about everything relevant to the chances* and not to be informed about the outcome except by way of information about the chances. Let us sum up the required constraint on information by using the term *admissible evidence*. Credence should match knowledge about chances provided that a person has only admissible evidence, evidence which bears on the outcome, epistemically, *only* by bearing on the chances of that outcome.

The link between well-informed credence and chances can be extended. Instead of considering a person's knowledge about chances, we may consider that person's beliefs about chances. In this case, his degree of belief in an outcome may not match the actual chances, but only the believed chances of that outcome. However, it does seem obvious that your reasonable degree of belief in an outcome should match what you believe to be the chances of that outcome, provided that your beliefs are all admissible ones – that is, provided that your beliefs all bear on the outcome only by way of bearing on the chances of the outcome. This is what Lewis calls the Principal Principle:

If a person believes (fully – to degree 1) that the chances of an outcome are *r*, and has only admissible evidence, then the reasonable degree of belief in that outcome is *r*.

We shall abbreviate this as the claim that credence must match believed chance.

In this formulation, we have supposed that a person believes the chances to be *r*. Implicitly, we are supposing that the person believes to degree 1, is certain, that the chances are *r*. Often a person may hold a range of hypotheses concerning chances and may hold each of these hypotheses with a different degree of belief. In that case, the reasonable degree of belief in an outcome is to be obtained by taking each of the possible chances of that outcome, weighting it by the credence for that chance hypothesis, and summing across all the possible chance hypotheses.

Lewis takes the Principal Principle to be obviously true. We agree. Lewis does not offer an explanation of why it is true. We wish to offer an explanation. And, as very often happens in philosophy and science, a desire for explanation leads to realism.

We hold that, when one proposition has a greater chance of being (or becoming) true than another, this difference in chances must be grounded in the natures of the propositions themselves. A similar situation holds for credences. Degrees of rational belief are relative to a person at a time. But given a person at a time, if one proposition has a greater credence than another then this difference must be grounded not only in the nature of the person, but also in the natures of the propositions.

Take a proposition to be a set of possible worlds. Take the chance of an outcome or an event to be the chance that a special sort of proposition is true: a proposition which asserts that that outcome will eventuate or that that event will occur. And we take a rational degree of belief to be a rational degree of belief in a proposition. (Some beliefs are, in fact, more complex than this: there are such things as *de re* beliefs, which relate a person not to a proposition, but to an object or an object under a description. These complexities are not, however, relevant to our present concerns.)

Whether a proposition is assessed for chance or for credence, there must be some feature of the proposition in virtue of which it has the chance or credence it does. We argue that the feature of a proposition which grounds either its chance or its credence is in either case the same: it is the proportion of logical space occupied by that proposition.

This is not to deny that truths about chances are, in a sense, empirical truths about the things within a given world. In a similar way, truths about what a person would do if offered a bribe are truths about the nature of that person in the actual world. Nevertheless, with probability, as also with counterfactuals, one way to give information about the nature of things within a world is to give information about how that world is located with respect to other worlds in logical space. Truths about chances for a proposition are, we argue, concerned with the proportions of logical space occupied by that proposition. This might appear to entail that truths about chances are not truths about the contents of the world we are in, but that appearance is illusory. Truths about logical space can give information about the worlds located in that logical space; and conversely, discoveries about the contents of the world we occupy will give us information about the logical space in our neighbourhood.

Credences too, like chances, are used to give information about the world we are in. But, again as with chances, information about logical space can be used to give information about the nature of the person whose credences we are describing; and conversely, information about a person will give information about the kind of logical space she thinks is in her neighbourhood.

The notion of a proportion of logical space is a generalization from a simple idea which arises naturally in the finite case. Imagine someone who is so extremely opinionated that only a finite number of possible worlds are compatible with that person's beliefs. In that case, a given proposition may be true in some of her belief worlds and not in others. There will be a *ratio* which represents the proportion of belief worlds in which the given proposition is true. This ratio is an objective (although relational) property of the proposition. It is plausible to suggest that a person's rational degree of belief in such a proposition will be proportional to the *ratio* of her belief worlds in which it is true.

When infinitely many worlds are compatible with your beliefs and you wish to estimate the proportion of belief worlds in which a given proposition holds, you cannot simply compare cardinalities. Often the set of all belief worlds will have the same cardinality as two of its disjoint proper subsets. If we measured probability by cardinalities, we would contradict the fundamental principles of the probability calculus, since the probability of the disjunction of two incompatible propositions would not be the sum of their probabilities taken singly. Hence, it would be a mistake to measure probabilities just by cardinalities.

And yet, we maintain, there is some objective sense in which a proposition may occupy a *proportion* of a given set of worlds. In a similar way, a region of space may occupy a proportion of some wider region. This proportion is not measured by the number of points it contains; nevertheless, it is an objective feature of a region that it occupies a specific proportion of a wider region. We can be confident that there is such a proportion, even when we are as yet unsure about how precisely this is to be analysed mathematically. Similarly in the case of propositions, we may be confident that one proposition is true in a greater proportion of worlds than another, even if we are as yet unsure about how precisely this is to be analysed mathematically. It is this proportion which is represented by the measure P, appealed to in the semantics for probability. This measure on a set of worlds is not a purely arbitrary imposition. We have argued that it is derived from the degrees of accessibility among the worlds which this proposition contains. Thus, the measure on a set of worlds represents a real property of that set of worlds, dependent on the intrinsic characteristics of the worlds it contains.

If we conjecture that there is an objective fact about the proportion of worlds in which a proposition is true, then we can explain the Principal Principle. The chance of an outcome is, we submit, determined by the proportion of accessible worlds in which it occurs. This property of proportion is the ground of chances. And it is also the ground of credence. The rational degree of belief in a proposition is, we submit, determined by the proportion of belief worlds in which it is true. The property of proportion is the same in both cases; what varies is only the reference class of worlds against which the proportion is measured. In the case of chances, we measure the proportion of *accessible worlds*. In the case of credence, we measure the proportion of *belief worlds*, of worlds compatible with a person's beliefs at a time. Both chance and credence are, as it were, *volume in logical space*. When belief worlds are based on all the admissible evidence, they come in line with the accessible worlds, and that is why there is an essential link between chance and credence, captured by the Principal Principle.

It is intrinsically plausible that chance and credence rest on proportions of worlds. By grounding both chance and credence in such proportions, we can explain why they are linked. This not only provides an explanation of the Principal Principle, but also provides confirmation of the underlying postulation of proportions among sets of worlds. In general, one of the strongest arguments for various brands of realism is furnished by their explanatory power. Realism about sets of worlds and about relations of proportion among them is no exception.

3.7 HIGHER-ORDER QUANTIFIERS AND METAPHYSICAL REALISM

We have allowed a name to be replaced by a pronoun (or variable) and then to be bound by a quantifier. In a sentence like

The stockmarket in Japan crashed,

we may replace the name 'Japan' by the pronoun 'it':

The stockmarket in it crashed.

We may then precede this by a quantifier to give, say,

For every country, the stockmarket in it collapsed.

This procedure is very powerful, but we may wonder why only one syntactic category should be singled out for this treatment. Why only names? Why not, for instance, predicates? It is, indeed, possible to extend quantification to cover predicates as well as names. The result is called *second-order logic*.[25]

The idea is this. We begin with a sentence, say

The wombat sleeps tonight,

25 See Boolos and Jeffrey (1980) for a sound presentation of second-order logic and Boolos (1975) for some suggestive explanation of its philosophical significance. A precursor to the present theory is sketched in Part III of Bigelow (1988a); the theory offered in this book is less ambivalent and much more fully articulated. Higher-order logic traces back, as we said earlier, through Whitehead and Russell (1910) to Frege (1879).

and we replace the predicate 'sleeps' by an analogue of a pronoun. We cannot use the same sort of pronoun that we used for names:

The wombat it tonight.

What we need is not a pronoun but a propredicate; here we will have to improvise:

The wombat thus-hows tonight.

We may then prefix a quantifier:

For somehow the wombat thus-hows tonight.

It is not easy to find everyday English to say what we need to say here. We are driven to symbols. What we do is to introduce a new category of variable, for which we still employ the Greek letters ψ, ψ_1, ψ_2, . . . ; these can replace any predicate in an atomic formula to yield an atomic formula. Thus,

$$Fa \equiv Fb$$

can become, say,

$$\psi a \equiv \psi b.$$

This can then be preceded by an appropriately second-order quantifier:

$$\forall\psi(\psi a \equiv \psi b),$$

'for anyhow, *a* that-hows iff *b* that-hows'. (We could use a special quantifier symbol $\forall^2$, but that seems unnecessary.)

Second-order logic has many intriguing virtues, but we shall mention just one here. Using second-order quantifiers, we can define the identity predicate. We do not have to add it as a new primitive, governed by extra axioms. We can introduce the identity predicate as a merely abbreviatory, defined term. The definition we need is called Leibniz's law:

$$\forall x \forall y((x = y) \equiv \forall\psi(\psi x \equiv \psi y)),$$

which says that any two things are identical just when, anyhow that one is, the other is too.

The syntax of higher-order logic is as sketched above. The tricky step, however, concerns the semantics. The question arises as to what language we should use in formulating our semantic theory.

The standard approach has framed a semantic theory for second-order logic within a first-order logic. More specifically, the theory is framed within a metalanguage containing first-order quantifiers and a set-membership predicate governed by set-theoretical axioms, usually those of ZF.

When our semantics proceeds in that manner, second-order logic becomes equivalent to a fragment of set theory. To illustrate, a sentence like

$$\exists \psi(\psi a),$$

'There is somehow that *a* is', asserts in effect that

$$\exists y(a \in y),$$

'There is some set to which *a* belongs'. The semantics for such a second-order sentence proceeds in the following manner. Corresponding to each predicate, we assign a set. Thus, a simple sentence,

$$Fa,$$

will be true provided that the referent of *a* is a member of the set corresponding to *F*. When *F* is replaced by a variable ψ, this variable will range over all the sets that a predicate like *F* might refer to. Hence,

$$\exists \psi(\psi a)$$

will be true iff there is a set to which *a* belongs.

It is interesting what Leibniz's law boils down to under this set-theoretical spotlight. Leibniz's law says that two things are identical just when, for anyhow one is, the other is that-how too. Set-theoretical semantics for this, however, makes it equivalent to something like

Two things are identical just when they belong to exactly the same sets.

This is true enough. But it is worrying, given that it is intended as a definition of identity. It defines identity, in effect, in terms of sets. Yet the set-theoretical axioms characterizing sets are framed using identity. There is a risk of vicious circularity. Technically, the charge of circularity can be rebutted by the observation that identity is being defined in one language, while set theory is being used in its metalanguage. Because the identity predicate and the set theory are on different language levels, there is no straightforward circularity.

Nevertheless, the circularity is not harmless. And there are further reasons for feeling discontented with the treatment of second-order logic as just a notational variant of set theory.

Consider, for instance, the assertion that x is a set:

$$\text{Set}\ (x).$$

We may then use second-order logic to assert that all sets have something in common, that there is somehow that all sets are:

$$\exists \psi \forall x\ (\text{Set}\ (x) \supset \psi\, x).$$

This is surely true: there *is* somehow that all sets are. They are all sets *because* they are all that-how. 'Being a set', for instance, is somehow that all sets are.

Yet if we try to interpret this assertion in the standard set-theoretical manner, we reach contradictions. Under a set-theoretical light, the claim that there is somehow that all sets are becomes

There is some set that contains all sets.

And this is provably false, as Cantor showed.

There can be no set of all sets. Yet there is somehow that all sets are. The moral is that second-order logic should not be conflated with set theory. If we are to take a second-order logic seriously, we should formulate our semantic theories in a second-order language. When framing the semantics for a second-order language, we should not do so in a first-order language. For present purposes, however, the main conclusion to draw concerns the nature of universals and the nature of the instantiation relation which binds instances to universals.

Universals, as we construe them, are things which can be referred to by names and which can be quantified over by first-order quantifiers. Many universals are distinctive primarily in their relationships to space and time: they can be in two places at once. They are not distinctive because they are especially closely linked with a predicate or open sentence. In the first place, not every predicate or open sentence corresponds to a universal. It is a mistake to infer from

There is somehow that these things are,
$\exists \psi(\psi x_1 \wedge \psi x_2 \wedge \cdots),$

without any extra premisses, to

There is something that these things have in common,
$\exists z(x_1 \text{ instantiates } z \wedge x_2 \text{ instantiates } z \wedge \cdots)$.

The latter identifies a universal, in our "recurrence" sense. The former does not. The second-order existence of *somehow* that things are does not, by itself, ensure the existence of *something* that stands somehow to each of those things. This reinforces and clarifies our earlier arguments to the effect that the primary arguments for universals are not semantic.

Theories of universals have always been bedevilled by the problem of instantiation. Suppose an individual has a property. What must there be in the world in order for this to be so? There must be the individual, and there must be the property: but this on its own is not enough. The individual and the property might exist without the one instantiating the other, so the mere *existence* of these two things is not enough to ensure that the world is as it has to be in order for it to be so, that the individual *has* the property. What else must there be in the world? Perhaps there must be the relation of instantiation. But the existence of that relation only raises the same problem over again: the world could contain the individual, the property, and the relation of instantiation, and yet it *still* not be true that this individual *stands* in the instantiation relation to that property.

The problem of instantiation arises out of a mistaken inference from second-order to first-order quantification. In order for an individual a to *have* a property F, there must be *somehow* that the individual stands to the property

$$\exists \psi(\psi(a, F)).$$

But this 'somehow' is not a 'something', so we should not infer that there is something (instantiation) which relates a to F. From

$$\exists \psi(\psi(\mathrm{a}, F))$$

we should not infer

$$\exists x \exists \psi_1(\psi_1(x, a, F)).$$

That leads to the infinite regress which has always harassed the theory of universals. On our diagnosis, the error arises from a failure to take second-order quantification seriously – from a mistaken conflation of the 'somehow' that instances stand to universals with a 'something' which links instance with universal.

A full theory of universals, then, requires a presemantic source for universals in the sense of recurrences – things of the sort which play fast and loose with space and time, and which instantiate without ever being instantiated. But a theory of universals also requires irreducible second-order existences. And it requires a recognition that first-order existences cannot be read off second-order existences without further premisses.

4

Modal ontology

4.1 TAXONOMY OF MODAL REALISMS

Modal realists should affirm some sort of correspondence theory of truth for modal language. A suitable sort of correspondence theory is furnished by the possible-worlds semantics found in Chapter 3 – provided that we can believe there really are the sorts of possible worlds and other possibilia which such semantics requires. We defended the view that possible worlds and other possibilia do exist and that possible-worlds semantics can be construed realistically, as a literally true account of what the things in the world are which make modal claims true.

In saying that possible worlds exist, however, we are so far saying very little about the nature of the truthmakers for modal claims. Very little has been said about what possible worlds are, what they are made of, how they are related to other things that exist, and so on. A modal realist must fill in such details about the nature of possible worlds before the formal semantics in Chapter 3 can provide an adequate explanation of the truthmakers for modalities.

Different realist theories will give different accounts of the nature of possible worlds. Yet it is convenient to express all such theories within a single framework. In all realist theories, something may be said to be *possible* just when there is a world which *represents* the actual world as being a certain way. Different possible worlds are different things, each of which represents the actual world as being a certain way. All but one of these worlds misrepresents the actual world – all except for the actual world, which represents how things actually are.

Possible worlds, then, will be described as representing the actual world; but for the time being, we ask you to take this word 'represents' as just technical jargon. It is not intended to carry any explanatory weight, at least not yet.

Not only do possible worlds represent or misrepresent the actual world, they will also be said to represent or misrepresent one another. We can then say that 'Possibly *p*' is true in a given possible world *w* just in case there is some world which represents or misrepresents world *w* as being such that *p*.

Having expressed modal realism within this terminology, we may map out the major distinctions that can be drawn among the varieties of realism. Different varieties of realism give different accounts of what possible worlds are like and of what features they have in virtue of which one world represents another world as being a certain way.

Three broad categories may be distinguished among the varieties of modal realism: *book* theories, *replica* theories, and *property* theories.

Books

The first category of realisms consists of theories which construe possible worlds as *maximal consistent sets of truthmakers* – or *complete books,* for short. Variations among such theories will arise from differing opinions on the nature and status of the truthmakers which band together to constitute possible worlds. Nominalists would like to take worlds to be maximal consistent sets of sentences (though for some of them, this theory can only be a stop-gap measure awaiting a decent theory which eliminates any appeal to set theory). Carnap spoke of "state descriptions", and more recently Richard Jeffrey has spoken of "completed novels". Plantinga takes worlds to be maximal consistent sets of propositions, which in turn he apparently takes to be unstructured simples. Alternatively, we might develop Russell's logical atomist conception of propositions. Take an atomic proposition to be a combination of particulars with universals: an ordered sequence consisting of an *n*-place universal followed by *n* individuals. Such a sequence may be taken to represent the possibility of those individuals instantiating that universal. Such a sequence will play the same role as a sentence with an *n*-place predicate and *n* referring terms: it will be a truthbearer. And a maximal consistent set of these may be taken to be a possible world. This way of constructing worlds is called *combinatorialism*. Alternative choices of truthbearers are possible. Ellis, for instance, builds functional equivalents for possible worlds under the guise of rational belief systems. These theories differ from one another over fairly important issues. But they all build worlds out of truthbearers. So they all treat worlds as, in a broad sense, books.[1]

Book theories of possible worlds also must give some account of what we have called the representation relation which holds between a book and the actual world or between one book and another. Suppose we want to say that it is true (in the actual world) that such-and-such is possible. The book-theorist will say that this is so because there exists a certain book. Why should the existence of this book have any bearing on the possibility in question? Only because this book *represents* the world as being one in which such-and-such is the case. If this book were one which represented the world as having, say, animals on the moon, its existence would be relevant to the possibility that there are animals on the moon. If it represented the world as having apes in the Alps, and no animals anywhere else, then the existence of that book would be completely irrelevant to the possibility that there are animals on the moon.

A book-theorist must give some account of what a book must be like in order that it should "represent" the world as being a certain way – of what a book must be like in order that its existence will be relevant to one given possibility rather than some other. In principle, there is room for book theories to subdivide into a variety of distinct theories, according to the different views they might take about the nature of the representation relation. In practice, however, book theories tend to be indistinguishable on this score. They all say equally little about the representation relation. Worlds are construed as sets of truthbearers; and it is assumed that they simply inherit a representational character from the representational character of the truthbearers out of which they are built. Presumably they represent in virtue of incorporating some sort of *reference* to things in the actual world.

Replicas

Book theories constitute the first of three broad categories which we will distinguish among realist theories. The second category

1 For the linguistic doctrine, see Carnap (1947) and Jeffrey (1965); for the abstractionist view, see Plantinga (1974, 1976, 1987); for the combinatorialist view, see Russell (1912) and Armstrong (1989); and for the conceptualist view, see Ellis (1979).

includes the modal realism of David Lewis.[2] Lewis does not construe worlds as sets of truthbearers, or as sets of anything else for that matter. He construes worlds as *replicas*. Each of these replicas "represents" any given world as being the way it is itself. A world which contains animals on the moon correctly represents the actual world as having a moon, but misrepresents it as having animals on the moon. Such a world contains something which straightforwardly resembles the actual moon – something which shares a vast number of properties with the actual moon. It represents the actual moon in virtue of its *resemblance* to the actual moon; Lewis calls it a "counterpart" of our moon. This otherworldly moon, however, differs from ours, because it is inhabited by things which resemble, and hence represent, actual-world animals, so that world represents these animals as being on the moon.

Lewis's replica theory is just as explicit as the book theories are about the nature of worlds; but it says much more about the relationships among worlds and about the features of worlds which explain the fact that they "represent" the actual world as being this way rather than that. The way they represent the world as being is simply the way *they are* themselves. Things in these worlds represent this or that actual individual in virtue of their resemblance to this or that actual individual. These things misrepresent an individual as being a certain way, in virtue of dissimilarities between them and actual individuals.

Replica theories other than Lewis's are possible. The key feature of Lewis's theory which invites variations is his counterpart theory. Lewis's replica of, say, the moon is an individual which resembles our moon but which is not numerically identical with our moon. Something in a world cannot count as a counterpart of our moon unless it passes a definite threshold of resemblance to our moon. Furthermore, it cannot count as a counterpart of our moon if it has a worldmate which resembles our moon more than it does. But if a thing in a world resembles our moon sufficiently, and at least as much as anything else in that world does, then Lewis calls it a counterpart of our moon. It *represents* our moon if it *resembles* our moon closely enough, provided that there are no better rival candidates, in the same world as itself, for the role of representing our moon.

2 Highlights from Lewis's output on modal realism include Lewis (1973a, 1986a).

Other modal realists dissent from counterpart theory: Hughes and Cresswell, Kripke, and Tichý, for instance.[3] They speak of other worlds not just as containing something which resembles our moon, but as containing our moon itself. Our moon may be inhabited in that world and uninhabited in ours; but it is one and the same moon which is inhabited in the one case and uninhabited in the other.

Lewis's counterpart theory does allow one sense in which one and the same thing can be present in several worlds and be inhabited in some worlds and not in the others. Lewis believes that there is, for instance, such a thing as the *aggregate* of all the counterparts of our moon. This aggregate may be said to be present in many different worlds, since different parts of it are present in each of many different worlds – each of these parts may be called a *world stage* of this aggregate of counterparts, and this aggregate is sometimes called a *hydra*.[4] The moon hydra may be said to be inhabited in some worlds and not in others, because some of its world stages are inhabited and others are not.

Some disputes between Lewis and opponents may therefore be based on a misunderstanding. What they call individuals he calls hydras; what he calls a (simple) individual (like our moon) they call an individual-in-a-world. Some disputes, then, may be merely verbal; but not all are.

There is another way in which realists can avoid counterpart theory. What is required is that we reconstrue what seemed to be *properties* as really being *relations*. We think of the shape of the moon as a property rather than a relation. But on reflection, it may be urged, we should note that the moon has slightly different shapes at different times. Consequently, we must construe shape-at-a-time to be a property but shape to be a relation between an individual *and a time*. This is a fairly orthodox view to take in response to the classical problem of change. So we should not be unduly alarmed at the idea that what we thought of as a property is really a relation.

Lewis, in fact, adopts a different yet equally orthodox view of the problem of change. He construes an individual, like our moon, to be an aggregate of "temporal stages", or "time slices". The moon

3 Hughes and Cresswell (1968), Kripke (1980), and Tichý (1971).
4 Lewis (1986a, pp. 96, 192).

has different shapes at different times, he says, because part of it is one shape and part of it is another – an early part of it may, for instance, be smoother than a later part of it. The moon is not an individual which is wholly present at distinct times. Lewis's solution to the problem of change is very closely analogous to his solution to the problem of transworld identity.

We here will adopt Lewis's account of the persistence of individuals across time. And by parity of reasoning, we favour the counterparts version of a replica theory of possible worlds. Yet it is worth noting that the arguments are not conclusive. There is still something to be said for the view that the very same individual may be wholly present at several distinct times. And the same holds for the view that the very same individual may be wholly present in several distinct worlds. The price to be paid is that we must reconstrue what we took to be a property, like the shape of the moon, as really being a three-place relation holding of an individual, a world, and a time.

Both the book theory and the replica theory construe worlds as representations; the difference between them is like the difference between a script for a movie and a remake of a movie. On a book theory, worlds are not literally similar to what they represent, any more than the word 'dog' is similar to a dog; on the replica theory, possible worlds are similar to the things they represent. Indeed, for realists other than Lewis, some of the things in other worlds may be said to represent actual individuals, not because they merely *resemble* those individuals, but rather because they are numerically identical with those individuals.

However, unlike actual scale models, other worldly replicas cannot be located at any distance from us, in any direction, either earlier or later than now. They are causally isolated from us. Anything which has an effect upon us, or which we can affect in any way, is automatically part of our world. This assumption of causal and spatiotemporal isolation is common ground among replica theories, both for Lewis's counterpart theory and for the rival transworld identity theories.

Replica theories have many advantages over book theories; and most of the objections raised against them are fairly feeble. We shall canvass some of these advantages and objections. But in the end we shall recommend neither a book theory nor a replica theory.

We shall argue that possible worlds cannot be taken to be books. There are, we believe, a great variety of books – that is, there are a great variety of sets of truthmakers. We believe that there are sets of sentences, and sets of propositions, and sets of beliefs. These things exist, sure enough, but they cannot bear the explanatory weight that possible worlds must bear if they are to deserve the name. In particular, construing worlds as books will prevent us from giving a satisfactory correspondence theory of truth for modal language.

Replicas stand a much better chance of furnishing a satisfactory correspondence theory of truth for modal language. Yet we shall argue that replicas fall victim to shortcomings which are analogous to the shortcomings of books. Both replicas and books represent the actual world as being a certain way, in virtue of their properties, the properties of the actual world, and consequent relationships between them and the actual world. We argue that all theories with that structure are mistaken. Fortunately, there is a third option for modal realists, a strategy recommended, for instance, by Robert Stalnaker and by Peter Forrest. Possible worlds can be construed as *universals*. Our theory is closer to Forrest's than to Stalnaker's, because Forrest is a much more robust modal realist than is Stalnaker.[5]

A world is not just something which represents the actual world as being a certain way. In order for anything to represent the world as being a certain way, *there must be* a certain "way" that the world is represented as being. A possible world, we will argue, is one of these ways: a world is identical with a way the world could be. And we construe this as meaning that a possible world is in fact a *property*. One of these properties is the way the world in fact is; but all the other properties are properties which the actual world does not instantiate. In fact, nothing instantiates these other properties. They are properties something could have had but nothing does actually have.

Any modal realism, we have said, must say something both

5 See, e.g., Stalnaker (1976) and Forrest (1986a). Although Stalnaker speaks suggestively of properties the actual world does not have but could have had, he is not a brazen Platonist about the ontological status of such properties. Forrest is much closer to our scientific Platonism.

about what possible worlds are and about how they represent or misrepresent the actual world. The property theory offers an account of both worlds and the representation relation. Worlds are properties; and a world represents the actual world correctly if it is instantiated by the actual world; it misrepresents the actual world if it is not instantiated by the actual world.

This theory, however, must be backed by an adequate theory of universals. We need to know more about what universals there are; whether they have parts, and if so, what parts they have; and so on. And we need to know something about how universals are related to particulars – and about what it is for something to "instantiate" a given property. In other chapters, we have attempted to do what we can to develop a theory of universals. Unfortunately, the theory of universals is in a state of considerable turbulence; there is no consensus view to draw upon, nor even any sizeable traditions within which there is a significant degree of consensus. Consequently, the property theory of possible worlds should be taken to be a schema for an adequate theory, waiting for details to be filled in gradually.

Before outlining our positive theory, we shall scout around some alternative strategies: first, the theory that no theory is needed, then the book theory, and then the replica theory.

4.2 AGAINST MODALISM

Book theories construe possible worlds as maximal consistent sets of truthbearers. The presence of the word 'consistent' in this thumbnail sketch signals the most important problem faced by book theories: the problem of circularity. The notion of consistency is essentially a modal notion: two truthbearers are consistent just in case it is possible for both to be true. If worlds are construed in terms of consistency, then indirectly, worlds are construed in terms of possibility.

Yet a crucial motive for introducing possible worlds arises from the desire to explain what makes modal claims true. Realism requires that, when certain things are possible, some account must be given of what things there are in virtue of which those things are possible. Possible-worlds theories suggest that something is possible just in case there exists a possible world of a certain sort. The existence of this possible world is supposed to explain what constitutes the fact that such-and-such is possible.

Keeping in mind the explanatory role which possible worlds are supposed to fill, consider again the nature of book theories. We are to explain the truth of a possibility claim in terms of the existence of a possible world; yet this world in turn is said to exist only because a certain possibility claim is true. This circle is too tight; it robs possible worlds of almost all their explanatory power. A modal realist must be able to define possible worlds without appealing to any modal primitives. We believe it can be done, but it is much harder than you might expect.

Because it is so hard to avoid modal primitives, it is natural to ask whether we really have to avoid them. What is so very wrong with taking one or another modal notion as primitive? The doctrine that some modal notions cannot ever be defined nonmodally is called *modalism*.[6] Echoing Hume's characterization of the fact–value dichotomy, we may sum up modalism in the slogan 'No *must* from an *is*'. When Hume said that we could not logically derive a moral 'ought' from nonmoral facts alone, he was presenting us with two options: either we say there are no moral truths, or we take some basic moral truths as primitive, and not analysable purely in terms of what nonmoral things there are in existence in the world. Hume's 'ought from an is' poses a choice between moral nihilism and moral primitivism. This problem concerning the moral 'must' carries over with equal force to the metaphysical 'must'. Modalism asserts that modal facts cannot be derived from facts solely about what (nonmodal) things exist in the world and how they are (nonmodally) related: No 'must' from a pure 'is'. This leaves us with a stark Humean choice, between modal nihilism and modal primitivism.

If you reject the correspondence theory of truth, perhaps for you there will be nothing wrong with modal primitives. Yet we have argued for a correspondence theory of truth. If we meet difficulties in defining possible worlds, it must be admitted that this provides some indirect reason for reconsidering the correspondence theory which led us into these difficulties. We think, however, that these doubts which can be cast on the correspondence theory have much less weight than the arguments in its favour. Consequently, we will persist in seeking an account of possible worlds which is compatible with the correspondence theory of truth.

6 Lycan (1979), Plantinga (1974, 1976, 1987), and van Inwagen (1985) deal sympathetically with the idea that some modalities do not have to be explained in terms of anything more basic.

The difficulty in avoiding modal primitives is severe enough, however, to prompt us to wonder whether we might be able to live with modal primitives without relinquishing the correspondence theory. We might argue, for instance, that a claim about possibilities is made true by a correspondence to things in the world, but that the things in the world to which it corresponds cannot be described without using modal terms.

Compare the case of modal terms with that of proper names. It might be thought that a name like 'Gough' refers to something nonlinguistic, something which is not itself a truthbearer. Yet when we are asked to specify which thing in the world corresponds to this name, we will have to refer to an individual. The natural way to refer to an individual is by using a name. And the name which is easiest to hand is 'Gough'. Thus, it is quite natural to find ourselves using a name in the process of giving a correspondence theory for that very name. We could, of course, avoid using that very name by using some other roundabout description. Yet, arguably, those other descriptions will always introduce *other* names. So we cannot state a correspondence theory for names without using names. There is no guarantee that these names are all noncircularly definable. Some may have to be taken as primitive. Yet admission of primitive names should not be seen as, by itself, constituting a threat to the correspondence theory.

Analogously, we might argue, modal primitives need not be seen as a threat to a correspondence theory. For instance, we might insist that the claim

Conchita can play the guitar

is true in virtue of a correspondence between this claim and things in the world, including Conchita and a certain property that she instantiates. This property is the property of being able to play the guitar. In describing the property, we use the modal word 'able'. Perhaps there is no way of describing this property without implicitly or explicitly drawing upon one modal notion or another. Yet, it may be argued, this is no threat to the correspondence theory of truth.

We agree. The case against modal primitives rests not on the correspondence theory alone, but rather on the thesis that modal properties must be supervenient on nonmodal properties. This is allied to the view that all dispositional properties (solubility, transparency, or whatever) must rest on a categorical basis – some conglomerate of nondispositional, nonmodal properties in virtue of which a thing possesses the relevant disposition.

To say that modal properties supervene on nonmodal properties is to say that we could not have any difference in modal properties *unless* there were a difference also in nonmodal properties. This supervenience claim has been called *Humean supervenience* by David Lewis[7] and forms a cornerstone of his metaphysics.

A supervenience thesis, if true, would guarantee the definability of modal properties in nonmodal terms: that is to say, it would guarantee definability provided that we allow for *infinitely complex* definitions. Supervenience of facts about one subject-matter, A say, on another class of facts, B, always ensures that the full story, perhaps infinitely complex, about B will *logically entail* the facts about A. This we have argued in Chapter 1. Such a logical entailment constitutes a definition of A-facts in terms of B-facts, not a finite or pragmatically useful definition perhaps, but enough like the more familiar definitions to warrant the extension of the term to these supervenience cases. There is no guarantee that supervenience of modal on existential facts will be accompanied by *finite* definability. Yet even if there is no finite definition of the modal in terms of the nonmodal, there should be some characterization to be found concerning the overall shape of the infinitary reduction, the constraints it obeys, and so forth. If the modal supervenes on the nonmodal, it should be possible to characterize the manner in which it does so. And this will yield, if not strictly a definition, nevertheless a characterization of the modal without appeal to modal primitives.

Hence, in what follows we shall restrict our attention to theories which aim to characterize possible worlds without appeal to modal primitives. Whenever a theory, on examination, proves to harbour a modal primitive, we will take that as grounds for setting that theory aside. This does not mean that the theory is indefensible; it may be thoroughly congenial if you are willing to abandon Humean supervenience. It does not mean that the possible worlds which are defined using modal primitives must be rejected as nonexistent. We put such definitions aside only because they cannot play the explanatory role which is required of them if they are to fill in the details for a Humean supervenience thesis.

7 Lewis (1986a, p. 14 and Intro.).

4.3 COMPLETE BOOKS

Some people have found mathematics philosophically perplexing and have been tempted to say that numbers are really just symbols – visible marks or audible noises. The trouble with symbols, however, is that in one way there are too few of them, yet in another way there are too many of them. There are too few in that, for some very large numbers, there would seem to be no corresponding visible or audible numerals. Yet there are too many in that, for smaller numbers, there are a great many more distinct visible or audible numerals than there are numbers represented. There are just two natural numbers between 3 and 6, yet there are very many more than two marks and noises corresponding to those numbers. For instance, there are '4', and 'IV', and 'four', and so on.

An analogous problem arises for the idea that possible worlds are books. On the one hand, there are not enough books, since there are lots of possibilities which no one has ever written about. On the other hand, there are too many books, since several different books may correspond to a single possibility. If any book were to correspond to a given possible world, that book could surely be translated into many different languages. Then there would be many different books, yet only a single corresponding possible world. Books, therefore, are both too numerous, on one front, and too scarce, on another. Hence, possible worlds cannot be books. So runs the initial argument against any identification of possible worlds with books.

The appropriate response for book-theorists is to specify more precisely exactly what sorts of books they have in mind. Possible worlds should not be construed as books in the sense of maximal consistent sets of *visible or audible* truthbearers. Yet that does not rule out the theory that possible worlds are maximal consistent sets of truthbearers of some other sort.

Not only are there fewer physical books than possibilities, but there are in fact no tangible books which describe just one possible world. No book ever written is *complete* in the sense that nothing could be added to it without introducing an inconsistency. No physical book contains a *maximal* consistent set of sentences.

And yet, for any physical book, we can imagine indefinitely many ways in which it would be possible to complete it. There are indefinitely many possible books consistent with any given physical book. It might be suggested that a possible world be identified with a *possible book* rather than a tangible book. An analogous strategy is sometimes used to save the thesis that numbers are just symbols: the higher numbers are construed as merely possible numerals.

We set this theory aside. It introduces a modal element into the definition of possible worlds. This is all right for a modalist; but we have given our reasons for restricting our attention to theories which aspire to a nonmodal characterization of possible worlds.

How can we construct the required sorts of books without falling into merely possible books? One way is by loosening our conception of the allowable ways of building a sentence. In spoken language, sentences are formed by uttering words one after the other. In written language, there is greater variation in the manner of constructing sentences. We anglophones place words one after another, left to right, top to bottom. Other language speakers do it differently. But the details of how sentences are constructed are extrinsic to most of our important concerns when we investigate sentences. It does not really matter how we write them. Indeed, we can even imagine alternative ways of constructing a spoken sentence. We could pause for a long time between one word and the next. We might even type or semaphore several sentences between one spoken word and the next. Or we might begin a sentence out loud and finish it in Morse code. And in Disney comics, Donald's nephews often share a sentence among the three of them.

It is convenient, therefore, to define the notion of a *sentence* in a way which abstracts from the details concerning conventionally accepted modes of aggregation that are used to create physical sentences. It is convenient to define a sentence as a *sequence* of words, in the mathematical, set-theoretical sense of the term 'sequence'. In this sense, a sequence can be formed – or rather, a sequence exists whether or not we do anything to "form" it – without any need for any sort of physical aggregation of the members of the sequence. We can take the word 'eppur', as written by Donald Davidson (call it a), 'si' as uttered by Al Capone (call it b), and 'muove' as muttered by Galileo (call it c), and there will be the sequence

$$\langle a, b, c \rangle.$$

This sequence will count as a sentence of Italian – even if no one uttered it. Legend has it that after affirming that the earth does not move, as he was instructed to do by church authorities, Galileo muttered, under his breath, '. . . and yet it does'. It makes a good yarn, though there is no evidence to support it. But the existence of the Italian sentence in question does not stand in need of a Galileo to utter those words in the right order, close together in space and time. The existence of the set-theoretical sequence of *a* and *b* and *c* is enough to guarantee the existence of that sentence *in the Italian language*.

A book can be defined as a sequence of such sequences.[8] And when defined in that way, there will be an unlimited number of books. There will even be an infinite number of maximal consistent books. It will not be easy to read them. Some of them will be too scattered. Some will not be scattered enough – the required sequences could all be built out of the physical words in a single copy of a dictionary. The resulting book would be in one sense entirely visible, being constructed entirely out of visible components. The trick would be, however, that of reading the words in the right order – the conventional order (left to right, top to bottom, page 1 onwards) would yield an ungrammatical sequence.

Applying set theory to physical symbols yields infinitely many "books". It might be argued that there are still not enough to serve our purposes. There are only countably many such books – that is to say, such books can be paired off with the natural numbers, with one book for each natural number and no books left unpaired. Yet, it may be objected, possible worlds cannot be paired off with the natural numbers. There are surely at least as many possible worlds as there are points in space. And it has been cogently argued that the points in space cannot be paired off with natural numbers – the Pythagoreans are said to have discovered this. Hence, there are more possibilities than there are books. We must conclude that possible worlds are not to be identified with books.

However, there are ways of extricating the book theory from this difficulty. We might, for instance, loosen up our notion of what counts as a *word*. After all, it does not matter whether we write a word in chalk or ink, or whether we write in red ink on yellow paper or blue ink on black paper or black ink on black paper. We might even doubt if it matters whether someone wrote it or

8 Lewis (1986a, p. 142) is our most immediate source for the present treatment of book theories, but the idea that sentences, theories, books, and so on are sequences of sequences is not idiosyncratic. Cresswell (1973), e.g., takes such a view more or less for granted.

whether it has been there through all time, as some Hindus believe to be true of some of their sacred scriptures. Loosening up to the limit, we can take any suitably shaped region of space to be a word, written in superinvisible ink, as it were. Then there will be no shortage of symbols. They will outnumber the points in space.

This profusion of symbols, however, will avail us nothing if they are all built from copies of a finite list of words in the vocabulary. The profusion of symbols will then be an embarrassment. For any one possible world there will be uncountably many books describing that world. Yet there will still be worlds which are not adequately described by any of these "books", because they contain things which are not describable within the vocabulary of the language in question.

Consider first the problem that there are many distinct books describing a single possibility. If we are to identify a world with anything, it should not be with an arbitrarily selected one of these books. Rather, we should identify a world with the set (actually an equivalence class) of *all* the books which describe it. To carry through such a reduction, however, we would need to give a noncircular, nonmodal account of how to demarcate the set of books which describe the same possible world. This is not easy. When establishing that two books describe the same world, we might begin by establishing that each sentence in one book can be matched by some equivalent sentence or set of sentences in the other book. Yet take care: what sort of "equivalence" should we be thinking of? If we are thinking of *logical* equivalence, we are drawing upon an unexplicated modal notion. Then we have left the game. Yet if we do not call upon logical equivalence, how are we supposed to delimit the set of *equivalent books,* the set of books which describe the same possible world?

Let us leave this problem for the time being. Suppose we could give a nonmodal definition of the set of books which we will identify with a possible world. We would still face the problem that, if all these books have a finite vocabulary, then there are not as many books as there are possibilities. The only way of solving this problem is by allowing a language to have an infinite vocabulary. We have to ensure that every individual, in any given world, has a name and that every property or relation corresponds to some predicate.

One way of ensuring that everything gets a name is by allowing, for instance, the following method of constructing names:

If *a* is any individual, then the sequence
⟨ this, *a* ⟩
is a name which refers to *a*.

If *all* the names and predicates in a language are built out of individuals and properties or relations, the language becomes what Lewis calls a Lagadonian language (in memory of an episode in Swift's *Gulliver's Travels*).[9] A Lagadonian language, then, is a set-theoretical structure built out of individuals, properties, and relations. Theories which build worlds that way are often called *combinatorial* theories, since they construe worlds as combinations and permutations of the things in the world we live in. What we arrive at are entities which are not normally thought of as *linguistic* at all. Instead of a subject–predicate sentence, we now deal with a sequence containing the individual referred to by that subject term and the property expressed by that predicate. This is structurally analogous to a sentence. But the notion of language has been stretched too far if you call this a sentence in a language. The combination of a property with an individual is, in many cases, an entity which would exist even if people never existed. It is misleading to call something a language if its existence is independent of the existence of anyone who could conceivably speak it. Yet there is still a point to highlighting the analogy between books and combinatorial structures. It draws attention to the fact that some of the central arguments against book theories do carry over into arguments against combinatorialism.

We can throw our net even farther afield, construing even more theories as fundamentally extensions of book theories. One crucial kind of theory we should acknowledge is the kind promoted by Plantinga. On this sort of theory, a possible world is a maximal consistent set of *propositions;* and propositions, in turn, are taken to be linguistic in only a very Pickwickian sense. Propositions, like combinations of individuals and universals, would exist even if people did not. But unlike combinations of individuals and universals, they are not construed as having any structure, either syntactic or set-theoretical.[10]

All these theories share a battery of serious problems. They all

9 Lewis (1986a, p. 145). Bigelow (1975) offers an extended implementation of the Lagadonian programme (and extended quotations from Swift).

10 Plantinga (1974, 1976, 1987) presents propositions as abstract and essentially representational entities. Lewis (1986a, sec. 3.4, pp. 174–91; 1986b; 1986c) criticizes all theories of the same sort as Plantinga's.

construe possible worlds as maximal consistent sets of *truthbearers* of one sort or another. Hence, they all face a question about what features these truthbearers have *in virtue of which* they represent what they do. The closer the truthbearers are to sentences in a natural language, spoken by a population, the easier it is to see a basis for the claim that they represent certain things and not others. The problem of explaining reference and so forth for a real, spoken language is hard enough; but it is much harder to explain what serves as a grounding for alleged representational properties in the case of "languages" which are entirely detached from any population which uses them. When we arrive at Plantinga's conception of truthmakers as "propositions" which lack all syntactic or set-theoretical structure, it becomes very obscure what justifies attribution of any representational relations. Lewis objects that the theory makes representation a "magical" relation. Plantinga counters, in effect, by taking the representation relation, for propositions, to be primitive and unanalysable but perfectly intelligible, and hence not magical.

The problem of analysing the representation relation (or justifying the claim that it is unanalysable) is closely allied to a problem mentioned earlier: the problem of giving an account of what makes two books *equivalent* – the problem of *translation,* if you like. At Plantinga's end of the spectrum, this problem vanishes, since he assumes that propositions which match representationally will always be numerically identical. But for other theories, defining some sort of equivalence is a genuine problem. If we identify a world with one specific book, we must justify choosing that book rather than some other, one which differs only in superficial syntax, for instance. Even combinatorialism faces the problem that there are rival set-theoretical constructions which can provide equally good candidates for identification with the same possible world. For these reasons, it is more rational to identify a world with a set of equivalent books. Two books are *equivalent* when they represent the same possible world – but the problem is to rephrase this without presupposing modality. This reduces to the question of what there is about each book in virtue of which it represents what it does.

Let us set aside, however, the problems of representation and of translation. There remains one blatant problem for all book theories, broadly construed. This is the problem of giving a noncircular account of what makes a set of truthbearers a *consistent* set. Even if book theories could solve all the other problems they face, we cannot see how they could solve this one, without being modified beyond recognition. We shall return to this issue later, when we discuss property theories. For the present, however, we note our broad agreement with Lewis about the shortcomings of book theories. We now move on to the analogous shortcomings of Lewis's own theory.

4.4 AGAINST REPLICAS

The big problem for book theories is to select which sets of truthbearers are to count as *consistent*. One way to guarantee that a description is consistent is by finding something which satisfies that description. If there is something which does have a given collection of characteristics, that collection cannot be a self-contradictory or inconsistent one. Displaying something which falls under a concept is one way of demonstrating the consistency of that concept. Call this the *instantiation principle*.

The importance of the instantiation principle can be illustrated by the history of mathematics. In mathematics, more clearly than in other subjects, consistency is of paramount interest. In other subjects, consistency is simply presupposed most of the time. When inconsistencies turn up, they must be eliminated; but this is an incidental inconvenience, like housekeeping, and not the central goal of the subject. In pure mathematics, in contrast, consistency seems to become an end in itself. Hilbert voiced this perspective on pure mathematics, saying that, once we have established the consistency of an axiomatic system, we have established all we have to establish for the purposes of pure mathematics.[11]

Frege objected to this consistency-based theory of pure mathematics. He held that existence precedes consistency. Consistency is not enough. Indeed, consistency presupposes existence. In order for a mathematical system to be consistent (or at least for it to be shown to be consistent) there must be a subject-matter which it accurately describes. If there exists something which satisfies a given definition, for instance, then that definition is obviously a consistent one. If, however, nothing satisfies a given definition, then, Frege asks, what warrant do we have for supposing the definition to be consistent?

11 For the debate between Hilbert and Frege, see Frege (1980).

Frege is thinking epistemically, in terms of "warrants". But Frege's question can be extended: if nothing satisfies a definition, what real difference is there between this definition and an inconsistent one? No difference at all, Frege assumes. Modern mathematics rests heavily on the Fregean assumption that there must be examples of structures satisfying any given definition if we are to be confident that the definition is consistent. The reason that set theory is important in pure mathematics is that it furnishes precisely what Frege asks for: examples which instantiate all the mathematical structures which any mathematician has ever wanted to study (at least until recently).

Lewis sees modality from Frege's standpoint rather than Hilbert's. If something exists which fits a description, obviously that description is consistent. Yet conversely too, Lewis claims, if a description is consistent, something will exist which fits that description. Lewis therefore offers an answer to the troublesome question which haunts all book theories of modality: the question of what it is that distinguishes consistent from inconsistent books. Complete books are just special cases of descriptions. Consequently, Lewis applies the Fregean litmus test to them. Consistent books are ones which truly describe some existing things; inconsistent books are simply ones which do not truly describe anything at all.

Lewis describes and defends his theory so well that he needs no help from us. So we shall pass over the positive arguments which can be mounted in favour of his theory. We shall also pass over many of the arguments against his theory; Lewis clearly states many such arguments, and in our view he does a good job of displaying their shortcomings.[12] We shall focus on only those objections which we find most illuminating.

The blank stare

Philosophical folklore parodies David Lewis by characterizing a Lewis-type proof that *p* as follows:

Most people find the claim that not-*p* completely obvious and when I assert *p* they give me an incredulous stare. But their confident claims are no proof

12 Lewis (1986a).

that not-*p*; and I do not know how to refute an incredulous stare. Therefore *p*.

But in fact, Lewis regards the incredulous stare, or, as we prefer to call it, the blank stare, as the strongest argument against his modal realism. In *On the Plurality of the World* Lewis says:

> I once complained that my modal realism met with many incredulous stares, but few argued objections. . . . The arguments were soon forthcoming. . . . I think they have been adequately countered. They lead at worst to standoffs. The incredulous stares remain. They remain unanswerable.[13]

Lewis believes that incredulous or blank stares are unanswerable but *inconclusive*. He argues that common sense is conservative, and it is this conservativeness that produces the blank stare. Lewis approves of such conservativeness. Yet conservativeness and common sense can be outweighed if the gains are sufficient. Modal realism offers many gains, and there is a lack of viable alternative theories. Nonetheless, the theory must pass the following test:

> A simple maxim of honesty: never put forward a philosophical theory that you yourself cannot believe in your least philosophical and most commonsensical moments.[14]

The blank stare indicates that, for the starer, the theory fails the test. Lewis believes that, while his denial of common sense is severe, it is not too severe given the gains. But he acknowledges that this judgement is contentious and that many will disagree. They go on staring. He cannot answer them – only stare back.

We believe that Lewis is too kind to the blank starers and that the blank stare should not be left as a stand-off. We shall argue that it is in fact at best a dubious argument – not one to worry the modal realist unduly.

There is a simplistic response to the blank stare – to deny steadfastly that it is an argument, to insist that to call the blank stare an argument is just a misnomer. After all, it does not present a body of evidence, or *a priori* premisses, and infer from such to a conclusion

13 Lewis (1986a, p. 133). We refer the term 'blank stare' in part from habit but also because it covers not only the case of incredulity, in the sense in which this implies understanding of what is said followed by a refusal to believe it, but also cases like that in which a person simply does not "get" a joke – the person understands what is said on the surface, but has the feeling that she is missing something. This involves elements other than mere incredulity, and these extra elements are very relevant to many philosophers' reception of Lewis's theory.

14 Lewis (1986a, p. 135).

that possible worlds do not exist. The characteristic features of genuine arguments seem to be missing. But such a pedantic appeal to a definition only succeeds in side-stepping a genuine question concerning rationality. An adequate theory of rationality must cater for such responses to a theory as that it is very implausible, or against common sense, or simply unbelievable, whether or not these technically constitute arguments. Such responses may be difficult to explicate within a theory of rationality, but some explication is required. And a blank stare is as good a way as any other to express such a response. Lewis, of course, does not give a merely pedantic response to the blank stare, nor do we. Our contention is that, in the case of the blank stare at concrete modal realism, this "argument" or response is hard to sustain within a viable theory of rationality. It does not present a significant worry for those sympathetic to modal realism.

Some might construe the blank stare as questioning the logical consistency of the theory, but that is not the kind of blank stare we have in mind. Allegations of inconsistency or incoherence are not communicated by mere expressions of incredulity. They play a role in a rational debate which is entirely different from that of a blank stare. They are settled by formulation of proofs of consistency or by derivation of a blatant contradiction.

Minimally, the incredulous stare would indicate a very low initial or prior probability. This is, incidentally, a low *epistemic* prior probability. Suppose we take an actualist to be someone who claims that the only things there are, are the things that are in this very world, the "actual" world. Such a person need not admit that although modal realism is false yet it *might have been* true. There is a sense, however, in which he should admit that it *might be* true: he should admit that he may be mistaken. Modal realism should be given some degree of epistemic probability. It is problematic to explain just how we can give nonzero epistemic probability to logical impossibilities, but it is obvious that there must be some way to do this. Stalnaker suggests, roughly, that what has nonzero probability is not the impossible proposition itself, but rather the contingent proposition that certain *sentences* expressing impossibilities *could have been* true.[15] This strategy may serve our purposes; but if not,

15 Stalnaker (1984). Another strategy for dealing with these and related problems can be found in Cresswell (1985b).

some other account must be found. It is clearly mistaken to assign an epistemic probability of zero to all logical impossibilities. Analogously, there must be some account of the rational means by which our confidence can be boosted, even when the theories under discussion are metaphysical ones which, if true, are necessary and, if false, impossible. We expect that some development of Stalnaker's strategies will extend to cover debates about metaphysical theories. And if not, some such account must be found which will permit us to assign epistemic probabilities (rational degrees of belief) to metaphysical doctrines.

It would seem to be against common sense to assign anything but a very low initial epistemic probability to a theory which contradicts many of our most confident existing beliefs. But if the blank stare reflects no more than this, it would not constitute a distinctive objection to Lewis. For there are so many cases of theories in both science and philosophy which have initially have been seen as contrary to common sense but that in the end have gained widespread acceptance. Given any finite initial probability, no matter how small, if there are enough gains with the theory, that is, enough cases in which the theory provides the best available explanation of some phenomena, then with the accumulation of such evidence the theory will eventually attain a significant level of probability. Lewis's theory has a low prior probability, so an incredulous stare may be an appropriate reaction to hearing the theory for the first time. But this is not an argument that can be used indefinitely, whatever the arguments Lewis presents for his theory. The blank stare that we object to is one which continues, undiminished, *after* hearing Lewis's supporting arguments.

Consider some analogous cases. When the mind–brain identity theory was suggested by Place and Smart,[16] the vast majority who first heard the theory were incredulous. But the theory was so powerful that, with its gains and as the evidence for it grew, it became widely accepted – and the blank stares faded. Those who continued to disagree were not thereby irrational, but they would have been irrational if they had done so merely by sustaining an incredulous stare. These cases are not unique to metaphysics. The Copernican world system and non-Euclidean geometries provide

16 For a historically illuminating picture of early discussions of the identity theory see Presley (1967).

further examples, and the list could go on and on. These cases demonstrate that very low initial probabilities can be boosted (rationally) until they reach near certainty. As we have said, there must be ways of doing this for metaphysical (and mathematical) doctrines that are analogous to the ways in which it is done for empirical ones.

Most agree that Lewis's "concrete" modal realism provides a great many analyses of phenomena and answers to problems in areas where there is a lack of viable alternative theories. Most agree with Lewis about the gains his theory would provide if it could be taken seriously. Why is it, then, that the blank stare has not faded away? There is nothing irrational about an initially low epistemic probability for Lewis's theory. What we question is the rationality of a blank stare which persists no matter what evidence Lewis presents in support of his theory.

To resist the gradual increase in probability for Lewis's theory as evidence mounts would require us either

(1) to assign a *zero* initial epistemic probability to the theory of concrete modal realism or
(2) to grant that while there is an increase in probability with the increase of evidence, the *total* increase is *infinitesimal*.

Let us look at each of these in turn.

Suppose the blank stare signifies the assignment of a zero initial probability to Lewis's theory. And if we assign it zero initial probability, it follows by standard confirmation theory that the probability of the theory remains zero no matter how much evidence is discovered. It is an interesting question as to whether it is ever rational to assign a zero probability to a theory in which one cannot detect a blatant self-contradiction.

But even if we suppose that one can sometimes assign a consistent theory zero probability, is it reasonable to assign *this* theory a zero probability? The problem is that we then have to make sense of the sincere believers – people like Lewis himself. Lewis presents blank starers with a trilemma:

(1) They may recognize that he is sincere, sane, thoughtful, intelligent, has thought longer and harder about these issues than they have, and means what he says; and that therefore what he says has a nonzero probability of being true.
(2) They may doubt his sincerity and so on.
(3) They may conclude that it is sometimes rational to give a zero probabil-

ity to some proposition which a sincere and so on person believes to be true.

No one who knows Lewis can accept option (2). So the trilemma reduces to a dilemma: either we give some credence to Lewis's theory, or we take option (3).

Option (3), however, is deeply disquieting. It is only a very short step away from a severe kind of relativism about rationality. Let us fill in that step a bit. Option (3) forces us to say that we may judge someone to be rational, we may understand all of that person's reasons for believing what she does, and we may have no reasons which she has overlooked, and yet we may nevertheless assign a zero *epistemic* probability to her hypothesis – we may (rationally) entertain not even the slightest doubt that she is wrong. We do not wish to pursue this in further detail: but we urge that it is a position on rationality that should not be accepted lightly.

The problem is worst if the blank stare expresses the assignment of a *zero* probability to Lewis's theory. Yet the problem is still severe even if the blank stare signals only a *very low* probability for Lewis's theory. When someone reaches a conclusion, after thinking longer and harder about it than you have, and when the subject-matter is one on which you have no epistemologically privileged access, then normally it is rational to give that conclusion quite a high probability. If we are to make exceptions, what principles would decide which cases may be treated as exceptions? Why should Lewis's testimony give a nearly zero probability to his theory? Was it rational, in an analogous case, to give the theory of relativity zero or nearly zero epistemic weight when it was proposed by Einstein? No, the theory is rationally held now, and so it was never rational to give it zero probability, at least after Einstein proposed it, no matter how counterintuitive it seemed at first.

We turn now to the suggestion that the blank stare continues because, although the low probability of Lewis's theory increases with the accumulation of evidence, the total increase in probability is infinitesimal. That is, we consider the suggestion that there is no finite change in the probability no matter how much evidence accumulates.

Our objection to this suggestion is that, from Bayes's theorem, this can be accepted only by agreeing that the prior probability of all the evidence that has been, or could in the future be, discovered or presented for concrete modal realism must be near certainty, and that no such evidence could surprise us. We believe that such a consequence is unacceptable.

We are not saying that it is *inconsistent* to resist any finite change in the epistemic probability of Lewis's theory as the evidence in its favour increases, evidence predicted by the theory or entailed by the theory and not predicted or entailed by contrary theories. But the consequence is again to relativize rationality. No matter what evidence Lewis mounts, the committed actualist will have to assign it a nearly certain prior probability (though it seems unlikely he will be able to recognize this until its relationship with Lewis's theory is pointed out). We would liken that sort of actualist belief to *faith;* and the consequence of allowing this way of defending a belief to count as rational would be to relativize rationality.

We leave, then, the persistent blank or incredulous stare. We feel there is good reason to be deeply suspicious of this persistent argument. There is a very strong prima facie case for saying that it marks nothing more than intellectual inertia, a lack of imagination, or groundless ignorance. We do not pretend to have refuted all possible accounts of rationality which might explain why the blank stare is rational when directed at Lewis. But until a suitable theory of rationality is offered, the presumption must surely be that the blank stare is a bad argument.

Isolated parts

According to Lewis, possible worlds are spatially, temporally, and causally isolated from one another. And whenever two things are spatially, temporally, and causally unrelated, according to Lewis, they belong to different possible worlds.[17]

It is thus fairly natural to ask whether a single world could not have an isolated part – a part not spatially, temporally, or causally related to other parts of the world. Lewis acknowledges that this is a prima facie plausible hypothesis, but feels in the end that it is a matter to be settled as "spoils to the victor":

17 This theory is not exclusively Lewis's, but his theory presents the clearest example of a theory which makes distinctions between one world and another depend on a lack of causal or spatiotemporal connections; see, e.g., Lewis (1986a, p. 17).

A first, and simplest, objection is that a world might possibly consist of two or more completely disconnected spacetimes. (Maybe *our* world does, if indeed such disconnection is possible.) But whatever way a world might be is a way that some world is; and one world with two disconnected spacetimes is a counter-example against my proposal. Against this objection, I must simply deny the premise. I would rather not; I admit some inclination to agree with it. But it seems to me that it is no central part of our modal thinking, and not a consequence of any interesting general principle about what is possible. So it is negotiable. Given a choice between rejecting the alleged possibility of disconnected spacetimes within a single world and (what I take to be the alternative) resorting to a primitive worldmate relation, I take the former to be more credible.[18]

Yet it is not really acceptable for Lewis to treat the alleged possibility of disconnected space-times as "negotiable". The matter cannot be treated so lightly – it matters too much. Being temporally related is the only plausible candidate on which the worldmate relationship can supervene.[19]

Other candidates which come to mind furnish extremely unsatisfactory analyses of the worldmate relation. Causation is, for Lewis, defined in terms of counterfactuals, and hence is modal. Thus, causation presupposes a principle of individuation for worlds, and so cannot be used to furnish such a principle. Space seems not to be modal; yet it, too, is of dubious value in individuating worlds. If things are said to be worldmates only when they are spatially related, it would seem that Cartesian egos could not be worldmates with anything. This would commit Lewis to saying there is no possible world with Cartesian egos, a view he should not be committed to, at least not merely as a consequence of his view on the individuation of worlds. Hence, causal and spatial relations do not by themselves seem adequate as a ground for the worldmate relation.

The worldmate relation is logically prior to, or at least independent of, causal and spatial relations. Perhaps, therefore, it is a mistake to seek to analyse it at all. Perhaps it should be taken as fundamental and in need of no deeper analysis. Yet to construe the worldmate relation as a primitive is far from attractive. For a concrete realist like Lewis, there would be very little plausibility for an objective primitive relation which holds between worldmates, even between those

18 Lewis (1986a, pp. 71–2).

19 See also Davidson and Pargetter (1980) for an earlier discussion of the importance of the relative isolation of possible worlds.

in different isolated parts of the one world, and yet does not hold between objects across world boundaries. And this plausibility objection rests on more than sheer intuition. An unanalysed worldmate relation would show all the features of a modal primitive, and hence would raise the spectre of a vicious circularity. The worldmate relation is equivalent to the modal relation of compossibility, and this is clearly a modal notion. Yet the essential motive of Lewis's theory is to explain modality. Thus, he should not fall back on taking the worldmate relation as primitive.

Thus, grant that the worldmate relation is not primitive and is not analysable causally or spatially. How is it to be analysed, if not by causal or spatial connectedness? No other candidates come to mind, except that of temporal connectedness. So the analysis of the worldmate relation in terms of temporal relations, or at least in terms of some disjunction incorporating temporal relations, had better work.

It is therefore important for Lewis's theory that all and only worldmates be temporally related. Yet a commitment to that claim is worrying for a variety of reasons. In the first place, there is a reasonably stubborn intuition that it is implausible to rule out the possibility of a world with temporally isolated parts. Lewis's theory, with a temporal analysis of the worldmate relation, does rule out temporally isolated parts – and this does count fairly strongly against Lewis's theory. What is particularly worrying about rejecting the possibility of a world with temporally isolated parts is that it is epistemically possible that the *actual* world is such a world. We would not merely be rejecting the existence of some esoteric world a long way away from what is actually possible; for to have a temporally isolated region of our world does seem to be compatible with the physical theories we accept concerning our world.

Einstein's general theory of relativity allows space-times of many different topologies, including ones with disconnected regions. We do accept, as would Lewis, worlds with nearly isolated regions, regions which could be reached only along a restricted number of world lines, and then perhaps only sometimes and even perhaps only with a certain objective chance. It seems stipulative to allow all these, but then rule out the case in which the final link is broken or the chances fall the wrong way.

We can certainly imagine worlds with branching time, worlds

where there is one past but different futures. By parity of reasoning, we should allow worlds with merging times, that is, worlds where different pasts merge into a single future. Thus, it seems we should have worlds with parts that share just a time segment – where their times merge and then branch. And finally there is the world where they come together for just a moment. A simple example would be a world with effectively two parts, temporally disconnected except for the sharing of one instant. And Lewis's theory should allow such a world.

We are then struck by the arbitrariness of requiring that the two parts of such a world must at least have one moment in common. What about worlds in which there is some objective chance that they will have that moment in common, a chance that, of course, might not result in any actual shared moment?

Suppose in that one moment we meet Jane from the other part of such a world. Consider the counterfactual: if we had not met Jane, then Jane would not have existed. We think Lewis must accept this counterfactual as true, but we think that it is clearly false. Since it is false, there must be at least one world in which we do not meet Jane and yet Jane does nevertheless exist. That world must contain both Jane and us, even though the sole point of temporal (and spatial and causal) contact between Jane and us in the actual world has been severed. A defender of the temporal criterion would therefore have to maintain that in that world the severing of contact between us and Jane is compensated by the appearance of some *other* point of contact. This commits the temporal theorist to the truth of another counterfactual of roughly the form: if we had not met Jane, then at least one of our "predecessors" or "successors" would have met one of her "predecessors" or "successors". (Take *X*'s predecessors and successors to include all entities temporally related to *X*.) We think that this counterfactual is false in the world in question, and indeed even less plausible than the initial counterfactual 'If we had not met Jane, she would not have existed'. The falsity of these counterfactuals is evidence against the temporal theory.

We end this reconsideration of the isolated-parts objection with a recapitulation. Lewis has a dilemma. Either he takes the worldmate relation as primitive and undermines his theory by admitting a modal primitive, or he has to legislate against certain plausible possibilities.

The worldmate relation, then, threatens to commit modal realism to an unanalysed modal primitive. But Lewis's full theory also rests on another important relation, called the *counterpart relation*. Like the worldmate relation, the counterpart relation is not a good candidate for an unanalysed primitive. And yet, again like the worldmate relation, the counterpart relation cannot plausibly be analysed without appeal to unanalysed modal primitives. Hence, we shall argue, the counterpart relation, or any substitute, will draw modal realism into the same kind of vicious circularity that undermines the explanatory power of many of its rivals.

Representation

Suppose that there are duplicates of our planet earth – planets very much like earth, except that some of them are causally and spatiotemporally isolated from us. For present purposes, their isolation is not important. So just consider a far-away planet, Twin Earth, which contains a being who shares a great many properties with you but who also differs in various respects – for instance, in being a little taller than you are.

Does such a replica of our earth represent the possibility that you could have been taller than you are? Perhaps. But we shall argue that certain conditions must be met in order that a "world" can represent a given possibility. And these conditions presuppose modalities. Hence, concrete realism falls into a kind of circularity analogous to that which afflicts rivals of Lewis's realism: particularly those versions of modal realism that characterize worlds as sets, or books, of some kind of truthmakers (sentences, beliefs, propositions, etc.). Concrete realism can be seen as taking "world books" and giving them a *live performance*. However, it is then hard to see what is to be gained by viewing a live performance rather than just reading the book.

The idea behind concrete realism is that worlds cannot be inconsistent in quite the way that books can. Hence, whereas the book theory falls immediately into blatant circularity, the concrete world theory does not. The book theory calls upon maximal *consistent* sets of truthmakers. The replica theory does not need to restrict attention to consistent concrete worlds. If something instantiates some nexus of properties, that nexus must be a possible one. There can be no inconsistent concrete worlds. The possibility that your car might have been at the airport simply *is* the existence of its counterpart at the counterpart of the airport. So concrete worlds cannot generate any inconsistent "representations".

Yet this masks a close analogy between the books and the worlds. In one sense, there can be no inconsistent books any more than there can be inconsistent concrete worlds. Books are inconsistent not in virtue of having inconsistent properties, but only in virtue of representing two or more things which cannot be true together. Concrete worlds cannot be inconsistent in the sense of themselves having inconsistent properties, but that is beside the point. The question is just whether a concrete world could represent two or more possibilities which cannot be true together. A concrete world represents a possibility *for your car,* say, in virtue of counterpart relations linking your actual car with part of that world, the actual airport with an otherworldly airport, and so on. The counterpart relation must be constrained in such a way as to ensure that a concrete world never *represents* inconsistent possibilities *for* actual entities. The consistency of the properties of the car in the other world is not by itself enough to ensure that it does not represent inconsistent possibilities *for your car.* If the otherworldly car were to be counted as a counterpart of both your car and, say, Gorbachev, that car would represent the possibility that Gorbachev could have been your car. And then, presumably, the other concrete world would not represent a consistent set of possibilities. In one sense, of course, it is in itself a consistent possibility. Yet it may not be consistent to describe it as the possibility that Gorbachev could have been your car. And other worlds are of interest here only insofar as they *can* be described as representing possibilities for such things as your car. The question of whether a concrete world represents consistent possibilities is not one we can evade by saying that, since it exists, it must be possible.

The trouble is that a representation, whether a book or a concrete world, may be taken to represent a variety of things. The fact that Twin Earth contains concrete beings rather than dead words does not, of itself, tell us anything about what possibilities it represents. Does the existence of Twin Earth entail that *you* could have been taller than you are? If so, why? Presumably because it contains a being who shares many properties with you but who is taller than you. Yet on Twin Earth, your counterpart has parents who are numerically distinct from your parents. Height is one respect in which your counterpart differs from you, parentage is another. Does this mean that the existence of Twin Earth entails that you could have had different parents from the ones you have? Or that you could have had parents from another galaxy? Surely not.

Your counterpart's parents have various properties, including being absent from our earth, but also such properties as being, say, richer than your parents. Twin Earth may be seen as representing the possibility that your parents could have been richer but not, say, that your parents might have had no common ancestors with earthlings. Why is this? The answer, we suggest, is because it *is* a possibility that your parents were richer but not that your parents had no common ancestors with earthlings. A modal realist can, of course, tailor the counterpart relations in such a way as to ensure that it generates the "right answers". Yet, we argue, this tailoring requires modal assumptions, and hence robs concrete realism of its potential to explain modality. If there is a modality-free definition of counterparts, this has to be demonstrated.

Even if we grant the existence of concrete worlds, there will have to be constraints which presuppose modalities. Hence, we argue, concrete worlds fall prey to the same kind of circularity that world books do. And, conversely, if Lewis can avoid circularity in his theory, there is no reason why world books could not be rescued by the same strategy. Concrete worlds have always appeared to have one decisive advantage over world books: a guarantee of consistency. That guarantee is, we argue, illusory. Concrete worlds run no more, and no less, risk of inconsistency than world books do.

Lewis's account of the counterpart relation has two components. First, something nonactual can be a counterpart of something actual only if it passes a threshold of sufficient resemblance to the actual thing. Second, the counterpart must resemble the actual thing at least as much as its worldmates do. The first place we suspect hidden modal presuppositions is in the definition of the threshold of sufficient resemblance. Having "enough" properties in common really just amounts to sharing all *essential* properties, all properties which the thing *must* have in order to exist.

In addition to worries about the threshold required for the counterpart relation, worries may be raised about Lewis's requirement that the counterpart of an actual thing cannot be less similar to the actual thing than any of its worldmates is. This rules out possibilities which, we argue, should not be ruled out.

It is possible that you could have had a duplicate, someone very, very similar to you, as similar as anyone could possibly be without ceasing to be a numerically distinct person. So there is a possible world with two people in it, one of them being your counterpart and the other being your duplicate. Lewis's theory allows this, provided that there is some feature your counterpart in that world has and your duplicate lacks, such that your counterpart resembles you more closely than the duplicate does. Your counterpart might, for instance, have originated by ordinary biological reproduction, whereas the duplicate occurred by some in vitro intervention, say.

Compare this possibility of the existence of a duplicate who is not you with the possibility that if history had been different you would have been a very different person in many ways. There is a possible world, then, which contains a counterpart of you, and that counterpart is very different from you, differing from you much more than the possible duplicate. This, too, is compatible with Lewis's theory, as long as the very different counterpart has no worldmates which are more like you than it (she or he) is. What Lewis cannot allow, however, is a world which contains *both* the duplicate who is extremely similar to you but is not your counterpart, *and* the counterpart who is very different from you, together as worldmates. Both would pass the threshold of sufficient resemblance, but the so-called duplicate would resemble the actual you more closely and hence would be your counterpart. And the so-called counterpart would fail Lewis's requirement that your counterpart never be less similar to you than any of its worldmates, and hence it would not be your counterpart.

Consequently, Lewis's theory does not allow the possibility that you could have been a very different sort of person and, at the same time, someone else could have been very similar to the way you actually are. This is, we maintain, an unacceptable consequence.

Lewis can block this unacceptable consequence, but only by putting more modal weight on the threshold required for the counterpart relation. In the world with the different you together with the duplicate, we could tailor the counterpart relation to ensure that the duplicate is not your counterpart. We could argue that, despite appearances, the duplicate does not pass the required threshold of resemblance. Perhaps the oddity of its nonnormal biological origin rules it out as a counterpart. Then the different you will, after all, be at least as similar to you as any of its "qualified" worldmates, as any of its worldmates that do pass the required threshold.

This defence of Lewis, however, places great strain on the threshold requirement. Some property which you have, and the duplicate lacks, will have to be essential to any thing's passing the threshold required for the counterpart relation. Why should this be so? Because this is a property which you could not possibly lack without failing to be the person you are. That is a modal presupposition. So again, the counterpart relation really conceals a hidden modal primitive.

We can summarize our complaint against Lewis's account of representation by giving a variant of our example. Suppose in the actual world we have identical twins, "identical" not only in their biology but in all the other ways in which twins can be found to be similar. We argue that each of the twins could have been different, so there are worlds where the twins have twin counterparts which differ in various ways from the actual twins but where the resemblance is sufficient to sustain the counterpart relation. Let the twins be Dum and Dee, and their counterparts be Tee Dum and Tee Dee. Presumably in some worlds the counterparts will be significantly different from one another, consistent with preserving their biological ties. Then it is possible for one counterpart, say Tee Dum, to be more closely similar to Dum than Tee Dee is to Dee. Yet what does Lewis's theory entail about such a case? It entails that of the nonactual twins Tee Dee is more closely similar to Dee, so is Dee's counterpart, which is what we would hope to find. Yet Tee Dee is also more closely similar to Dum than any of his worldmates are, so he also counts as the counterpart of Dum. Tee Dee is the counterpart of both Dee and Dum, and Tee Dum is the counterpart of neither. Yet in some other worlds, possible individuals exactly matching Tee Dum are counterparts of Dum. The failure of Tee Dum to be a counterpart of Dum is a consequence of his twin brother's closer resemblance to both Dee and Dum and has nothing to do with his own intrinsic character.

Yet it is not plausible to say that Tee Dee is the counterpart of both Dee and Dum and that Tee Dum is the counterpart of neither. That is equivalent to the modal claim that one of the twins could not have been different unless the other had been different too. That is unacceptable. The only way Lewis could block that consequence would be by appeal to essential properties distinguishing between

Dee and Dum. That is unacceptable to Lewis not only in its appeal to individual essences, but also in its appeal to modal presuppositions in a theory which is intended to explain modality.

There is yet another way in which replica theories can be argued to rest on a modal primitive, when we look at their representational role. Consider the way that a story may turn out to be inconsistent. It may contain a sentence saying that, when Toula drives to work by one route, she crosses the railway line just once. Another sentence may say that, when she drives to work by an alternative route, she crosses the same railway line just twice. It may also be implicit or explicit in the story that neither route goes around the end of the line, that space is near enough to being Euclidean that no spooky translocation or time warps are occurring, and so forth. The story then entails that Toula's place of work is on the opposite side of the tracks from her home and that it is on the same side as her home.

What makes it possible for an inconsistency like this to occur in a book? Words in the book map onto individuals and properties, and then a name may be syntactically linked to two predicates, even though the individual referred to could not have both of the corresponding properties.

What prevents a replica from falling into exactly the same inconsistency? Why could we not find that the *representation* relations, linking the replica to the actual world, yield exactly the same inconsistency that the *reference* relations yielded for a book? To avoid inconsistencies, we must ensure an intricate harmony in the assignment of counterpart relations. We must, for instance, ensure a harmony between the assignment of counterparts for Toula, for Toula's place of work, for the train lines, for her trips to work, and so on. Suppose Toula's counterpart goes on a drive, in a possible world, crosses the tracks twice, and arrives at a place of work. Does this show that there could be a route to her actual place of work which crosses the tracks twice? It would, if that drive were the counterpart of one of her actual drives to work, which crosses the tracks just once, say.

But could the twice-crossing drive be the counterpart of the actual once-crossing drive? Yes, but only if Toula's place of work had been on the same side of the tracks as her home. Either that, or the twice-crossing drive must represent Toula's drive as being a drive not to work, but to somewhere else. You could not allow the counterpart relation to hold between the actual and the possible drives unless you also allowed a change in the counterpart relation for her place of work or in the identification of which properties the counterpart represents the actual drive as having. Changing one counterpart would require changing other counterparts too, in order to ensure that the world in question represents a *consistent* set of possibilities. Books represent possibilities only when they are consistent, when their referential relations are adequately harmonized, as it were; similarly, replicas represent possibilities only when their representational, counterpart relations are adequately harmonized. The harmonization of the counterpart relations is every bit as modally loaded as the required harmonization of referential relations for world books.

If we are right that the counterpart relation conceals modal primitives, then rival replica theories begin to seem more and more seductive to a modal realist. Rival replica theories circumvent Lewis's arguments for his counterpart theory. Lewis imagines an entity, in another world, which is numerically distinct from the real you and which qualifies as your counterpart. His rivals say that the other world contains you yourself, not someone numerically distinct from you. In place of Lewis's counterpart relation, the rival theory employs the *numerical identity* relation.

The identity relation does not harbour any modal primitives, at least not in any sense which would threaten the replica theory with vicious circularity. Hence, it might appear that, by replacing the counterpart relation with transworld identity, the replica theory can eliminate modal primitives and evade vicious circularity.

This appearance is illusory. There is an isomorphism between counterpart theories and transworld identity theories. Pushing the modal primitives out of the counterpart relation will avail us nothing: they will pop up again in different dress, in a different place, but playing the same role.

Suppose that another possible world contains not an entity which is numerically distinct from Toula, but Toula herself. Then consider a property Toula might or might not have, say anger. Lewis might say that the actual Toula is angry and her counterpart is not. The transworld identity theory cannot say that. Either she is angry or she is not. In order for her to be angry in one world and not in another, given that it is the very same Toula in both, we must say that anger is, in fact, not an intrinsic property of Toula, but a relation in which she stands to a context. In the first place, she may be angry at one time and not at another. In the second place, she may be angry in one world and not in another. So anger is a relation in which Toula stands to a time and a world. Attributions of anger take the fundamental form not of

Angry (Toula),

but of

Angry (Toula, t, w),

meaning that Toula is angry at time t in world w.

Under this version of the replica theory, what becomes of the problem of consistency? Lewis's theory requires constraints on the counterpart relation. Those same constraints will now have to apply to the relations in which a "transworld" individual stands to various distinct worlds.

Consider Toula's problematic trip to work. A story can misrepresent Toula as going to one and the same place of work by two routes, one crossing the tracks twice, the other just once. Nothing *in the book* itself crosses the tracks, but things in the book *represent* two distinct and incompossible situations. Likewise, a replica of our world cannot contain two routes to the same destination, one crossing the tracks once and the other twice. And Toula's actual place of work cannot be on both one side, and the other side, in the replica world.

But why not? Why can't Toula's place of work be located on the same side of the tracks as her home, and also be on the other side of the tracks, and yet not be in two places at once? On the transworld identity theory, that would require her place of work to stand in several distinct relationships to one and the same world. It would have to stand in the "this side of the tracks" relation to that world and, at the same time, in the "that side of the tracks" relation to the very same world. There is no world to which her place of work stands in both those relations at once. That is, there is no world which *represents* her place of actual work as being on this side of the tracks and as being on that side of the tracks at the same time. Why not? Because these two properties, or relations to world–time pairs, are incompatible. Yet that is a modal fact. This modal incompatibility between properties (or relations) has been smuggled into the theory in disguise.

We can add a step to the argument here. There is *no* world w to which Toula's place of work stands in *both* the relation R_1, 'being this side of the tracks in w' *and* the relation R_2, 'being that side of the tracks in w'. We should ask, however, what features of Toula's place of work and of world w ground the fact that at least one of the relations R_1 and R_2 always fails to hold. Do these relations hold or fail to hold quite independently of the intrinsic characters of the related terms? That is a possible position one could take. But we would reject that position for precisely the reason that Lewis rejects Plantinga's sort of book theory, as resting on "modal magic". We cannot believe that such relations hold or fail to hold of a world independently of the intrinsic character of the world.

Suppose, then, that relations of things in our world to world w hold or fail to hold depending on the intrinsic character of world w. Then we are owed an account of what that intrinsic character amounts to. This takes us back irresistibly to Lewis's counterpart theory. We then repeat our arguments to show that this theory hides modal primitives no more, and no less, than book theories do.

There is one further way of construing transworld identity theory. The properties of 'being this side of the tracks' and of 'being that side of the tracks' are *incompatible*. There is a relationship between these properties. Suppose we want to encode this relationship between properties set-theoretically. Suppose we represent properties by set-theoretical relations, namely, by conveniently chosen sets of sequences. Then it may be convenient to choose sequences in such a way that, when properties are incompatible, no sequence corresponding to one has a last member which also features as a last member of a sequence corresponding to the other.

On this construal, possible worlds feature merely as ways of set-theoretically encoding modal relations like incompatibility. That makes the theory unexceptionable for the formal purposes it is well suited to meet. But there is a kind of thirst for an explanation of modalities, which such a theory cannot begin to quench. Such a theory gives set-theoretical models of relations among properties, such as incompatibility. This is worthwhile, but incomplete – it leaves us still lacking an explanation of what incompatibility and other modal relations amount to. We hope that these can be analysed in terms of a part–whole relation among universals, ultimately explicated in terms of two distinct levels of quantification (as we urged in the final sections of Chapters 2 and 3). Any appeal to such relations among universals, however, will take the theory of modality well outside the scope of the replica theory of possible worlds.

Let us sum up the argument so far. In counterpart theory, a replica of our world represents possibilities for a thing in our world only if it contains a counterpart of that thing in our world. And we have argued that constraints on the counterpart relation are needed, and they presuppose modalities. In a transworld identity theory the picture is different, but also the same. A replica of our world represents possibilities for a thing in our world only if that thing stands in several distinct relationships to the replica world. And we have argued that constraints are needed on such relationships to those replica worlds, and those constraints presuppose modalities.

Hence, the replica theory does harbour modal primitives after all, whether it takes the form of a counterpart theory or a transworld identity theory. A replica theory is, after all, in the same boat as a book theory. The supposed advantage of replicas over books rests on the supposition that replicas cannot be contradictory, whereas books can. Yet this overlooks the representational role which replicas must play in the theory and which they share with books. A replica cannot have contradictory properties, but neither can a book. The important thing is not whether it can have contradictory properties in itself, but whether it can represent contradictory properties *for* things in our world. To ensure that a replica represents only compossible possibilities for things in our world, constraints must be placed on the representational relations in which it stands to things in our world. Those constraints are no more and no less modally loaded than the required constraint in a book theory, namely, that a world book must be a consistent set of sentences. Changing a highly *conventional* representation (a book) into an *iconic* representation (a replica) does not escape the modal consistency constraints which apply to all representations. And possible worlds can be relevant to modalities only if they are construed, in one way or another, as representations (or misrepresentations) of the actual world we live in.

If the existence of something, whether a book or a concrete world, is to be relevant to everyday modalities, it must represent some way that things could have been here on earth. Some of the properties and relations instantiated by its components must represent properties and relations that things in our world could have had. Others must not. The question remains whether we can give a modality-free account of the representational role of either books or concrete worlds. We do not believe that concrete modal realism has yet shown that it does.

We have argued against concrete modal realism; yet we wish to stress that many of the things we wish to do with possible worlds can be done whether or not you accept concrete modal realism. Much of what we say later about possible worlds will not depend on the details of our own world-property theory or on our rejection of book and replica theories.[20]

4.5 WORLD PROPERTIES

The actual world instantiates various properties. There are other ways the world could have been. If the world had been one of those ways, it would have instantiated different properties than the ones it does instantiate. Those other properties are uninstantiated. But, we claim, those uninstantiated properties do exist. They are what constitute the alternative possible ways the world could have been. So said Stalnaker and then Forrest (with a more congenially realist intention), and so, too, say we. Possible worlds are universals. More specifically, possible worlds are complex structural universals of the sort described in Chapter 2.

It may be objected that not every uninstantiated world property constitutes a genuine possibility. It may be argued that only *consistent* properties can be counted as constituting possible worlds. And so a modal primitive will creep into the property theory, just as it does into book theories and replica theories. We maintain, however, that there are no inconsistent properties. There are inconsistent predicates, of course, but they do not correspond to any universals. Only consistent predicates correspond to universals.

By positing the existence of the right number of universals, not too many and not too few, we can give a realist account of modalities. We can explain modalities in terms of existence. We can derive a 'must' from an 'is': what is possible depends on what there is, particularly on what universals there are and how they stand to one

20 This section is based on Bigelow and Pargetter (1987b). David Lewis provided useful comments on this material.

another. In this respect, the property theory is like the book or replica theory. But those theories presuppose the property theory. In order for a book or replica to be relevant to modality, there must be some way which it represents the world being. We take the phrase 'there must be' here to require the *existence* of a universal, a world property. Book and replica theories presuppose world properties; but world properties do not stand in any explanatory need of either books or replicas. Books and replicas may serve a useful purpose in picking out world properties. But it is the existence of those world properties which really does the explanatory work. There can be replicas or books which represent those properties only on the prior assumption that there exist such properties for things to have or represent.

In the world-property theory, we do not have to deny the existence of books (or replicas). In fact, we believe there is good reason to accept the existence of books, particularly books of the combinatorialist sort, built set-theoretically out of the individuals, properties, and relations in the world. We deny, not the existence, but only the explanatory force of such things. A satisfactory realist correspondence theory for modalities must draw on further resources to separate the consistent from the inconsistent combinations. We urge that the consistent combinations are those which correspond to appropriate sorts of world properties. The inconsistent combinations, like inconsistent predicates, do not correspond to any appropriate universals.

Consider an example. Suppose that 'red' and 'round' refer to universals and that *a* is some particular apple. Then does a combination including

$$\{\langle \text{red}, a\rangle, \langle \text{round}, a\rangle, \ldots\}$$

represent a possibility? Yes, but this must not be simply assumed without scrutiny. Suppose that 'cubic' represents a universal too. Then does a combination including

$$\{\langle \text{round}, a\rangle, \langle \text{cubic}, a\rangle, \ldots\}$$

represent a possibility? No. The reason is that there is such a property as being red and round, but there is no such property as being round and cubic.

Consider another example. Suppose *F* and *G* are universals, and *a* and b are individuals. Then does a combination which pairs F with a and G with b,

$$\{\langle F, a\rangle, \langle G, b\rangle, \ldots\},$$

represent a possibility? Perhaps not. It could be that a is an individual with an essence that rules out being F – an individual such that, if it had had property F, then it could not have been the individual it is. Likewise for b and property G. And even if a could be F and b be could be G, it is conceivable that these two possibilities could not be true together. *Internal relations,* as we use the term, are defined to be relations which are such that the related things could not have failed to be thus related. It might be a consequence of such internal relations that a could be F only if b fails to be G.

The existence of internal relations is sometimes doubted. A plausible example is sometimes proffered by referring to points in space. Under one conception of space, it consists of points, and these points are without intrinsic properties. Their essence lies entirely in their mutual relationships. A point could not have been the point that it is if it were related in a different way to the other points.

Another model for internal relations might be provided by quantum mechanics. The Pauli exclusion principle lays constraints on the properties a pair of fundamental particles may have. One particle may have a spin of $+½$, say, but only provided that another has a spin of $-½$.

The most comprehensive, all-purpose method of coping with internal relations is to consider higher-order properties and to consider properties not just of things in the world, but of the world as a whole. A combination including

$$\{\langle F, a\rangle, \langle G, b\rangle, \ldots\}$$

is a possible one provided that there exists such a property as

containing something with the property of being a and being F, and something with the property of being b and being G.

If we consider larger and larger aggregates, we can encompass more and more such properties. Some of these are instantiated; others are not. When we reach the largest possible aggregate of worldmates, a whole possible world, we then have available the broadest properties of that sort. A maximal consistent property of that sort is a possible world, a way the world could be.

Let us recount some of the ways in which a theory of this sort might be articulated. To begin with, consider combinatorial theories of the sort descending from Wittgenstein's *Tractatus*.[21] Consider, for instance, some of Carnap's proposals for the underlying structure of the world. Suppose we take the ultimate, simple individuals in the world to be the points of space-time. Each of these points may either have or lack certain given qualities. And the whole state of the world, we are to suppose, will be determined entirely by the ways in which qualities are distributed through space-time. Everything, we are to suppose, will supervene on this distribution of qualities. For purposes of illustration, we may simplify further. Suppose there is only one relevant quality which points may either have or lack: that of being occupied by matter. Each point is either occupied or unoccupied. We may then consider the following set of ordered pairs,

$$\langle \text{occupied}, x \rangle,$$

such that x is a space-time point which is occupied by matter (is *actually* occupied, is occupied in the actual world). On our supposition that everything supervenes on the distribution of matter, it follows that such a set of ordered pairs entirely determines all truths about the actual world. A different set of ordered pairs of that form, however, will not represent the actual world, but will *mis*represent the actual world. Sets of such ordered pairs determine what Cresswell has aptly called Democritean possible worlds.[22]

A pair of the form

$$\langle \text{occupied}, x \rangle$$

is a proposition in roughly the sense of Russell's logical atomism. It is a sequence of an n-place universal (a one-place universal, or property, in this case) followed by n (in this case, one) individuals. It represents the individual space-time point x as being occupied. We may suppose that any two such propositions are logically independent. That is, a given point may be either occupied or unoccupied,

21 Combinatorial theories include those of Wittgenstein (1922, 1929), Carnap (1947), Armstrong (1989), Skyrms (1981), Barwise and Perry (1983), and Bigelow (1988b).

22 Cresswell (1972, 1973).

whether or not another such point is occupied or unoccupied. In that case, any set of such propositions will be consistent. Any such set determines a possible world.

We do not say, however, that such a set of (Russellian or Carnapian) propositions "constitutes" a possible world. A possible world, we have argued, is not a book. It is not a set of propositions. A possible world is a universal – in fact, a structural universal. The "complete book of the world" is not itself a world, but only represents a world – it represents a way the world could have been. What we must do is identify the world property which corresponds to a Democritean world book.

The way to turn a world book into a world property is as follows. Begin with a basic combination of particular with universal, an ordered pair,

$$\langle \text{occupied}, x \rangle.$$

This corresponds to the conjunctive property 'being x and being occupied'. The same correspondence will hold for another such pair,

$$\langle \text{occupied}, y \rangle,$$

which will map onto the property 'being y and being occupied'. These two properties can then be combined in one of a number of ways. Most notably, we may combine them by first forming the aggregate of the two points x and y and then by considering properties of this aggregate. The aggregate of x and y is a miniature region of space; and it has the structural property of

having a part which is x and occupied and a part which is y and occupied, and being exhausted by the parts x and y.

This structural property then corresponds to the miniature book formed by taking the two propositions together:

$$\{\langle \text{occupied}, x \rangle, \langle \text{occupied}, y \rangle\}.$$

This technique of mapping miniature books onto structural properties of aggregates can be repeated with longer and longer books, until we finally arrive at a mapping of each complete Democritean book onto a maximal world property.

The Democritean model can be modified and elaborated in many ways. For a start, we can add more individuals. In addition to (or even instead of) taking points of space-time as individuals, we may posit a large number of fundamental particles (or waves, or whatever they are): protons, neutrons, electrons, photons, and so forth; or quarks and gluons, or something else of that sort. These individuals may be assigned a variety of properties. Again, the choice is to be made in tentative stages by the progress of physical science: for illustrative purposes we may suggest such properties as charge, position, momentum, and mass. For present purposes it does not matter exactly what the individuals or the properties are. The point is that, in place of Democritean propositions, we may construct propositions like

$$\langle P_1, x \rangle,$$
$$\langle P_2, x \rangle,$$
$$\text{etc.},$$

where x is a fundamental particle, and P_1, P_2, and so on are physical properties. We may parcel these propositions into a single sequence,

$$\langle P_1, P_2, \ldots, x \rangle,$$

which represents x as having the properties P_1 and P_2 and so on – in other words as having the conjunctive property of being P_1 and P_2 and. . . .

Note, next, that many of the properties P_1, P_2, and so on may be determinate properties which fall under some determinable. That is, they may be quantities; they may stand in relations of proportion to others of their kind. It will be convenient, then, to designate some specific determinate, under a given determinable, to serve as a unit. Each determinate under that determinable will stand in a distinctive proportion to the unit. Thus, each determinate will be identifiable by reference to the real number which constitutes the proportion in which that determinate stands to the unit. Hence, the sequence linking an individual x to properties P_1, P_2, and so forth can be represented by a sequence

$$\langle r_1, r_2, \ldots, x \rangle,$$

where each r_i is a real number representing the proportion in which P_i stands to the chosen unit.

A set of such sequences,

$$\{\langle r_1, r_2, \ldots, x \rangle, \langle r'_1, r'_2, \ldots, y \rangle, \ldots\},$$

will then represent a possible world – a much richer possible world than a Democritean world.

Another variation is possible. Suppose one of the things we wish to specify about an individual is the sequence of positions it occupies during its existence. We could specify this by a long (infinitely long) list of four-tuples, each one giving a single space-time location. Yet there may be other, more economical ways of representing the odysseys of particles. We can sum up all the positions of a particle in a single function. A similar treatment can be given for the other quantities we may assign to a particle: we may pair the particle, not with separate numbers describing the proportions in which its various quantitative properties stand to their units, but rather with functions of various sorts. These functions may characterize variations in the particle's properties across time. They may also characterize variations in the "particle's" properties across a region of space. (Remember that fundamental "particles" now seem to be more like waves, which collapse from time to time, than like little billiard balls.)

By pursuing elaborations such as these, of the basic Democritean world books, we can arrive at world books of extreme mathematical complexity. We can arrive at the Hilbert spaces in which current fundamental physics is often formulated.

The world book for the actual world, then, may be a Hilbert space. But a world book is not a world. A world book for the actual world corresponds to two things: to a world property, or maximally specific structural universal, and to something, the actual world itself, which instantiates that universal. Other world books correspond not to two such things, but only to one. They correspond to world properties which are uninstantiated.

Although the world books for our world and its near neighbours will be much more complicated than simple Democritean books, many philosophically significant features are common to both. When faced with the ferocity of the mathematical jungles of Hilbert spaces, it is easy to lose sight of what they represent. The numbers which pervade the mathematics seem to be abstract and remote from any tangible reality. They may "represent" physical things, but only as a mirror does, by creating unreal, intangible counterparts of real things. That is how it seems. But that is not really how it is. The numbers which feature in world books are not as abstract as they seem. They are the proportions in which the properties of particles stand to the properties selected as units.

Just as Democritean world books can be matched with world properties, so too can the more complex world books of modern physics. Exactly the same technique carries over. Pairing a particle with a property P corresponds to a world property of the form 'having a part which is x and has P'. Pairing x with a *function* is just a way of pairing it with an infinite conjunction of such properties. And a sequence of functions attributes a conjunction of conjunctions of such properties. The complexity is rising, but the principle stays the same.

The construction of world books and world properties using quantities at the fundamental level has one metaphysically crucial consequence. It promises to ensure the existence of a nonarbitrary measure of the distances between worlds. The closer the numbers assigned to corresponding particles in two worlds, the "closer" those worlds are to one another: the more accessible one is from the other. This promises to secure a firm grounding for the kind of accessibility which we take to underpin the physical modalities, as we shall see in later chapters.

It is also worth noting one more specific feature of the way that a distance measure emerges from a quantitative construction of worlds. When two worlds contain the same individuals and the same quantitative properties, a measure of the distance between them will emerge very naturally. When two worlds contain different individuals, a distance measure will not emerge so easily. And when two worlds contain different qualitative properties, when one contains entirely different *kinds* of properties, ones which stand in no proportions at all to the properties in the other world, there will probably be no nonarbitrary measure of distance between them. On our semantics, that means that two such worlds will be inaccessible from one another. One is not nomically possible from the standpoint of the other. This will support the theories we offer about physical modalities, as we shall see in the next chapter.

All attempts to give a reductive account of modality face a common threat: that of circularity. Even Lewis's concrete modal realism, we have argued, faces such a threat. It is no surprise that our theory, too, faces such a threat. Yet the prospects for avoiding circularity, we claim, are much better for our property theory than for its rivals. As a last resort, we could fall back on a modal primitive as an ingredient in our theory. That would be disappointing. But it would not set our theory at a disadvantage with respect to its rivals.

Lycan compared combinatorial theories favourably with respect to concrete modal realism.[23] The combinatorial theories he was speaking about were, in our terms, versions of the book theory (with what Lewis called Lagadonian books). Like all book theories, the combinatorial theory must draw on a distinction between consistent and inconsistent combinations: only the former represent possibilities. Lycan urges that there is nothing wrong with taking such a notion of consistency as an unanalysed input into the theory. Our reply is, yes, there is nothing wrong with it, if you really cannot find any better option.

A better option, however, is offered by Forrest in his property theory of possible worlds.[24] Forrest does appeal, in the end, to a modal primitive. Yet this modal primitive lies at a different level than the ones which are called upon by book theories. Lycan, for instance, envisages a modal primitive relation between representations. Forrest offers a theory which grounds that intralinguistic relation on an extralinguistic relation. The inconsistency between propositions (truthbearers) is grounded on a relation between universals (loosely, truthmakers). The required relation between universals may be called *incompatibility.*

Given that one property, *F* say, is incompatible with another, *G* say, this will entail an inconsistency between two propositions. The proposition which ascribes *F* to an individual *a*, for example, will be inconsistent with the proposition which ascribes *G* to that same individual *a*. The inconsistency between these propositions is explained by the incompatibility between the properties *F* and *G*.

Forrest's strategy is disappointing in its admission of a modal primitive. Yet this modal primitive is better than others. The reduction of modal relations among truthbearers to modal relations among truthmakers does stand as a contribution of considerable importance. The presence of a modal primitive does not remove all explanatory significance from the theory. Light is cast on modality if modal characters of representations can be grounded in real properties and relations in the world. Forrest at least offers the realist a

23 Lycan (1979).
24 Forrest (1986a, b).

way of giving a correspondence theory of truth for modal claims. Nevertheless, an unexplained modal relation between properties does strike us as worryingly magical. As magic goes, it is just a bit too far off white. A satisfying theory should be able to say more about modal relations among universals.

Our property theory does aim to give a reductive analysis of modality. Possibility claims are reduced, ultimately, to existence claims. Something is possible when *there is* a corresponding property, a corresponding way for a world to be. The required properties will be structural universals, in the sense explained in Chapter 2.

If a threat of circularity still lurks in this theory, it is to be found in the theory of structural universals. There is a general problem about how a structural universal of a more complex sort is related to the simpler universals which in some sense serve as its constituents. There are problems in construing a structural universal as simply the mereological sum of its constituents. Yet if we admit a "nonmereological" mode of composition, there is a severe threat that modal primitives will creep in. We will then fall back on a theory like Forrest's, which is better than nothing, but not what we hoped for.

There is still the possibility that an adequate theory of structural universals may be able to avoid modal primitives altogether. Our discussion in Chapter 2 gave some indication of how this might be done. In Chapter 3 we explained at greater length the way in which existence claims, quantification, must be allowed at two quite distinct levels. There is first-order quantification over *somethings;* and there is second-order quantification over *somehows* that things stand to one another. The key to understanding structural universals lies in the interaction between the somethings that a structural universal contains and the somehow that it contains them. The interaction of 'something' and 'somehow' is the key to structural universals. Thereby, it is also the key to modality.

A reductive theory of modality aims to derive a *must* from an *is*. This is hard to do. Indeed, it may be impossible, as long as we fail to recognize an ambiguity in the 'is'. We cannot derive a 'must' from only a first-order 'is'. But we can derive a 'must' from a combination of a first-order and a second-order 'is'. A way things are is a way things must be when *there is no* alternative somehow for things to be. Modality is grounded in structural universals, and these in turn are grounded in two distinct kinds of existence. Realism about possibilities and about truthmakers for modal truths rests on the existence of structural universals; and these in turn require the existence not only of various somethings, but also of various somehows for those somethings to stand to one another. The key to modality is to be found in an adequate theory of universals.

5

Laws of nature

5.1 LAWS AS NECESSITIES

Among the things we seek in science are the laws of nature. Various proposals have been made in the history of science as to the laws which govern the events in our world. What have been proposed as laws of nature have generally turned out to be false. But many of them have been, we trust, reasonably near the truth. The actual laws, whatever they are, are somewhat different from the laws which have been proposed. Yet in some sense, these alleged laws resemble the actual laws – or at least some of them are not as far from the truth as others.

Consider some examples. It has been proposed as a law that what goes up must come down. This has been refined in a variety of ways. Greek atomists, such as Lucretius, believed that there was a particular direction ("down") such that anything composed predominantly of earth or water must always move in that direction unless prevented from doing so by an obstruction. Objects moving in that direction would be moving along parallel lines.

The more astronomically informed Greeks, however, from the Pythagoreans through Aristotle to Ptolemy, believed that the earth was round. Aristotle and Ptolemy believed that earthy or watery things must always move, not in parallel lines, but towards one particular point, the centre of the cosmos, unless prevented from doing so by the presence of an object between them and that point. Because of the operation of this law, the centre of the earth is at the centre of the cosmos. As things fall towards the centre of the earth, they do not move in the same direction in the sense of moving along parallel lines. Rather, they move closer to one another as they get nearer to the centre of the earth. Over short distances it may appear as though they move along parallel lines, but in fact they will be closer when they land than when they began falling, even though the variation in distance between them is generally too small to observe.

In some respects, Aristotle is closer to the truth than Lucretius, though in other respects Lucretius was arguably closer to the truth than Aristotle. It is true that falling things move towards the centre of the earth, as Aristotle thought, but Lucretius was right to deny that the centre of the earth is always in a privileged place, the centre of the cosmos.

Between Newton and Einstein, what was accepted as a law of nature was that masses must always move in such a way that their centres of gravity approach one another at an accelerating rate, unless they are prevented from doing so by the presence of an object between them or by some other interference.

Many philosophical disorders can be prevented by taking care to feed the imagination with a sufficiently varied diet of examples. Wittgenstein perhaps overestimated the benefits of a healthy diet, but in the case of the philosophy of science there is good reason for thinking that misunderstandings have arisen from a diet too heavily biased towards physics. It is thus worth recounting briefly another sequence of proposals about the laws of nature, this time from biology.

There is a significant generalization, one which we would be blind to overlook, roughly to the effect that each living animal has a mother and a father of the same species as itself.[1] This generalization has some exceptions that we know of today. But through the history of biology the predominant trend has been one of finding that the law is even more nearly true than we thought. Many alleged exceptions have turned out not to be exceptions at all.

Consider, for instance, the theory of spontaneous generation – the theory that some creatures, like flies and eels, can be formed from a mixture of inanimate materials, without any parents. All of the apparent cases of spontaneous generation turned out to involve ordinary reproduction by male and female. In the early seventeenth century, Harvey speculated that this was so, and over the next

1 For an excellent discussion of the development of scientific views about generation in the seventeenth to nineteenth centuries, see Gasking (1967).

century and a half experimental verification became more and more compelling.

In the case of some insects, it appears as though creatures of one species give birth to creatures of another species, which in turn give birth to creatures of the first species. Butterflies produce eggs which grow into caterpillars; caterpillars produce "eggs", chrysalides, which give birth to butterflies. In the late seventeenth century, the metamorphous life-cycles of many insects came to be more and more clearly understood. A caterpillar becomes a butterfly – the butterfly is not only of the same species but the same individual as the caterpillar. And each such individual has a mother and father of the same species.

The Greek hypothesis that all (or most) animals have two parents was never intended to apply to plants. It was a surprise to find that reproduction in plants is much more closely homologous to animal reproduction than anyone expected.

It was also discovered that, among animals, the details of the reproductive process are more universal than expected. It was known that birds came from eggs, which originated in ovaries. Yet it was thought that animals did not. Harvey, in the early seventeenth century, speculated that mammals too originated from eggs. And later de Graaf presented evidence to suggest that there were mammalian eggs and that they originated in the "female testes", which he renamed the ovaries. Although de Graaf was right, his evidence was far from conclusive, and no consensus was possible on the matter until early in the nineteenth century, when von Baer first saw mammalian ova. By this time it was apparent that not only animals, but also most plants reproduced by way of an ovum from one parent and a sperm from another parent of the same species.

The generalization that, as the Bible says, each creature reproduces "according to its kind", by copulation of male and female, has always been believed to have exceptions. Traditionally, it was thought, for instance, that not all humans had a mother and a father of the same biological species as themselves. Jesus, it was thought, had a mother but no human father. It was also thought that neither Adam nor Eve had a mother and a father. Current evolutionary theory does not postulate two initial parentless creatures of each kind. Complex creatures, like humans or aphids, *always* have at least one parent of the same species as themselves. Very seldom do creatures have just one parent, like aphids, which sometimes reproduce parthenogenically (as Bonnet showed in the mid-eighteenth century). And never (yet) has a *complex* organism arisen without any parents at all. Presumably in the distant past, a *simple* living thing originated from nonliving chemicals; but even this exception is still just speculation.

It was also long thought that an offspring need not be of the same species as the parents – witness the case of insect metamorphosis, but also the case of hybridization, as, for instance, in the production of mules by crossing a mare with a jack or of hinnies by crossing a stallion with a jenny. Modern theory has dealt with these apparent exceptions to the biblical rule of reproduction, "each according to its kind", as follows. If interbreeding is possible, producing fertile offspring, the individuals are by definition of the same species. If interbreeding produces only sterile (or nearly sterile) offspring, the individuals are *to some degree* of the same kind, but not to a sufficient degree to count as of exactly the same species. Every individual is born of parents which are to a high degree the same kind as itself.

The difficulties we have faced in trying to frame an exceptionless generalization about generation are really not very much more severe than they have been in physics. In laws of motion, a clause is often introduced which excludes "interfering factors". If we restrict our biological law in an analogous way, it is not that hard to find exceptionless generalizations. 'All mammals have a mother' might be a truly exceptionless generalization.

Yet it is also important not to overestimate the importance of attempts to eliminate all counterexamples to a law. Even amid a sea of exceptions to the rule, it is obvious that there is something important to be taken into account concerning the roles of mothers and fathers. In searching for a law of nature, we are in one respect looking for *more* than an exceptionless regularity. Not every regularity is a law. But in another respect we are looking for something *less* than an exceptionless regularity. Even if we cannot formulate a law carefully enough to cover all exceptions, we may often be sure that we have uncovered some very important features of the cases which our law *does* cover. The features we have uncovered are, we claim, *modal* features. We are uncovering facts not just about what does occur, but also about what could not have occurred, about what would have occurred under different circumstances, about why things happen as they do. We do not have to formulate a strict law which makes explicit allowance for parthenogenesis, to know *why*, say, our cat is pregnant. We know that, if there had been no loose toms around, she would not have been pregnant. Exceptionless generalizations would be nice; but we do not have to wait until we have found one before we start wondering whether we are faced with a law of nature.

Ideally, it would be desirable to find exceptionless generalizations of the form 'All *F*s are *G*s':

$$\forall x(Fx \supset Gx).$$

When such a generalization has the modal status required to be a law, we will write it as

$$\boxed{N}\forall x(Fx \supset Gx).$$

But sometimes it may be good enough to find facts of the form

Most *F*s and *G*s.

Such statements may, in certain cases, have the modal status required to give them explanatory power. Sometimes it may be possible to find truths, not of the form

$$\boxed{N}(\text{Most } Fs \text{ are } Gs),$$

but of the form

Most *F*s are necessarily *G*s,

where the term 'necessarily' modifies only the consequent.

Even if it is not true of all animals that they would not exist if there had been no parents who produced them, it may nevertheless be true of our kittens that *they* needed parents in order to exist. Hence, it is true of our kittens that, if they exist, necessarily they have parents. If a variable x refers to one of our kittens, then

$$(\text{Exists}(x) \supset \boxed{N} \text{ Has parents}(x))$$

is true. And this is true of most animals. The necessity operator here governs only the consequent and not the whole conditional, yet its presence may nevertheless be of crucial importance. It may, for instance, provide an adequate ground for a range of explanations.

It is tempting to regiment 'Most *F*s are necessarily *G*s' by a formula like

$$(\text{Most } x)(Fx \supset \boxed{N}Gx).$$

This, however, does not capture the interpretation we have in mind. This formula is equivalent to

$$(\text{Most } x)(\sim Fx \vee \boxed{\text{N}}Gx).$$

And this will be true provided that most things are not *F*s – *even* if all the things which *are* *F*s are also necessarily *G*s. The formalization of 'Most *F*s are necessarily *G*s' is, in fact, quite tricky. The problem is not solved simply by replacing the material conditional in the above formula by some other conditional, say the counterfactual $\Box\!\!\rightarrow$.

$$(\text{Most } x)(Fx \Box\!\!\rightarrow \boxed{\text{N}}Gx)$$

could be true for the wrong reasons, if, for instance, following Lewis we take it to be vacuously true for most *x*, simply because for most *x* the antecedent *Fx* is true in no possible worlds at all.

The logical form of 'Most *F*s are necessarily *G*s' must incorporate some means of allowing the predicate *F* to limit the scope over which the quantifier 'Most' ranges. The resulting formula will have to have a combination of scopes of modifiers, as reflected in a formula like

$$((Fx)(\text{Most } x))\boxed{\text{N}}Gx.$$

The languages we outlined in Chapter 3 do not lend themselves easily to a formulation of this sort; and we refrain here from developing the linguistic resources we would need for a rigorous treatment. Nevertheless, the intended interpretation can, we think, be grasped without the need for rigorous formulization.

There are some crucial features to notice about proposed laws of the form 'Most *F*s are necessarily *G*s', even though we have given only the scantiest sketch of them. Note that they aim to describe ways in which events occur over and over again. Such laws are concerned with regularities, with recurrences, with repeatable phenomena, patterns which occur everywhere or almost everywhere, at least throughout our epoch and our neighbourhood. Generalizations of this rough and ready sort can often be found, even if we cannot find any *strictly* true, *thoroughly* universal generalizations and even, we urge, if there are none to be found.

The first notable feature of laws is that they often involve generalizations of some kind. A second notable feature is that they often attribute some sort of necessity to a generalization. Not every correlation counts as a law. If it is a law that things move towards the centre of the earth, then not only must it be true that things do move in this way, it must also be true that things *must* move in this way. Some generalizations are true only because each of their instances is true. These generalizations are not laws. In the case of other generalizations, the order of explanation is reversed. The generalization is not true because each of its instances is true, but rather each of those instances is true because it is an instance of that law. The law explains its instances: each of the instances has to be true because there is a law which requires this.

For instance, if it is a law that masses move towards a given point unless obstructed, things move towards that point because in some sense they *must* move in that way. They do not move in that way by chance. The instances of the law, each instance of a falling body, all have the character they do because of the truth of the law. But the nature of this necessity is not straightforward. It is not logical necessity, since it is possible, in some sense, that masses not move in the way they do move. It is for this reason we use the symbol $\boxed{N}$ for this kind of necessity – nomic necessity.

To illustrate, compare Aristotle's laws of motion with a near relative of Newton's, a theory close to Copernicus's sketchy comments on the topic of falling bodies. On Aristotle's laws, earthy objects move towards a specific point unless obstructed, and this point happens to be the centre of gravity of the earth. On a quasi-Copernican theory, we may suppose, the law is that earthy objects move towards the centre of gravity of the earth. Lunar material naturally moves towards lunar material, and that is why the moon is round. Yet Copernicus had no reason to think that lunar material would naturally move towards the earth; quite the contrary, since the moon does not fall.

Notice that Aristotle's theory entails that every motion of earthy things fits the quasi-Copernican theory. Aristotle says that earthy (or watery) things move towards a given point *and* that that point is (nonaccidentally) the centre of gravity of the earth; hence things move towards the centre of gravity of the earth, unless obstructed. Yet Aristotle's law differs from that of Copernicus. The difference is not due simply to the fact that the earth moves. Even if the earth did not move, there would be a difference between the laws. There would be a difference between the laws even if there were, in fact, no difference at all in the motions of bodies. Even if all actual motions did fit the description Aristotle gives (being directed towards a specified point), Aristotle might still have got the law wrong. Things might move in the direction Aristotle specifies, towards a certain point which is the centre of gravity of the earth, but not for the reason Aristotle gives. Aristotle says that they move towards that point not because it is the centre of gravity of the earth, but because it is the centre of the cosmos. Yet it might be objected that the reason is rather that they move towards that point only because it is the centre of gravity of the earth. That is, it may be said that, although things do, in fact, always fall towards a given point, it is not the case that they *must* fall towards that point. Rather, they *must* fall towards the centre of gravity of the earth. The centre of gravity of the earth could have been elsewhere; hence, things could have fallen towards a different point. They could have fallen towards a different point, so Aristotle was wrong. He was wrong not because he said things *never do* fall towards any other point, but because his theory required that things in some sense *could not* fall towards another point.

Putatively, then, we have two crucial features of laws: they describe regularities of some sort, and they ascribe some sort of necessity to these regularities.[2] We should note a third feature of laws. When we come to state a law as a generalization, however carefully we choose our words, the generalization we set down will almost inevitably be either strictly false or else vacuous.[3] It will be strictly false because there are always, or at least most often, interfering factors which we have not foreseen and which modify the course of events in ways which we have not explicitly allowed for in our formulation of the law.

For instance, suppose an atomist says that earthy things move in a given direction, "down", unless they strike something "below" them. This is strictly false, since it overlooks the possibility that a thing may be struck from the side as it falls. If the atomist adds further specifications ruling out all impacts from the side, for instance, some previously falsifying instances will cease to fall under the generalization, and so cease to falsify it. There is a risk, however, that we have overlooked other sources of interference. We can achieve greater security by rephrasing the law: 'things move thus

2 Dretske (1977), Tooley (1977), and Armstrong (1978, 1983), for example, offer theories of laws which satisfy these requirements.

3 See, e.g., Cartwright (1983) and Hacking (1983).

and so unless something interferes'. Yet we then secure the truth of our generalization at a considerable cost. There is something unsatisfactory about the rider 'unless something interferes'. It threatens to reduce our law to an empty tautology with the same content as 'things move thus and so unless they don't'. And even setting this worry aside, our alleged law, once qualified, may have secured truth only at the cost of so restricting its scope that nothing falls under it any more. Consider an atomist law of downward motion with just two qualifications: 'earthy things move down unless they strike something below *or* are struck from the side'. It may be argued that as things move downwards they are always continually struck from the side, by atoms of air, for instance, and hence that falling bodies are almost never moving precisely in the direction designated as "down". So if stated carefully enough to count as strictly true, there will then be no actual motions which fall under the conditions specified by the atomist's law, even by the atomist's own standards. The law secures truth at the cost of vacuity. Yet this is puzzling. A law, it seems, will have no positive instances, that is, no actual instances which meet all the added auxiliary conditions. Why, then, are laws so valued in science?

We will argue that we can resolve this puzzle by understanding the way in which laws involve modalities. Laws are truths about possibilia. They help us to understand actualia precisely because a full understanding of actualia cannot be achieved without an understanding of nonactual possibilia. We understand the actual world only when we can locate it accurately in logical space.

We shall argue that a law of nature cannot be adequately described merely in a nonmodal language such as the predicate calculus or its propositional fragment. A law is not merely a regularity and cannot be described just by a universal generalization of the form

$$\forall x(Fx \supset Gx)$$

or any variant of this. Rather, there is a law of nature only when a *modal* statement of some sort is true. The relevant modal truth will often be a universal generalization, but we should not rule out the possibility that it may often be some other kind of generalization or sentence of some other logical character. Nevertheless, we shall assume here for simplicity that the relevant law involves universal generalization, and so will have the form

$$\boxed{N}\forall x(Fx \supset Gx).$$

The natural necessity possessed by laws, we shall argue, is such as to entail that laws of nature involve counterfactuals. Laws of nature are concerned not only with what does happen, but also with what *would* happen if things were different in various ways. For instance, a law of free fall concerns how objects would move if there were no obstructions or other interferences. When a law holds, we argue, not only will there be some regularity of the form 'All *F*s are *G*s',

$$\forall x(Fx \supset Gx),$$

but in addition, it will be true that any *F would be* a *G*:

$$\forall x(Fx \mathrel{\Box\!\!\rightarrow} Gx).$$

Both the actual regularity and the subjunctive regularity follow from the law, which we take to be the truthmaker for the sentence

$$\boxed{N}\forall x(Fx \supset Gx).$$

For the reasons we have given, therefore, there is a strong case for taking laws of nature to be modal claims. It is natural for modal realists to take these claims to have truthmakers of the sort set out in the framework of possible-worlds semantics outlined in Chapter 3. A law of the form

$$\boxed{N}\forall x(Fx \supset Gx),$$

would then be true just when, in all accessible worlds, if anything is an *F* then it is a *G*. And the accessibility here will be of a kind which is appropriate for nomic necessity.

This is the theory of laws that we shall defend. Yet many philosophers – even a committed modal realist like David Lewis – have been persuaded that such a theory must be rejected. It has seemed to many that such a theory embodies a vicious circularity.[4]

5.2 RELATIVE NECESSITIES

Once we have secured a notion of logical entailment, as we did in Chapter 3, it is easy to generate any number of necessity operators, in a manner we shall explain below. These derivative necessities are

4 See especially Lewis (1979).

called *relative necessities.* The orthodox theory of the necessity of laws of nature, of natural or nomic necessity, has been that natural necessity is merely one salient sort of relative necessity.

Within the semantic framework we have described, a class A of sentences may be said to *entail* a given sentence α just when sentence α is true in *every* possible world for which all members of the class A are true. When α is entailed by a class of sentences A, then α is true in every possible world within a certain specific class C. The relevant class of worlds is specified by the class A of sentences. The class C of worlds is the class of all worlds in which all the sentences of A are true. When A entails α this means that α is true in *every* world in set C. 'Truth in every world' is the idea behind the notion of *necessity.* Hence, whenever there is an entailment between A and α, we may say that α possesses a kind of necessity: it is necessary relative to A and relative to C.

Suppose, then, that we single out a class A of sentences and designate them (for one reason or another) as laws of nature. This class of sentences demarcates a class C of worlds, the worlds in which all these laws are true. Then we may say that any sentence α is *nomically necessary* just when it is true in all worlds in class C. Nomic necessity is then explained as being simply entailment by the laws.

The question then arises, how are we to single out the set A of sentences which count as laws? Clearly, which sentences count as laws in a world will depend on what that world is like. A sentence may be a law in one possible world, but not a law in another. It may be a law in one world, true but only accidentally so in another, and false in yet another. For instance, there are possible worlds which are governed by Aristotle's laws of motion. In those worlds, some of the Copernican descriptions of motion are accurate descriptions of the ways objects move, but these descriptions do not have the status of laws. In other possible worlds again, neither Aristotle's nor Copernicus's descriptions are either true or lawlike. Hence, if nomic necessity is to be relativized to a class of laws, it must be doubly relativized, since the class of laws must itself be relativized to a possible world.

Corresponding to each possible world w, then, there will be a set A_w of the sentences which are laws in w. This will determine a set C_w of worlds whose events fit all the laws in w. Any sentence p then counts as nomically necessary in w just when it is true in all worlds in C_w – in all worlds compatible with the laws in w.

We define a notion of accessibility as follows. Any world u is accessible from a given world w just when

$$u \in C_w,$$

that is, when u is compatible with all the laws of w. Then we can say that a sentence α is nomically necessary in a world w if and only if α is true in all worlds accessible from w. Representing nomic necessity by $\boxed{N}$, we may say that

$$\boxed{N}\alpha$$

is true in world w if and only if α is true in all worlds accessible from w. This regiments nomic necessity so that it falls smoothly into the pattern given in the semantics outlined in Chapter 3.

Note that, in thus construing nomic necessity as a kind of relativized necessity, we define nomic necessity by appeal to a prior designation of a class of sentences which count as *laws* for a given world. Laws of nature are used to define natural necessity. Hence, we cannot then define a law of nature to be a generalization which is nomically necessary. We cannot say that a generalization

$$\forall x(Fx \supset Gx)$$

is a law in a world w just when it is nomically necessary in w, when

$$\boxed{N}\forall x(Fx \supset Gx)$$

is true in w. To define a law in that way would be viciously circular if the nomic necessity we appeal to is explained, in turn, on the assumption that the generalization in question is in fact a law.

It has been widely believed that the notion of natural laws cannot be explained in terms of a modal concept of natural necessity. It has been assumed that natural necessity must be explained as relativized necessity – necessity relative to laws. If that were so, laws would have to be explained nonmodally.

The argument for this view runs as follows. A modal analysis of a law would have to appeal, in effect, to a notion of accessibility: to be a law, a generalization must be true in all accessible worlds. (We agree with this step in the argument.) But a world is accessible, in the relevant sense, just in case it is compatible with the same laws. (True enough.) There is no other available account of the required sort of accessibility; and it is not reasonable to take this sort of accessibility as primitive. (We question at least one of these two premisses.) Therefore, it is unacceptably circular to define laws of nature modally.

This circularity objection to a modal theory of laws is essentially a negative one: it is claimed that there is no possible account of nomic necessity except one which presupposes laws. We must settle for a construal of nomic necessity as a relative necessity because there is no viable option.

There is also, however, a more positive argument in favour of the relative necessity theory. It is argued that there is in existence a positive theory of laws which makes no prior appeal to nomic necessity. Or rather there is a galaxy of nonmodal theories to choose from, most of them being descendants of David Hume's celebrated discussion of causation. Most nonmodal theories of laws are aptly called Humean theories of laws. We will not explore these theories in great detail, but we will give a brief outline of what sorts of Humean theories are available. For the present, however, we are concerned with arguments against the modal theory of laws, and from this perspective all that matters about Humean theories is their resolutely nonmodal status. They explain what makes something a law in entirely hygienic, nonmodal terms.[5]

Thus, it is possible for Humeans to argue that there is no need for a modal theory of laws, since there are adequate nonmodal theories of laws already available. And once we have such a theory of laws, we can account for nomic necessity as a relative necessity, without vicious circularity. Hence, it is argued, the account of nomic necessity as a relative necessity is a good one which rests on an independently plausible account of laws. It is reasonable, therefore, to use laws to explain nomic necessity, rather than the reverse. It is reasonable, since it forms part of a reasonable overall package deal. This point may be reinforced by an unfavourable comparison with the modal theory of laws. The relative necessity theory explains the obscure (nomic necessity) in terms of the familiar (laws of nature); the modal theory, in contrast, explains the familiar (laws of nature) in terms of the obscure (nomic necessity and accessibility relations among possible worlds).

5 Hume's account is given in Hume (1739/1960). For other Humean accounts see note 8.

We shall reply to these arguments against our modal theory. We claim that an accessibility relation among possible worlds is not unacceptably mysterious, especially if possible worlds are construed as structural universals, world properties, as we recommend. The nonmodal Humean theories of laws face problems of their own. They distinguish laws on the basis of nonmodal features which laws do often, we grant, possess; but Humean theories are unable to *explain* those features. The greater theoretical commitment of modal realism is simply the price to be paid for the explanatory power which realist theories provide.

Of necessity, our replies will be somewhat programmatic. The first argument against a modal theory of laws alleges that there is no adequate theory of nomic necessity except the relative necessity account. An acceptable reply to this challenge would have to present a satisfactory rival account of nomic necessity, without prior appeal to laws of nature. We shall outline a modal realism which we believe to be satisfactory, but it is impossible to give a conclusive proof of a theory of this sort. All that can reasonably be asked is that we give reasons for believing that theories of this sort have enough explanatory power to make them worth taking seriously.

Consider also the Humeans' other argument against the modal theory of laws. They argue that there are perfectly adequate nonmodal theories readily available, so the modal theory of laws is redundant. A conclusive reply to this argument would require us to show that no Humean theory will be satisfactory. There is, however, no limit to the possible variations which could be attempted under the general Humean umbrella. Given any objections to extant Humean theories, there remains the possibility that these objections might be side-stepped by some future revisions of the Humean programme. Hence, all that can reasonably be asked of us is that we uncover some serious problems facing Humean theories – enough problems to make it worth investing some time in an exploration of alternatives. Let us look briefly, therefore, at the range of Humean theories available.

5.3 REGULARITY THEORIES

Lewis gives very detailed reasons for thinking that natural necessity must be explained in terms of laws rather than the other way around. He maintains, first of all, that degrees of accessibility between worlds (i.e., the relation which supplies the semantics for counterfactuals) should be construed as degrees of *similarity*. But then he notes that we must be more specific. We must ask what respects of similarity are relevant to counterfactuals. And then he argues that similarity *with respect to laws* must be given a higher priority than similarity with respect to matters of particular fact.[6]

This can be illustrated by the following example. Consider the actual world, in which Darwin asks his father for permission to sail on the *Beagle,* and his father grants permission. Darwin sails and writes a book about his voyage, many people read it, and many more hear about it. Call this world *w*. Compare this world with world *u*, which is exactly like *w* up to the time when Darwin asks permission, but in which, at that time, permission is denied, and Darwin does not sail, he does not publish a book about the voyage of the *Beagle,* no one reads such a book, and no one hears about it. Compare worlds *w* and *u* with a third world *v*. In world *v*, Darwin's father utters words which express a refusal of permission, but as they travel along the son's auditory nerves they are transformed into words which grant permission; and Darwin's father instantly forgets what he has just said and forms the false memory of having given permission. From then on, *v* is just like *w*.

In world *w* it is the case, we may suppose, that *if* Darwin's father had refused permission, Darwin would not have sailed, we would not have read or heard of the book of his voyage, and so on. Our semantics, and Lewis's, tells us that this counterfactual is true in *w* only if worlds like *u* are the most accessible from *w*. Lewis then adds that these degrees of accessibility should be construed as degrees of similarity. Hence, world *u* must resemble *w* more closely than *v* does. If *v* resembled *w* more closely, then we would have to say that the following counterfactual would be true in *w*:

If Darwin's father had refused permission, then Darwin would have misheard, and his father would have misremembered what he had said.

That counterfactual is not true in world *w*. Consequently, world *u* resembles *w* more closely than *v* does – there is a greater degree of similarity in those respects which are relevant to degrees of accessibility.

What respects of similarity, then, *are* relevant to degrees of accessi-

6 See Lewis (1979, 1986a). In contrast, see Jackson (1977a), who focusses on causation rather than laws in his account of similarity.

bility? It cannot be simply similarity in respect of matters of particular fact – in such matters as who sailed on which voyage, who wrote which books, who read or heard about which books, and so forth. In those respects, there is much greater similarity between *w* and *v* than between *w* and *u*; yet *v* must be counted as much less accessible from *w* than *u* is. Hence, the degree of similarity in those respects alone does not determine the degree of accessibility.

Yet Lewis still maintains that degree of accessibility is degree of similarity. He therefore construes it as degree of similarity with respect to features other than merely a matching in matters of particular fact. World *v* is very similar to *w* in matters of particular fact. But *u* is closer to *w* in another respect: world *u* has fewer exceptions to the *laws* which govern *w*. World *v* requires events which, from the standpoint of *w*, would be miraculous: a transformation of auditory information to Charles from his father and a matching lapse in his father's memory. World *u* does not contain those miracles. In world *u*, after Charles's father utters the sounds, their effects follow exactly as required by the laws governing *w*; and after he resolves to refuse permission, his memory operates according to the same general rules that govern its operation in *w*. The content of his memory differs between worlds *w* and *u*, but the *kinds* of circumstances under which his memory fails remain the same between *w* and *u*.

Lewis notes, however, that although a world like *u* has fewer miracles than *v* does, *u* may not be entirely devoid of miracles. World *u* is exactly like *w* up to the time when Darwin asks permission to set sail, so in *u* Charles's father is in exactly the same mental state as in *w* when Darwin asks permission. Yet in *u*, although we have exactly the same causes as in *w*, these are followed by a different outcome – a refusal instead of a granting of permission. Does world *u* obey the same laws that govern *w*? Or is it a minor miracle for those precise causes to produce a different effect? It depends on further details about the psychology of Charles's father, and perhaps it depends on the solution to the problem of free will. Yet it may have been, for all we know about the details of the events in the actual world, that given *all* the background to the decision by Charles's father, it would have required a minor miracle for him to have decided to refuse permission. At any rate, it might have been *as much* of a miracle if he had refused permission as it would have been if he had misremembered what he had decided and if Charles had misheard. Consequently, both world *u* and world *v* may depart from the laws which govern *w*. World *u* is not entirely obedient to the laws of *w*, although it may be said to be more obedient than *v* is.

Nevertheless, there may be worlds which are more obedient than *u* is, worlds which avoid the minor miracle that generates what we are supposing (for the sake of the argument) to be an unexpected and unreasonable refusal. Consider a world *z* in which, prior to Charles's request for permission, there is a network of events which make it entirely to be expected that his father would refuse permission. Suppose, for instance, that in *z* Darwin's father has just been told a few hours earlier that in Whitehall war is being planned with France, and that the *Beagle* is to be sunk and France to be blamed, thereby giving a pretext for declaring war. A fanciful story, perhaps, but there surely is such a possible world, and one moreover which has no exceptions to any of the natural laws governing world *w*.

Suppose that similarity with respect to laws were the only thing which determined degrees of accessibility. We must agree that world *z* resembles *w* more closely, in that respect, than *u* does. It would then follow that *z* must be more accessible than *u* is. Our semantics, and Lewis's, would then force the conclusion that the following sort of counterfactual will be true in world *w*:

> If Charles's father were to refuse permission, then war would be on the verge of breaking out between England and France, or there would be some other such factor which would motivate the father's refusal.

This counterfactual is false in *w* on at least one reading. Hence, Lewis concludes that similarity in laws is not the only respect of similarity which is relevant to degrees of accessibility.

Lewis's conclusion is that degrees of accessibility are to be explained in terms of similarity with respect to *both* laws *and* matters of particular fact, with a specific schedule of weightings assigned to each. He argues that exact match of particular fact over long stretches of time has priority over exact obedience to laws. That is why world *u*, which is exactly like *w* up to the time Charles asks permission, is more accessible than *z*. But obedience to laws has priority over approximate match in particular facts, and that is why *u* is more accessible than *v*, in which the added miracles produce a convergence in matters of particular fact after an initial deviation.

Lewis's theory is impressive. Insofar as it succeeds, it presents a problem for our modal theory of laws. We explain laws in terms of degrees of accessibility. But Lewis argues that degrees of accessibility must be explained by appeal to laws. If he is right, our theory of laws is viciously circular. He is then obliged to offer an alternative, nonmodal account of laws – which is precisely what he does. But that is a topic that we shall consider in a moment.

Our reply to Lewis's charge of circularity is that we deny that accessibility is to be defined in terms of similarity. The examples discussed earlier may be taken as showing that it is not, in fact, the case that the most accessible world is the most similar world. Of course, there will always be some respects in which it will be most similar. But unless tight constraints are placed on what respects we are allowed to appeal to, the mere fact that *some* respects of similarity may be found is too vacuous to bear any explanatory weight. Even if the most accessible worlds are most similar in the respects that Lewis outlines, this does not disprove the modal theory of laws. It may indeed fall out as a corollary of the modal theory of laws.

It is incumbent upon us, however, to provide a rival analysis of degrees of accessibility, one which does not presuppose laws of nature. Or if we cannot find such an analysis, we must defend the proposal that we take degree of accessibility as an unanalysed primitive. Lewis presents us with a challenge, the challenge to come up with a better theory. But there are enough problems facing Lewis, and enough attractive features of the modal theory of laws, to make it worth taking up Lewis's challenge and developing a rival modal theory.

Hume and regularity theories

In one sense, David Hume did advance a modal theory of laws. He analysed causation as including two essential elements: *constant conjunction* and *necessary connection*. He did not deny that our idea of causation includes a modal element, that of necessary connection; on the contrary, he held it to be an essential component of our idea of causation that the effect *must* follow, given the cause. Furthermore, a case of causation must always, he said, be an instance of a generalization or constant conjunction. This causal generalization will therefore be a generalization whose instances all contain a modal element. It will be a modal generalization – a law of nature. Laws of nature, for Hume, do involve a kind of modality.

Nevertheless, we see Hume as an opponent of our modal theory of laws. The reason is that, although superficially his theory is not an *eliminativist* one with regard to modality, it is nevertheless a severely *reductionist* one. Furthermore, the kind of reduction he offers is diametrically opposed to the kind of modal realism we are defending. Hume's reductionist account is, from our perspective, indistinguishable from an eliminativist theory. His account says that laws are necessities, but the necessity of a law amounts only to the possession of a certain family of characteristics which we will outline later, characteristics involving its relationship to people and their customs, habits, and expectations. He might just as well say that there is no necessity to a so-called law of nature and that what we call laws are really just generalizations which stand in specified relationships to people, their customs, and so forth.

A Humean can pick out a set of generalizations and designate them as laws; and then a concept of *relative necessity* may be defined, as outlined in the preceding section. But this relative necessity cannot then be used to explain the nature of laws, on pain of vicious circularity. Hence, Humeans' ability to define a concept of necessity should not be interpreted as permitting them to hold a modal theory of laws.

Humeans may define a substitute for possible worlds and for an accessibility relation between worlds. Then they may wheel in the possible-worlds semantics outlined in earlier sections. This will result in superficial agreement between our theory and theirs. The difference is that we construe this possible-worlds semantics as explaining the fundamental truthmakers for laws of nature. We construe degrees of accessibility as explaining the nature of laws. We give a *realist* theory of laws of nature and a *correspondence* theory of truth for the linguistic or mental representations of laws of nature. Humeans do not. Given that Humean theories are barred from using modal semantics to explain the nature of laws, how then can they explain laws? The standard strategy is to split the task into two stages. First, laws are construed as generalizations. Their truthmakers are provided by the semantics for quantifiers. A Humean is perfectly free to take a realist view of generalizations and to construe a Tarski-style semantics as providing a correspondence theory of truth for laws, insofar as they involve generalizations.

A Humean may also appeal to further standard semantic features of the generalization which is alleged to be a law. It is sometimes urged, for instance, that if it is to be a law a generalization must contain no reference to any individual – it must be "entirely general". This requirement is both too strong and too weak. It is too strong because it excludes a great many of the laws proposed in the history of science. Kepler's laws of planetary motion furnish one famous example – they refer to all planets in orbit around the sun, and hence they do refer to an individual (the sun). More important for our present purposes, however, is the objection that a "pure generality" requirement is too weak to define laws of nature. It is possible to cook up pure generalizations which are, despite their generality, not laws. An example was furnished earlier by the Aristotelian possible world in which it is a true generalization, yet not a law, that earthy things move towards the centre of gravity of the aggregate of all earthy things. This generalization is not a law in the possible world in question, even though it is true and "entirely general".[7]

The second stage of the Humean account is then to explain the special status of those generalizations which count as laws – the status which gives us the (illusory) idea that they involve some sort of "necessary connection". In this phase of the Humean theory, realism and the correspondence theory of truth are firmly excluded.

At this point, a Humean appeals to one or more features of a law which give it a special status; but not by way of correspondence relations between the law and the world. Presumably intrinsic properties of the law are also implausible candidates for the law's distinctive features. Consequently, it seems, the distinctive features of laws must lie in their relationships to one or more of the following:

(1) psychological relationships (such as expectations);
(2) social relationships (such as customs);
(3) technological relationships (such as practical utility);
(4) logical relationships (such as the relationships of axioms to theorems);
(5) epistemic relationships (such as evidential relationships of laws to sensory input).

7 See, e.g., the discussion in Pap (1963, chap. 16).

We do not claim that these categories are mutually exclusive or that they exhaust the possibilities. They do, however, give some idea of the range of Humean theories on the market.

Hume, of course, stressed (1). Others have shifted the emphasis more heavily to (2)–(5). Wittgenstein, in the *Tractatus,* presented a dramatic denial of natural necessities, and on this issue the later Wittgenstein remained in sympathy with his earlier self. A great many philosophers of science this century have been deeply influenced by Wittgenstein. Various mixtures of (2)–(5) have been advanced as accounts of laws by philosophers including Ramsey, Goodman, Ayer, Nagel, Hempel, Braithwaite, Quine, Lewis, van Fraassen, and others.[8]

All these theories are reminiscent of the various rivals of the correspondence theories of truth. The coherence theory of truth, for instance, appeals to relationships of truthbearers to one another, relationships like those mentioned in (4) and (5). Pragmatism defines truth to be dependent on relationships of truthbearers to people and their practical projects and interests, relationships like those mentioned in (2) and (3).

Humean theorists, therefore, offer a kind of coherence or pragmatist theory of natural necessity. We urge a correspondence theory of natural necessity. We have already argued for a correspondence theory of truth. Our reasons for holding a correspondence theory of truth carry over into analogous arguments in favour of a correspondence theory of necessity.

We argued, for instance, that realism originates in an extroverted form with people who have as yet no self-conscious concerns with the properties of, or relationships among, truthbearers. Initially, they may have precious few truthbearers to worry about. They have yet to generate a theory as a result of their attempts to understand things other than theories.

The same considerations apply in the case of laws. We wonder what laws govern some range of events – animal reproduction and inheritance, for example. Initially, we are not thereby wondering which theories have which features. We may not be searching for a theory at all; we are trying to understand something, and a theory

8 For example, Wittgenstein (1922), Ramsey (1929, 1931), Goodman (1955), Ayer (1956), Quine (1960), Nagel (1961), Hempel (1965, 1966), Braithwaite (1968), Lewis (1973a), Mellor (1980), and van Fraassen (1980). See also Armstrong (1983) for a critical review of such theories.

emerges in the process. As the theory emerges, we may reflect on its logical and epistemic structure, its practical and social relevance, and so forth.

Notice also that considerations (1)–(5), the psychological, social, practical, logical, and epistemic status of a law, may vary strikingly as we progress from initial speculations to mature theories. Initially, we may speculate that some law governs certain events, but not be the least bit surprised when it turns out that we are wrong. We may be not at all reluctant to abandon our proposed law if conflicting arguments or data are presented. We may have only very vague ideas about the logical relations of this proposed law to other propositions. And we may have no idea whether this proposed law will serve any practical purposes. As time goes by, if things go well for us, our proposed law may become entrenched psychologically, socially, practically, logically, and epistemically. We may then be surprised by conflicting arguments or data; we may become reluctant to abandon it and willing to make serious revisions to other beliefs if that will save the law. We may work out a detailed logical structure for our theory, in which our proposed law serves as an axiom and a great many consequences are derived as theorems. And we may work out practical applications for many of these consequences.

Yet all these accretions emerge only with time. We wonder whether something is a law long before it has become thus entrenched. And in wondering whether it is a law, we are not speculating about the future fortunes of a linguistic item among theorists. We might even expect a reversion to wholesale delusion and superstition; and yet we may still wonder whether something which has just come to mind is, in fact, one of the laws of nature. In so wondering we are not wondering about the actual future role of our proposed laws. We might be persuaded that, if it is a law, it would come to be entrenched if scientists were to continue to be rational. That is, we might be persuaded of some *counterfactual* entrenchment of the proposed law. Yet this counterfactual is as badly in need of a truthmaker as the proposed law is. An analysis of laws in terms of counterfactual entrenchment seems a dangerously circular strategy for a Humean. Besides, when we wonder whether something is a law, we are not thereby wondering about counterfactual entrenchment. We may be persuaded of the counterfactual only *after* deciding whether it is a law.

Philosophers who try to analyse the nature of laws think too exclusively of the well-entrenched laws which have been the successful survivors in the history of science. It is too easy to think that they are laws because they are so entrenched. In fact, quite the reverse is the case. They are entrenched because they are laws. We should not take the anthropological fact of entrenchment as an unexplainable datum. We should ask why the successful laws have the logical, sociological, epistemic, and other such features that they do. In finding an explanation of these features, we will have to take account of the nature of theorists, no doubt. But we will also have to take account of the world in which these theorists live. In particular, we will have to take account of the laws which govern the world in which the theorists live. Laws cannot adequately be explained in terms of entrenchment or other such introverted considerations, because these in turn can be explained only by appeal to laws.

Humeans argue that our modal theory of laws is viciously circular. We deny this, and we return the compliment. Most Humean theories are themselves viciously circular. There is just one broadly Humean theory which apparently evades our charge of circularity. It is the theory which Lewis extracts from Ramsey and endorses himself – a theory which rests on (5) above, the logical relationships of laws to other truths.[9] Let us take a theory to be an abstract entity – a set of propositions, organized into axioms and theorems which are logical consequences of those axioms. Given a world, there will be a class of theories which are true of that world. These theories may never be propounded by anyone, but they are true nonetheless. Some of these theories maximize the number of truths concerning matters of particular fact. Of these, some are like enormous almanacs which simply list facts one after another. Other theories, in contrast, maximize the simplicity of truths which are taken as unproved axioms and include only such further truths as can be proved as theorems from those chosen axioms. The best theories, however, are not found at either of those extremes. The best theories are ones with an optimal balance between simplicity of axioms and richness of information on matters of particular fact which can be derived from those axioms. Laws of nature, Lewis says, are statements which are axioms within an optimal theory of that sort.

9 Ramsey (1929, 1931), Lewis (1973a), and Mellor (1980).

This theory bypasses many of our objections to other Humean theories, or at least it seems to do so. On further investigation, old problems might emerge. It is hard to tell, in the absence of more details about the nature of theories and about the criteria for optimal balance of simplicity of axioms and richness of particular detail. The Lewis theory does not make sense if theories are construed as sets of propositions and if propositions in turn are construed as sets of worlds. Then the best theory for a world will have just one axiom, the unit set of that world. Every truth about this world, including every bit of information whatsoever concerning matters of particular fact, will be entailed as a theorem derivable from that single axiom. This would mean that there could only be a single law for any given world and that no two worlds could ever have a law in common. Clearly, therefore, Lewis cannot construe a theory as a set of propositions in the sense of sets of worlds. A theory must presumably be a set of sentences. This requires specification of a syntax and a semantics. It is then not clear where this will lead us. Constraints on the range of possibilities for the language's syntax and semantics will be called for. Otherwise, the shift from propositions to sentences will have been in vain – without any constraints, a language-based theory could collapse into something isomorphic to the propositions theory. It would be tempting to introduce constraints relating to human purposes and so forth. Yet that would bring Lewis's theory back into the same territory as other Humean theories, which we have been criticizing. If, however, the relevant constraints reflect only objective, metaphysical structures, we escape Humean problems but move outside the framework of Humean regularity theories altogether.

Hence, we suspect that there will be something unsatisfactory about Lewis's theory of laws (taken as a Humean theory), even if we grant that all and only laws are axioms in an optimal theory. Yet we have further doubts. It is not clear that all axioms in such optimal theories will be laws. Consider again the world which fits Aristotle's cosmology. In that world it is a law that earthy things move towards the centre. This may perhaps be used as an axiom in an optimal theory. Now substitute for that axiom the Copernican generalization that earthy things move towards the centre of gravity of the aggregate of earthy things. It is quite conceivable that this substitution might cause no reduction at all in the richness of theorems concerning matters of particular fact in that world. So the new theory, resulting from the substitution, will be equally optimal. Hence, the Copernican generalization would seem to be a law in the Aristotelian world, according to Lewis's theory. Yet it is not, we have claimed, a law in that world. And so, we argue, Lewis's theory is mistaken.

Lewis's theory, along with other Humean theories, fails to do justice to the modal content of laws. Natural necessity is not definable by reference to laws: it is what makes laws laws rather than mere generalizations.

The Hume world

Laws are generalizations which involve natural necessities; and natural necessities are truths which hold not only in this world but also in all appropriately accessible worlds. Natural necessities differ from logical necessities because logical necessities hold in all worlds without restriction, whereas natural necessities are only required to hold in all accessible worlds. Logical necessities count as a special case of natural necessities, since whatever is true in all worlds must also be true in all accessible worlds. But not all natural necessities count as logical necessities. (Often the term 'natural necessity' is used to mean *merely* natural necessity, a natural necessity which is not logical necessity. But it is tidier if we allow that, strictly, natural necessities include logical necessities as a proper subclass.)

If this account of laws is to play the explanatory role which we intend, it must not collapse into a vicious circularity. Hence, the relationship of *accessibility* must be explainable without appeal to laws of nature. How, then, should accessibility be explained? Among the possible theories, we may draw an important distinction between what we may call supervenience theories and magical theories.

Consider three worlds *w*, *u*, and *v*, where *v* is different from *w* and *u*, but *w* and *u* exactly match one another in all respects *other* than their external relationships to other possible worlds (so that *w* is *u*'s "double"). Suppose *w* and *u* contain exactly the same individuals, instantiating exactly the same properties and relations. Suppose also that in *w* and *u* all the properties and relations themselves stand in the same properties and relations. Suppose, that is, that *w* and *u* exactly match at *all* levels of higher-order universals, as well as at the level of individuals, in all respects which are internal to the worlds themselves.

We are inclined to adopt the view that the two worlds *w* and *u*, thus described, must really be a single world. Being indiscernible in all internal respects, they must be indiscernible in all respects, and hence must be numerically identical. This position amounts to a supervenience theory of accessibility. The accessibility relations among worlds supervene on the contents of those worlds. That is, there could not be a distinction between the accessibility relations holding for one world, and those holding for another, unless there were a difference in which things instantiate which universals in those two worlds, at some level of the hierarchy of universals.

The opposing viewpoint would take accessibility to be a primitive relation, not supervenient on the intrinsic natures of the related worlds. We call this the magical theory of accessibility. On this theory, two worlds may differ in their relational properties even if they exactly match each other in all nonrelational properties.

This magical theory cannot be dismissed simply on the basis of the principle which is known as the *identity of indiscernibles,* at least in its minimally controversial form. This principle requires that when things are indiscernible in all respects, that is, when they exactly match in *all* their properties and relations (both internal and external), they are numerically identical. We endorse this principle: it is not beyond dispute, but neither is it highly controversial. The magical theory of accessibility does not violate identity of indiscernibles. Worlds *w* and *u* may exactly match in their intrinsic character, but they might nevertheless differ in some of their relational properties. The accessibility relation might hold between *w* and *v* but not between *u* and *v*; so *w* has a relational property that *u* lacks. Hence, *w* and *u* are not identical.

The distinctive feature of the magical theory is not that it violates the identity of indiscernibles. Rather, what distinguishes the magical theory is its treatment of relational properties. It allows some relations which hold among things to be independent of the intrinsic characters of the related things. That is, relationships among things are allowed which do not supervene on the intrinsic characters of those things.

Leibniz maintained that all relationships supervene on intrinsic properties. Yet there is now ample precedent for allowing relationships which do not supervene on intrinsic properties. Bertrand Russell diagnosed the fundamental error underlying Leibniz's metaphysics to be the assumption that all relations among individuals supervene on intrinsic qualities of the individuals. An individual, or "monad", stands in no irreducible relations to other monads; according to Leibniz, "A monad has no windows".[10] And this leads Leibniz to some of the most bizarre among the strange consequences of his monadology.

We believe, with Russell, that there are irreducible relations. Consider first a controversial case: the relation of nonidentity. An argument can be mounted for the view that nonidentity fails to supervene on the intrinsic properties of the related things. One powerful argument of this sort begins from a premiss concerning the possibility of certain sorts of global symmetries in the world. Usually, the symmetries appealed to are spatial or temporal symmetries; but let us save space and time for a moment and consider a case which focusses purely on nonidentity, without interference from problems concerning space and time.

Imagine two spirits, say Thomas and René, who are extremely similar in their perceptions and thoughts. Suppose for simplicity that they are as similar as they possibly could be, except that Thomas has one property that René lacks – suppose Thomas is doubtful. At a specific moment, God reveals to each the presence of the other, and each is equally surprised. In the world described (call it w), it is arguable that René could have been doubtful. It is implausible to say that, if René had been doubtful Thomas would have been opinionated. Rather, Thomas would (or might) still have been doubtful even if René had been doubtful too. Yet doubt is the only property which distinguishes between them in world w.

Given these modal claims about world w, we are committed to the existence of another world, w^* say, in which

$$\text{Thomas} \neq \text{René},$$

even though the one nonrelational property which distinguished them has been altered, so that now they exactly match in all nonrelational aspects. Hence, the nonidentity relation fails to supervene on qualitative differences.[11]

10 See discussion in Russell (1900).

11 This kind of argumentation owes much to Adams (1979).

This argument is not beyond debate; but let us set it aside for the present. The existence of irreducible relations is normally taken to rest not on nonidentity relations but on spatial and temporal relations. It is arguable that the existence of a spatial or temporal relation between two objects does not supervene on the intrinsic properties of those objects. Suppose two objects are a certain distance apart; then, in some cases at least, it would be possible for those two objects to be farther apart without changing any of their nonrelational properties. Similarly, one event may occur before another; but they could have occurred in the opposite order without any of their nonrelational properties changing. The fact that one event is earlier than another is not reducible to their intrinsic, nonrelational properties.

There are, therefore, precedents for the positing of relations which fail to supervene on nonrelational properties. Hence, we cannot rule out the magical theory of accessibility on entirely general grounds. It construes the accessibility relation as analogous to spatial or temporal relations. Formally, the set of worlds together with the accessibility relation counts as a "space" in the mathematical sense – it is aptly called *logical space*. The magical theory of accessibility construes logical space as more closely analogous to physical space than we had expected.

Nevertheless, if it were possible to avoid the magical theory, it would be desirable to do so. It is one thing to admit the irreducibility of spatial or temporal relations, but quite another matter to add more irreducible relations to the list. Such a new primitive would face a heavy burden of justification. It would have to explain an immense amount before its theoretical support would rival that of irreducible spatiotemporal relations. It would be much easier to defend the introduction of an accessibility relation if it were supervenient on the contents of the related worlds.

For this reason, we urge that a world's accessibility relations should be supervenient on which things in the world instantiate which properties and relations. If two worlds differ in their accessibility relations, they must also differ with regard to the instantiations of some of the properties or relations of things which exist within those worlds.

A world-property theory of possible worlds is especially well suited to meet such a supervenience thesis for the accessibility relation. In fact, a world-property theory of possible worlds, or a combinatorial theory, will entail a supervenience theory for accessibility, unless some fairly radical changes are made in the manner in which worlds are combinatorially generated. We cannot impose accessibility relations on worlds until we have already generated the worlds which are to stand in those relations. We cannot use a relation between worlds as one of the ingredients in the construction of those worlds. If we model the combinatorial construction set-theoretically, an accessibility relation between worlds will be modelled by a set of pairs of worlds. This set of pairs of worlds cannot itself be used in the construction of one of the worlds in those pairs. Hence, worlds must be constructed from properties and relations *prior* to the accessibility relations between worlds. This conclusion follows not only for combinatorial theories, but also for world-property theories which correspond to the combinatorialist's world books in the manner described in Chapter 4.

Given that possible worlds are constructed from properties and relations, excluding the accessibility relation itself, it follows that accessibility relations *must* supervene on the contents of the related worlds. If two worlds are constructed from properties and relations, and there is *no* difference between them with regard to those properties and relations, then they are not, after all, two distinct worlds, but are numerically identical. Hence, their accessibility relations cannot differ. (This follows from the *indiscernibility of identicals:* two things which are numerically identical must have all their properties and relations in common.) We conclude that accessibility relations among worlds do supervene on the contents of those worlds – on which things instantiate which properties and relations.

Yet we offer a warning. We do not believe that accessibility relations supervene on the *first-order* properties and relations of individuals in those worlds. On the contrary, we believe that two worlds may exactly match one another at the level of first-order universals and anything which supervenes on these, and yet may differ in their accessibility relations. Their differences in accessibility relations will indicate a difference in higher-order universals, at some level, but it need not indicate any first-order difference between them.

There is a striking illustration of the failure of first-order supervenience of the accessibility relation: namely, the Hume world. Consider any generalization which we believe to be a law, say 'All *F*s are *G*s', in which *F* and *G* involve only first-order universals and higher-order universals which supervene on them. Then we assume that certain counterfactuals hold – for instance, that *if* this object or that object had been an *F*, it would have been a *G*. However, we may acknowledge that we could be mistaken about whether or not this is a law. Even if we are sure that the generalization has no exceptions, even if we are sure that all *F*s are *G*s, we may nevertheless be unsure whether *it is a law.* Uncertainty about whether it is a law will be reflected in uncertainty about the truth of associated counterfactuals. Although we believe that we are in a world in which the generalization is a law, we allow that Hume *might* have been right in this case. The generalization might have been just a generalization, with no objective necessity to it at all. And this commits us to the existence of two worlds: the one we believe we are in and another one which is (at the first-order level) *exactly like* our world, except that what is a law in our world is not a law in that other world. A world in which it is true, but not a law, that all *F*s are *G*s may be called a Hume world with respect to the law 'All *F*s are *G*s'. A world which is a Hume world with respect to *all* the laws of our world may be called simply *the* Hume world. Strictly speaking, this is only the Hume world corresponding to *our* world, or at least to the world we believe to be our world. It is not the only Hume world. Other law-abiding worlds abide by laws quite different from our own; and there will be a Hume world for each of those worlds, just as there is a Hume world corresponding to our world. Given any law-abiding world, there will be another world in which all the necessary first-order generalizations are still exceptionlessly true, but are not necessary.

We claim that there is a Hume world for each law-abiding world. We have three grounds for making this claim. First, there is a powerful intuition that what we take to be true as a matter of law *could* be true simply as an accidental generalization. Second, positing Hume worlds brings some theoretical, explanatory benefits, which we will indicate later. And third, a "slippery slope" argument like that advanced by Frank Jackson[12] can be used to support the conclusion that the Hume world does exist. This slippery slope argument runs as follows.

Grant first that there is a world in which the first-order laws governing our world not only are not necessary, but also have exceptions. There is a world in which it is entirely random whether

12 Jackson (1977a).

an *F* will be a *G* or a non-*G*. We would expect roughly half the *F*s in that world to be *G*s. But in that world, each of the non-*G*s may be a contingent being, a being which could have entirely failed to exist. Or, even if it did exist, it could have been a *G* without affecting any of the other *F*s. So we can infer that there is a world in which it is still random whether any given *F* will be a *G* but in which there is one fewer *F* which is a non-*G*. From that world, we can infer the existence of a further world in which there is yet one fewer *F* which is a non-*G*. Repetition of this argument will, in the end, prove the existence of a world in which each *F* has a 0.5 chance of being a *G*, and yet all *F*s are *G*s.

This argument is a cousin of a theorem of probability theory. If it is random for each given *F* whether it will be a *G*, then it is highly probable that the frequency of *F*s which are *G*s will be about 0.5. But there will be a finite probability that all *F*s are *G*s. So there will be possible worlds in which all *F*s are *G*s, conditional on the assumption that it is random whether any given *F* will be a *G*.

We conclude that there is a powerful case for the existence of a Hume world corresponding to each law-abiding world. Each of the propositions which is a law in the given world will be true but not a law in its Hume world. This assumes that laws involve a necessitation of a first-order proposition which does not involve any such necessitation. The simple paradigm is that of a regularity or correlation which is asserted to be necessary. We have argued that not all laws need have such a simple form. This entails that the definition of Hume worlds will become a matter of some delicacy. For present purposes, however, we will bypass complications and think mainly of simple laws of the form 'Necessarily, . . .'. In this chapter, we also assume that what is said to be necessary, in a law, is something which entirely supervenes on first-order universals. In Chapter 6, when we discuss causation and forces, we shall argue that some laws concern things which do not supervene on first-order universals. For the present, however, we limit our scope to laws which have a simple logical form and which concern simple first-order regularities. For any world in which all the laws have this simple form, we argue, there will be a corresponding Hume world.

But admission of a Hume world commits us to denying that accessibility supervenes on *first-order* properties and relations within a world. A lawful world and its Hume world exactly match one another with respect to all correlations of first-order properties and relations. Yet correlations which are necessary in the lawful world are not necessary in its Humean counterpart. Counterfactuals which hold in the lawful world fail to hold in its Humean counterpart. This entails that there are worlds which are accessible from the Hume world but are not accessible from the lawful world. Hence, the lawful world and the Hume world differ in their accessibility relations, without differing in their first-order contents. Accessibility does not supervene on first-order properties and relations.

Yet there are compelling reasons for believing that accessibility does supervene on the contents of worlds. It follows, therefore, that the first-order contents of a world do not always exhaust the contents of that world. In particular, law-abiding worlds must contain more than just first-order properties and relations of individuals. The details of the contents of worlds have yet to be hammered out. But we can now lay down a constraint on the combinatorial programme for constructing possible worlds. Combinatorialism must countenance higher-order universals. And so, of course, must the world-property theory of possible worlds. These higher-order universals must hold the key to the nature of natural laws.

5.4 THE MODAL THEORY

A statement is logically necessary in any given world if it is true in all worlds. It is physically or nomically necessary, or a law of nature, in a given world if it is true in all worlds which are accessible from that world.

Terminology here is somewhat ambiguous. We can say that a generalization of the form

$$\forall x(Fx \supset Gx)$$

is a law in a world, provided that it is true in all worlds accessible from that world. Or we can say that the above generalization is not itself the law, but rather the law is a modalization of it; the law is a statement of the form

$$\boxed{N}\forall x(Fx \supset Gx).$$

Alternatively, we sometimes say that a law is the truthmaker for a statement of the above form, so that the law is not the statement itself, but rather the law is what such a statement asserts. We will vary our terminology whenever this aids smooth exposition, but there will be no serious risk of fallacies due to such equivocation.

As mentioned earlier, the schema $\boxed{N}\forall x(Fx \supset Gx)$ is chosen as a representative sample only. In fact, we believe that schemata of a range of forms can qualify as laws. The key requirements are that a law involve some sort of generality and that it involve nomic necessity. So we will expect a law to contain the symbol $\boxed{N}$ (or some equivalent) and a quantifier such as $\forall x$. But they need not always appear in the order

$$\boxed{N}\forall x$$

as the outermost operators in the law. However, we set aside the problem of giving a taxonomy of laws. We shall focus only on the representative case,

$$\boxed{N}\forall x(Fx \supset Gx).$$

If such a law holds in a given world, then a correlation, a universal generalization, also holds in that world. Furthermore, if a law holds in a given world, a generalization holds in all accessible worlds. The law requires that a regularity *holds* in all accessible worlds. But the law does not require that this regularity *be a law* in all accessible worlds. In order to require that a correlation be a law in all accessible worlds, we would have to assert,

$$\boxed{N}\boxed{N}\forall x(Fx \supset Gx).$$

Yet this is not entailed by

$$\boxed{N}\forall x(Fx \supset Gx);$$

the latter could be true and yet the former false. Or so we maintain. We do not require the accessibility relation to be transitive.

It is sometimes the case, we maintain, that

$$\boxed{N}\forall x(Fx \supset Gx)$$

is true in w, and

$$\boxed{N}\boxed{N}\forall x(Fx \supset Gx)$$

is false in w. The latter may be false in w because there is a world w^* accessible from w such that it is false in w^* that

$$\boxed{N}\forall x(Fx \supset Gx).$$

If this law is false in w^*, there must be a world w^{**} in which the generalization

$$\forall x(Fx \supset Gx)$$

is false. So in w^{**} there are exceptions to the correlation between F and G. If w^{**} were accessible from w, the correlation between F and G would not hold in all worlds accessible from w. So

$$\boxed{N}\forall x(Fx \supset Gx)$$

would not be true in w, whereas we have stipulated on the contrary that it is true in w. Hence, w^{**} cannot be accessible from w. We therefore have a chain of worlds: w can see w^*, and w^* can see w^{**}, yet w cannot see w^{**}. (We say one world can "see" another when the latter is accessible from the former.)

Consider the mediating world w^*. In this world the regularity which is a law in w still holds as a regularity but is not a law. So exceptions to the regularity are not possible from the perspective of w, but they are possible from the perspective of w^*. Hence, w^* functions as a Hume world with respect to w. It is like w with respect to generalizations, but not like w with respect to the nomic status of those generalizations. Because w^* differs from w with respect to its accessibility relations, it also differs in which modal claims hold in it and which counterfactuals hold in it.

From the perspective of the law-governed world w, consider the status of exceptions to lawlike correlations such as

$$\forall x(Fx \supset Gx).$$

Exceptions would be cases of the form

$$Fa \wedge {\sim}Ga.$$

These are not possible with respect to w; in w,

$$\Diamond_{N}\,(Fa \wedge {\sim}Ga)$$

is false. But in w^* it is true. Furthermore, w^* is possible with respect to w, so in w,

$$\Diamond_{N}\Diamond_{N}\,(Fa \wedge {\sim}Ga)$$

is true. An exception to the law is *not possible,* but it is *possibly possible.*

Given the existence of a Hume world for a given lawful world and assuming that this Hume world is accessible from the lawful world, it follows that anything which is logically possible is possibly possible. Under the assumption, therefore, of both the existence and the accessibility of a Hume world for each lawful world, it will follow that

$$\Box\alpha \equiv \boxed{N}\boxed{N}\alpha,$$
$$\Diamond\alpha \equiv \Diamond_{N}\Diamond_{N}\,\alpha.$$

Thus, although we have argued that it is not possible to give a reduction of the nomic concepts in terms of logical ones, it may be possible to give a reduction the other way around. Hence, the theory is not committed to a multiplicity of independent, primitive modal concepts, only to the acceptance of a single core modality.

The reduction of the logical to the nomic modalities would work smoothly under the assumption that there is an accessible Hume world for each lawful world. We have argued that there must *be* a Hume world for each lawful world. But we have not argued that this Hume world must be *accessible* from the lawful world. In fact, there are reasons for thinking that, for some worlds at least, the corresponding Hume worlds cannot be accessible.

For example, consider a Newtonian world in which there are fundamental laws of nature which govern *forces*. There will be, for instance, a law requiring that any two masses will exert a gravitational force on one another. In Chapter 6 we will argue that forces are higher-order relations. We will also argue that forces will be absent from Hume worlds. Thus, we have a dilemma. If the Hume world is accessible from the Newtonian one, the force laws are not true in all accessible worlds. If the force laws are true in all accessible worlds, as the modal theory requires, the Hume world is not accessible from the Newtonian one. We conclude that the Hume world is not accessible.

The reduction of logical to nomic modalities does not, however, require the accessibility of a Hume world. What is required is something weaker. The regularities which are laws in a world need not all be first-order regularities. But whatever their level, they must be regularities which hold in every accessible world. They need not be laws in every accessible world, but they must be regularities in every accessible world. Consider, then, any world with laws and compare it with a world where there are regularities that are not laws, yet which otherwise exactly match the regularities that are laws in the lawful world. Call this lawless world the *Heimson* world corresponding to the lawful world. It may not be a Hume world, because it may have lawless yet regular correlations among higher-order things like forces. Yet it can play exactly the same role in reducing logical modalities to nomic modalities.

This concept of nomic possibility is closer to the prephilosophical concept of possibility than is abstract logical possibility. According to this prephilosophical concept of possibility, it is no more possible for a cow to jump over the moon or for a rocket to go faster than the speed of light than it is to construct a five-sided square. Consistency may be a necessary condition for possibility, but it is not a sufficient one. If nomic possibility were to be taken as providing an analysis of this possibility concept, consistency or logical possibility could in turn be provided with a reductive analysis in terms of nomic possibility. Logical possibility becomes what is possibly possible, a result which itself does not lack intuitive preanalytic appeal.

In Chapter 3 we argued that Leibnizian logical possibility is truth in all worlds, not truth in all accessible worlds. This calls for a different semantic treatment of the modal operator, one which makes no reference at all to the accessibility relation. We therefore have two semantically distinct kinds of necessity: logical and nomic necessity. We have just argued that there is an equivalence between logical necessity and iterated nomic necessity. From a formal point of view, it is tempting to construe this equivalence as a definition and to define logical necessity in terms of nomic necessity. For the reasons given in Chapter 3, however, this would be a mistake. Logical and nomic necessity are semantically quite distinct. The equivalence between them is not a purely logical or definitional matter. It requires a metaphysical lemma, namely, the accessibility of a lawless Hume or Heimson world corresponding to each lawful world. The presence of a metaphysical commitment here is an inevitable consequence of the interpreted status of nomic accessibility, with this kind of accessibility supervenient on the intrinsic characters of possible worlds. Leibnizian logical necessity, however, is purely "logical", and it should therefore not be definitionally tied to nomic necessity.

Consideration of the Heimson world, or Heimson worlds, allows us to see the relationship between logical necessity and possibility, on the one hand, and natural or nomic possibility and necessity, on the other. But another key use for the Heimson world is in connection with counterfactuals.

On our theory, when something is a law, exceptions are not possible. There are *no* accessible worlds in which exceptions occur. In evaluating a counterfactual, we recall, we look to the *most* accessible worlds meeting certain conditions. We take it that counterfactuals involve the same nomic accessibility that is involved with natural laws, that is, that the counterfactuals are evaluated using the nomic accessibility relation. Therefore, in evaluating counterfactuals we cannot look at any worlds in which there are exceptions to the laws of nature. On this point, we align ourselves with Pollock, against Lewis, Jackson, and others.[13]

Yet there are some very persuasive counterexamples to the theory that there are no accessible worlds in which our laws of nature are violated. There are many counterfactuals whose truth seems to require the existence of accessible worlds in which events occur which violate the laws governing our world. First, there are so-called *counterlegals,* counterfactuals which explicitly state something about what would be the case *if the laws of nature were different* in specified ways. To evaluate these, it would seem, we must consider the most accessible worlds in which our laws are violated. Yet this would mean that we must allow *some* accessible worlds in which our laws are violated. And our theory says that there are no such worlds.

Even if we set aside counterlegals, we will find an endemic problem arising for everyday counterfactuals. We often wonder what would have occurred if things had happened differently at a given moment of time. Sometimes we engage in what Lewis called "backtracking". We imagine a hypothetical change in an event at a given time, and we then go on to describe ways in which prior events *would have to have been* different in order for the hypothetical event to have occurred. These backtracking counterfactuals pose no problem for our theory. Yet a problem does arise for our theory, because not all counterfactuals are backtrackers. Often we consider a hypothetical change in an event at a time, and we wonder which future events would have followed that hypothetical change. When we do this, we do not simultaneously wonder how the past would have to have been different too. In fact, we assume that the past remains

13 As we shall explain, our theory allows "backtracking" for counterfactuals. A paradigm backtracker is Pollock (1976). See also Bennett (1974). For non-backtrackers see Stalnaker (1968), Lewis (1973a), and Jackson (1977a).

completely fixed. We imagine a world exactly matching ours, up to a time very close to that of the hypothetical change in question. We take the history of the world as given and tack on an alternative event. We then consider which future consequences would have followed. The reason these forward-tracking counterfactuals pose a problem for our theory is that often it is not possible to leave history unaltered and to tack on a counterfactual alternative event *without violating laws of nature.* Often, in order to bring about the counterfactual supposition we must, given history, introduce what Lewis calls "minor miracles".[14]

For instance, you may say that, if there were a drink here beside you now, you would drink it. But how could that drink have arrived there beside you? By many routes; but which would be the one in the most accessible world? Suppose you are alone in the house. Then the least far-fetched way a drink could have arrived would have been by your having brought it from the kitchen. So the most accessible world is one in which you got the drink. This means that, if there were a drink beside you *now,* you would have had to have brought it yourself. To have brought it, you would have to have gone to the kitchen. But if you had gone to the kitchen, you would have drunk it there. So you would not have drunk it here now. Hence, it is *not* true to say that, if there were a drink here now, you would drink it – or so it seems.

And yet, despite the argument against it, it does seem plausible to say that, if there were a drink here now, you would drink it. How can we explain this? We just imagine a world in which there is a drink here now – but we must *not* consider how it could have got here. We must imagine that the drink is here now – by magic, as it were. Then among worlds just like ours up to now, except for the magical presence of the drink, we must look for the *most accessible* of such worlds. We must judge whether, in the most accessible worlds in which a drink is magically here now, you drink that drink. If you do, the counterfactual is true; if you do not, it is false. Yet this requires us to consider the most accessible worlds which match the actual world up to now but which magically contain a drink here and now. How can there be any such accessible worlds? Such worlds violate the laws of nature. A drink is here now, in such worlds, without any of the changes in historical antecedents which

14 Lewis (1979).

would have been required by the laws of nature. It would seem to follow that our theory cannot explain the (nonvacuous) truth of ordinary, nonbacktracking counterfactuals.

Our reply to this objection rests on a distinction between the theoretically fundamental, regimented counterfactuals, such as

$$(Fa \mathrel{\Box\!\!\rightarrow} Ga),$$

and ordinary-language counterfactuals. We should not presume that ordinary-language counterfactuals carry over uniformly into simple regimented counterfactuals. We suggest that ordinary-language counterfactuals may often be construed as *iterations* of regimented counterfactuals.

For instance, the counterfactual

If there were a drink here now, you would drink it

requires us to consider a world in which a drink is present here and now by magic. No such world is accessible from our world. But such a world is accessible from any Heimson world. Thus, one way to introduce consideration of inaccessible worlds is by first considering a suitable Heimson world. Instead of boldly saying,

If there were a drink here now . . .

we may first say,

If this were a Heimson world. . . .

If we assume that there is a Heimson world accessible from our world, such a counterfactual will be nonvacuously true or false. To evaluate it, we must shift attention to the Heimson world, a world in which no things happen, which are miracles from the perspective of our world, but in which such things would be possible. Then, having shifted to a Heimson world, we may consider what would be true if there were a drink in such a world:

If this were a Heimson world, then if there were a drink here now, you would drink it.

In fact, the Heimson world may not be quite the stepping-stone we need. The Heimson world wipes out *all* laws. This certainly makes space for minor miracles. But it does so in a heavy-handed way. We *need* to make space for only a few, local miracles; and the Heimson world allows that these miracles are possible. Yet everything else is possible, too, from the perspective of the Heimson world. If we evaluate counterfactuals by mediation of the Heimson world, there will be *no* carryover of laws from our world. Yet this is unsatisfactory. When we evaluate a counterfactual, we often allow a bit of magic to set up a hypothetical state of affairs. But we then assume strict compliance with the laws of nature when we work out the future consequences of that hypothetical state of affairs. Strictly speaking, what is required in such cases is only compliance with the regularities which are laws in our world; and this does not automatically require that these regularities have the status of laws in these worlds. The worlds we envisage when evaluating a counterfactual must *comply* with our laws, from the time of the miraculous antecedent event onwards. But a further argument is called for, to show that they must also *share* those laws.

However, in some cases at least, such an argument may be forthcoming. Consider, for instance, counterfactuals of the form

Even if there were a drink here now, it would still be a law of nature that . . . .

Some of these are intuitively true; but the consequent is not true in any Heimson world. Thus, we cannot explain their plausibility with respect to our world by appealing to our Heimson world, where there are no laws at all. Hence, when evaluating forward-tracking counterfactuals, we must take up the perspective of a world which allows a miracle just before now, but for which *future* exceptions to our laws are not possible.

We can gain the required effect by construing a counterfactual like

If there were a drink here now . . .

as involving an initial step not to the entirely lawless Heimson world. Rather, we should judge from the perspective of a world in which our laws hold universally *except for the minimal violations* required to permit the counterfactual's antecedent to hold. We must judge the counterfactual from the perspective of a world which is *locally* Heimsonian, but elsewhere law abiding.

Hence, our counterfactual should be construed as some sort of iterative counterfactual, of something like the following form:

If it were possible for a drink to be here now, then if a drink were here now, you would drink it.

The natural-language counterfactual

If there were a drink here now, you would drink it

is, we may suppose, intuitively true. So, plausibly, is the nested counterfactual

(It is possible for a drink to be here now $\Box\!\!\rightarrow$ (A drink is here now $\Box\!\!\rightarrow$ You drink it)).

This nested counterfactual can, in turn, be made true by our semantic theory. And it can be made true even if there is *no* world accessible from ours in which there is a drink here now. All that is required for the truth of the nested counterfactual is that there be an accessible world in which it is *possible* for there to be a drink here now. And there can be an accessible world where this is *possible* even if there is no accessible world in which it is *true*.

Forward-tracking counterfactuals seem to be true, in some cases, even though their surface antecedents are true in no worlds accessible from ours. Yet we can allow them to be true, provided that we make two postulations. First, we must construe forward-tracking natural-language counterfactuals as equivalent to a nesting of our formalized counterfactuals. And second, we must assume that, although the antecedent in question is true in no accessible world, it is nevertheless possible in some accessible world. These two assumptions are plausible enough to undermine significantly the argument from forward-tracking counterfactuals to the existence of accessible worlds which violate laws of nature.

If we deny the existence of accessible worlds which violate natural laws, there are important theoretical benefits we can secure. We thereby bind counterfactuals to laws in a way which resists any proliferation of accessibility relations. The very same measure of degrees of accessibility suffices for both natural necessities and counterfactuals. The costs are that we must assume the accessibility of the Heimson world, and that we must construe forward-tracking natural-language counterfactuals as nestings of regimented counterfactuals. These costs are not so exorbitant as to justify summary dismissal of the theory. Nor are they so negligible as to justify complacency; later we shall explain some promising variations of the theory advanced so far. For the moment, however, we shall continue to pursue the theory which takes *all* degrees of accessibility to presuppose *conformity* with all the laws of nature.

A closely related problem for our theory of counterfactuals is that of counterlegal conditionals. These counterfactuals have antecedents which explicitly assert the truth of laws incompatible with the laws of the actual world. What is the truth value of such conditionals in the actual world? Consider an example:

If the gravitational attraction between two bodies had been inversely proportional to the *cube* of distance between them (instead of the square), then bodies would have accelerated towards each other faster than they actually do.

It is plausible to claim that this is true. The usual Lewis account of similarity allows this result, and this must be admitted to be a point in its favour.

And yet we recommend that counterlegals be treated in the same way as counterlogicals: both involve *impossible* antecedents, antecedents which are true in no accessible possible worlds. Hence, we urge, in neither case can the semantics for counterfactuals be applied in a way which discriminates between the plausible and implausible instances. The semantic rule we have given (in Chapter 3) for counterfactuals, V9, entails that all counterfactuals with impossible antecedents are trivially false. It would be easy to rewrite the rule in such a way as to reverse this verdict on all counterlegals and counterlogicals, while leaving the truth values of all other cases unchanged. This would ensure that all counterfactuals with impossible antecedents would be trivially true. Lewis adopts this formulation, but rightly stresses that the choice of formulations is inconsequential. The point remains that these semantics for counterfactuals will give an undiscriminating verdict on all counterfactuals whose antecedents are true in no accessible worlds. And counterlegals are, we claim, in this respect on the same footing as counterlogicals.

There are two ways we can accommodate our intuitions that some counterlegals have nontrivial truth values. In our theory, counterlegals are not quite on a par with counterlogicals. The antecedent of a counterlegal is true in some possible worlds, even if those worlds are not accessible ones. The existence of these inaccessible worlds, in which a counterlegal's antecedent is true, can be used to explain some of our intuitions about counterlegals.

Consider the conditional

If it had been the case that the laws were $p_1, \ldots, p_n$, and the matters of particular fact were $q_1, \ldots, q_m$, then it would have been the case that . . . .

Our intuitions on the truth of this conditional will generally match our intuitions on the truth of the nonconditional

In a world with laws $p_1, \ldots, p_n$, and matters of particular fact $q_1, \ldots, q_m$, it is the case that . . . .

That is, such a conditional can be taken as elliptical for a description of some possible world the nonaccessibility of which is irrelevant. The counterfactual formulation is misleading because it suggests that degrees of some accessibility relation are relevant when in fact in this case there is no role for any such accessibility relation.

Alternatively we could represent such a natural-language conditional by nested counterfactuals with the semantics we have endorsed. A natural-language counterlegal can plausibly be construed as logically equivalent to the iterated counterfactual

If this world had been a world where there were only accidental generalizations, then, if it had been the case that the generalizations were $p_1, \ldots, p_n$ and the matters of particular fact were $q_1, \ldots, q_m$, it would have been the case that . . . .

Now this nested conditional can be treated in the standard way, for we can consider the most accessible Heimson world (most similar to the actual world) and then from that world consider the most accessible Heimson world with $p_1, \ldots, p_n$ as true generalizations and $q_1, \ldots, q_m$ as matters of particular fact.

We have theoretical grounds for defending backtracking counterfactuals as the most fundamental sort. Initially, this seems to run afoul of intuition. Yet, we have argued, on reflection our backtracking theory can do at least as much justice to intuition as do its rivals.

We conclude the presentation of our modal theory of laws by discussing so-called *derived laws.* Many laws fail to be true in all accessible worlds. This must be shown to be compatible with the modal theory of laws.

We commonly make a distinction between fundamental laws, together with the entailments of fundamental laws, and derived laws. The account of laws so far developed is an account of fundamental laws. Derived laws are universal generalizations which are entailed by the fundamental laws conjoined with accidental facts. The law of universal gravitational attraction (Newton's law) is a fundamental law, while a law which expresses the force of gravitational attraction on a mass near the surface of the earth (Galileo's law) is a derived law. It is entailed by the law of universal gravitational attraction together with certain facts about the earth.

Derived laws will not be true in all accessible worlds, for the accidental facts on which (along with the fundamental laws) their truth depends will vary from one accessible world to another. In fact, as any matter of particular fact is changed, there may be a corresponding change in the derived laws. However, there is an obvious explanation of the lawlike nature of derived laws, since any proposition which is a derived law in world w_0 will be entailed by the fundamental laws of w_0 together with the particular facts $q_1, \ldots, q_m$. Hence, it will be true in w_0 and in all worlds accessible to w_0 in which the particular facts $q_1, \ldots, q_m$ are true. Truth throughout this subclass of accessible worlds will amount to a kind of necessity, truth in *all* worlds within a specified class. This restricted kind of necessity will be robust enough to support a range of relevant counterfactuals. Hence, it will be possible to explain why derived laws do have a lawlike status which is closely analogous to that of fundamental laws.

An interesting application of the distinction between fundamental laws and derived laws involves the notorious controversy concerning the difference in the status of the laws of physics and the laws of biology. It has even been suggested that biology does not have any generalizations which are truly laws. This is because the laws of biology seem to depend on certain contingent facts, such as the fact that living organisms are carbon based. Were they silicon based, living things would be very different.[15]

But it is wrong to challenge the lawlike status of the generalizations of biology, though it may be appropriate to deny that some of the more common laws of biology are fundamental. Often the laws we know and use do depend on accidental facts about the actual world. Thus, they are better seen as derived laws which hold in all accessible worlds where these accidental facts are true. In other accessible worlds, often accessible to a much lesser degree, where these accidental facts do not hold, there will be different (derived) laws of biology. Nevertheless, our theory does grant a lawlike status to the laws of biology, as well as to the laws of physics. Our theory of laws is more charitable to derived laws than are theories

15 Smart (1968) argues that so-called laws in the biological sciences are of a very different nature than those of physics and chemistry.

which demand strict generality. And yet, at the same time, our theory requires laws to have a modal strength beyond that of mere generalities. This strengthening of modal force, combined with a weakening of generality, yields an account of derived laws explaining both their similarities to and their differences from more fundamental laws.[16]

5.5 THE LOGIC OF NOMIC NECESSITY

Formal possible-world semantics defines one kind of possibility in terms of a set of worlds and a binary relation on those worlds called accessibility. This semantics construes possibility as a relative notion. A proposition is not possible absolutely; rather, it is possible relative to some world. It is possible in a world if and only if it is true in some world accessible from that world. Similarly, a proposition is necessary in a world if and only if it is true in all worlds accessible from that world.

We have argued, however, that for the most fundamental kinds of necessity and possibility, consideration of accessibility is not relevant. Logical possibility and necessity should not be taken to be relative notions. As we argued in Chapter 3, the intended interpretation of logical necessity and possibility should be the Leibnizian one, an interpretation not involving accessibility.

Thus we distinguish two modal realist kinds of possibility:

(1) A proposition is logically possible if it is true in some world.
(2) A proposition is nomically possible in world w_0 if it is true in some world accessible from world w_0.

Correspondingly, with necessity:

(1) A proposition is logically necessary if it is true in all worlds.
(2) A proposition is nomically necessary in world w_0 if it is true in all worlds accessible from world w_0.

This commits us to system S5 as a logic for logical possibilities and necessities. But what logic should we accept for nomic possibilities and necessities?

Take a proposition, 'Every *A* is a *B*', which is true in world w_0. It is a law in w_0 if and only if it is true in all worlds accessible

16 This discussion of the modal theory is based on Pargetter (1984) and reflects useful comments by Frank Jackson, John Pollock, and David Lewis.

from w_0. This does not mean that all accessible worlds have the same laws, only that all propositions which are laws in w_0 will be at least exceptionless regularities in all worlds accessible from w_0. It is a core commitment of our theory that although accessibility is not *defined* in terms of laws, nevertheless a regularity which is a law in one world will hold *as a regularity* in any accessible world. Having matching nomic regularities is a *necessary* condition for accessibility. This leaves it an open question, however, whether match in these regularities will yield a *sufficient* condition for accessibility.

Consider the postulate that it is a *sufficient* condition for the accessibility of a possible world that it has regularities which match all those regularities that are *laws*. Then it will follow that, for any law-abiding world, its Heimson world will be accessible. The resulting logic of nomic necessity is intriguing. The accessibility relation which results will be reflexive, but not symmetric or transitive, as we shall now show.

Under the supposition that matching of regularities is sufficient for accessibility, the accessibility relation will not be symmetric. In a Heimson world, there are no laws, so any world is accessible. The accessibility of any world from the Heimson world follows automatically from the supposition that obedience to all (zero) of its laws is a *sufficient* condition for accessibility. Consider, then, a regularity which does *not* hold in the Heimson world. Suppose there is a world where that regularity is a law. The Heimson world, then, will fail to conform to a regularity which is a law in that other world. Such conformity would be a necessary condition for the accessibility of the Heimson world from that world, and this necessary condition is not met. Hence the Heimson world is not accessible from that lawful world. And yet that lawful world is, as every world is, accessible from the Heimson world. Hence, the accessibility relation is not symmetric.

An extension of this argument will show that accessibility is not transitive. Suppose there are two lawful worlds with different and incompatible laws, neither being accessible from the other. Then from the first lawful world, its Heimson world will be accessible; and from that Heimson world, the other lawful world will be accessible. Transitivity would then entail that the second lawful world would have to be accessible from the first, and this we know not to be so.

Although accessibility fails to be transitive, it does meet an analogous formal condition. We can connect any world with any other world, even one with different laws, in "steps". The easiest way is via a Heimson world, that is, a world with no laws but only accidental generalizations. For any world w_0 there will be an accessible Heimson world in which there are no laws and where the regularities which are laws in w_0 are merely true accidental generalizations. All worlds are accessible from any Heimson world. Thus, we can get from w_0 to any world, even one with different laws than those of w_0, in two steps. From w_0 we may move to a Heimson world accessible from w_0, and then from this Heimson world we may move to any world including worlds with laws different from those of w_0.

As so far characterized, the logic of nomic possibility and nomic necessity is a system containing the system T. Besides accessibility being reflexive, we have accessibility being "two-step" transitive: every world has access to some world from which all worlds are accessible. This can be captured axiomatically if to the axioms of T for nomic modality (see axioms A7 and A9 in Section 3.2),

A27. $\boxed{N}(\alpha \supset \beta) \supset (\boxed{N}\alpha \supset \boxed{N}\beta)$,
A28. $(\boxed{N}\alpha \supset \alpha)$,

we add

A29. $(\diamondsuit_N \boxed{N}\boxed{N}\alpha \supset \boxed{N}\boxed{N}\alpha)$,
A30. $(\diamondsuit_N \diamondsuit_N \alpha_1 \wedge \cdots \wedge \diamondsuit_N \diamondsuit_N \alpha_n) \supset \diamondsuit_N(\diamondsuit_N \alpha_1 \wedge \cdots \wedge \diamondsuit_N \alpha_n)$,
for every $n > 0$.

Axiom A29 requires, in effect, that for any world there be an accessible Heimson world. Axiom 30 requires, in effect, that any world be accessible from a Heimson world. This system is complete (and sound).[17]

In summary, the modal system we have arrived at is one which we might call HW (for 'Heimson world') and which is constituted by the following:

Definitions

D1. $(\alpha \wedge \beta) =_{df} \sim(\sim\alpha \vee \sim\beta)$,
D2. $(\alpha \supset \beta) =_{df} \sim\alpha \vee \beta$,
D3. $(\alpha \equiv \beta) =_{df} ((\alpha \supset \beta) \wedge (\beta \supset \alpha))$,

17 George Hughes has shown this in a personal communication.

D4. $\forall x\alpha =_{df} \sim\exists x\sim\alpha$,
D5. $\Box\alpha =_{df} \sim\Diamond\sim\alpha$,
D6. $(\alpha \Diamond\!\!\rightarrow \beta) =_{df} \sim(\alpha \Box\!\!\rightarrow \sim\beta)$,
D7. $\boxed{N}\alpha =_{df} \sim\Diamond_{N}\sim\alpha$.

Rules of inference

R1. The result of uniformly replacing any atomic sentence in a theorem by any given sentence will also be a theorem,
R2. If $(\alpha \supset \beta)$ and α are theorems, then β is a theorem,
R3. If α is a theorem, so is $\forall x\alpha$; and so is the result of replacing x by any other variable,
R4. If α is a theorem, so are $\Box\alpha$ and $\boxed{N}\alpha$,
R5. If $(\beta_1 \wedge \cdots \wedge \beta_n) \supset \gamma$ is a theorem, so is $((\alpha \Box\!\!\rightarrow \beta_1) \wedge \cdots \wedge (\alpha \Box\!\!\rightarrow \beta_n)) \supset (\alpha \Box\!\!\rightarrow \gamma)$.

Axioms

A1. $(Fa \vee Fa) \supset Fa$,
A2. $Fb \supset (Fa \vee Fb)$,
A3. $(Fa \vee Fb) \supset (Fb \vee Fa)$,
A4. $(Fb \supset Fc) \supset ((Fa \vee Fb) \supset (Fa \vee Fc))$,
A5. If β is a sentence with an individual constant or free variable and $\forall x\alpha$ generalizes β by uniformly replacing that constant or variable in β by x and then binding it with $\forall x$, then $\forall x(\alpha \supset \beta)$ is an axiom,
A6. For any sentences α, β, any sentence of the form

$$\forall x(\alpha \supset \beta) \supset (\alpha \supset \forall x\beta)$$

is an axiom, provided that α contains no occurrences of x.

The following are the modal axioms:

A7. $\Box(\alpha \supset \beta) \supset (\Box\alpha \supset \Box\beta)$,
A9. $(\Box\alpha \supset \alpha)$,
A12. $(\Diamond\alpha \supset \Box\Diamond\alpha)$,
A13. $(\forall x\Box\alpha \supset \Box\forall x\alpha)$.

(A8 follows from A9; and A10 and A11 follow from A12 – hence, their omission.) To these we add the counterfactual axioms:

A14. $(\alpha \Box\!\!\rightarrow \alpha)$,
A15. $(\sim\alpha \Box\!\!\rightarrow \alpha) \supset (\beta \Box\!\!\rightarrow \alpha)$,
A16. $(\alpha \Box\!\!\rightarrow \beta) \supset (\alpha \supset \beta)$,
A17. $(\alpha \Diamond\!\!\rightarrow (\beta \wedge \gamma)) \supset ((\alpha \wedge \beta) \Diamond\!\!\rightarrow \gamma)$,
A18. $((\alpha \Diamond\!\!\rightarrow \beta) \vee ((\alpha \wedge \beta) \Diamond\!\!\rightarrow \gamma)) \supset (\alpha \Diamond\!\!\rightarrow (\beta \wedge \gamma))$.

The identity axioms are

A19. $\forall x(x = x)$,

A20. $(\alpha \wedge \sim\alpha_{[\sigma/\lambda]}) \supset (\sigma \neq \lambda)$,
where σ and λ are any names or variables, and λ occurs in α and $\alpha_{[\sigma/\lambda]}$ is the formula which results from replacing every occurrence of λ in α by σ.

The probability axioms are

A21. $(\boxed{N}\alpha \supset \textcircled{1}\alpha)$,

A22. $(\sim \Diamond_{N}\alpha \supset \textcircled{0}\alpha)$,

A23. $(\sim \Diamond_{N}(\alpha \wedge \beta) \wedge \textcircled{p}\alpha \wedge \textcircled{q}\beta) \supset \textcircled{r}(\alpha \vee \beta)$, where $r = p + q$,

A24. $(\alpha \mathrel{\Box\!\!\rightarrow} \beta) \supset (\alpha \mathrel{\textcircled{1}\!\!\rightarrow} \beta)$,

A25. $(\alpha \mathrel{\Box\!\!\rightarrow} \sim\beta) \supset (\alpha \mathrel{\textcircled{0}\!\!\rightarrow} \beta)$,

A26. $((\alpha \mathrel{\Box\!\!\rightarrow} \sim \Diamond_{N} (\beta \wedge \gamma)) \wedge (\alpha \mathrel{\textcircled{p}\!\!\rightarrow} \beta) \wedge (\alpha \mathrel{\textcircled{q}\!\!\rightarrow} \gamma)) \supset (\alpha \mathrel{\textcircled{r}\!\!\rightarrow} (\beta \vee \gamma))$,
where $r = p + q$.

Nomic necessity requires at least the axioms for system T:

A27. $\boxed{N}(\alpha \supset \beta) \supset (\boxed{N}\alpha \supset \boxed{N}\beta)$,

A28. $(\boxed{N}\alpha \supset \alpha)$.

Nomic necessity will not satisfy axioms of the form of A10, A11, or A12, since the relevant accessibility relation is neither symmetric nor transitive. It is debatable whether nomic necessity would satisfy the Barcan formula A13. It does, however, satisfy two further axioms, under the assumption that for any world there is an accessible Heimson world:

A29. $(\Diamond_{N} \boxed{N}\boxed{N}\alpha \supset \boxed{N}\boxed{N}\alpha)$,

A30. $(\Diamond_{N}\Diamond_{N}\, \alpha_1 \wedge \cdots \wedge \Diamond_{N}\Diamond_{N}\, \alpha_n) \supset \Diamond_{N}(\Diamond_{N}\, \alpha_1 \wedge \cdots \wedge \Diamond_{N}\, \alpha_n)$,
for every $n > 0$.

Under semantic rules V1–V11 in Chapter 3, the axiomatic system HW, containing A1–A30 is both sound and complete.

6

Causation

6.1 ACCESSIBILITY, CAUSATION, AND NECESSITY

Laws of nature, we have argued, should be explicated in terms of a particular accessibility relation among possible worlds, a relation we call nomic accessibility. Counterfactuals should be explicated in terms of degrees of the same accessibility relation among possible worlds.

How, then, are we to explicate degrees of accessibility? We have urged a supervenience thesis for accessibility. The degree of accessibility between two worlds should be determined by the intrinsic natures of those worlds. We recommend that some sort of combinatorial world-property theory offers the best chance of providing a satisfactory theory both of the worlds and of the degree of accessibility between them. Each world arises from a recombination of individuals in the actual world with the various quantitative properties and relations in the actual world, such as mass, charge, relative velocity, and force. More specifically, each world is a structural universal, standing in a host of internal relations to its own constituent universals and to other possible worlds. The degree of accessibility between worlds will be a function of the proportions holding between the different quantities assigned to the same individuals in these different worlds. This holds not only for the combinatorialist's world books, but also for the world properties which, we have argued, correspond to them. Thus, accessibility will be a quantity, a determinable, and each determinate degree of accessibility will be a structural universal standing in a host of internal relations to the worlds it relates and to various other universals.

When we begin to flesh out this combinatorial story, a great deal will depend on our decisions about which quantitative properties and relations are parts of the basic furniture of nature. A good case can be made for admitting *forces* to this status. And, as we shall argue, there are essential connections between fundamental forces and the basic causal relation. Hence, the causal relation must be admitted as part of the basic furniture of nature. The causal relation therefore enters into the explanation of degrees of accessibility and thereby contributes to the explication of laws, counterfactuals, and probability.

It follows that we are committed to the rejection of a substantial number of attractive theories of causation. On pain of circularity, we cannot explain causation in terms of laws, or counterfactuals, or probability.

Theories which link causation with laws, counterfactuals, or probability often have a very strong intuitive appeal. The reason for this is that there are necessary connections among these concepts, connections which should emerge as theorems to be derived from any adequate theory. If necessity and counterfactuals are to be analysed, as we suggest, in terms of causation, then there will, of course, be some important necessary interconnections among causation, necessity, and counterfactuals. We certainly do not claim that *all* such alleged necessary connections are invalid. We claim only that they cannot play the explanatory role for which they are intended.

Causation, we argue, must be construed as a basic relation, a universal. It is, in fact, a relation which can be reduced to the action of fundamental forces. We urge a strong thesis: that all causation is reducible to the action of forces or to some complex processes involving the action of forces. Basic causation is a *structural universal* the constituents of which are forces. And forces, in turn, are *vectors*. Our views on the metaphysics of vectors are explained in Chapter 2. So are our views on structural universals. What we say about forces and causes does not presuppose all the details of our account of vectors, quantities, structural universals, combinatorialism, and so on. Yet vectors and structural universals are important in science and do have to be explained in any case. Thus, our theory of causation reduces one problem to other, pre-existing problems, and it does not introduce any new problem which other theories can avoid.

Before looking at our own account of causation, however, we give our reasons for rejecting some of the more plausible among the rival theories. Modal and probabilistic analyses, in particular, are extremely plausible, and our first task is to look at these.

There is a widely held and seductive, yet mistaken, view which defines causation as some sort of "necessary connection". This is conceived as entailing either that the cause is a *necessary condition* for the effect, or that, given the cause, the effect was a *necessary consequence* of that cause. That is to say, the cause is conceived as either a necessary or a sufficient condition for the effect – or else it is both a necessary and a sufficient condition for the effect.[1]

Some recent theorists have preserved this general structure, but have weakened the notion of necessary and sufficient conditions. Instead of the traditional construal in terms of *impossibility* of finding cause without effect, or effect without cause, it has been suggested that we should content ourselves with some sort of relative *improbability* of finding cause without effect, or effect without cause. These theories are called probabilistic accounts of causation, but they are close enough in spirit to a modal account of causation to be included, for our purposes, with necessary connection theories.[2]

Of course, a necessary connection theory of causation owes us some account of what kind of necessity it rests upon. Either one of the realist or one of the nonrealist theories of necessity (or probability) may be added to some necessitarian (or probabilistic) account of causation to give the full theory. But regardless of how the theory of necessity (or probability) is spelt out, we believe that both necessitarian and probabilistic theories of causation should be rejected. A cause need be neither a necessary nor a sufficient condition for an effect. The effect could have come about without the cause, either from some other cause or by no cause at all, and consequently the cause is not a necessary condition for the effect. Nor can the occurrence of the effect always be taken to ensure a high probability that it was preceded by the cause, nor even to increase the conditional probability of the cause above what it would otherwise have been. Neither necessity nor its probabilistic weakening is essential to causes. Similarly, neither sufficiency nor its probabilistic weakening is essential to causes. A cause need not be a sufficient condition for the effect and may not even ensure an increased probability for that effect.

1 For a classic collection of important papers and further references, see Sosa (1975).
2 We have in mind particularly the defence of probabilitistic theories by Lewis (1979); see also Tooley (1987).

We have learned this largely from Hume.[3] Hume's contributions to the theory of causation have a theological background. French theologians, notably Descartes and Malebranche, belonged to a tradition which insisted that God could not be fettered by any constraints whatever upon His freedom. Hence, given a cause, any cause, God could not be thereby compelled to permit the effect to follow. If cause was followed by effect, this could be only by the grace of God, by God's entirely free choice to permit the effect to follow. God could intervene and present us with a miracle whenever He chose. Hence, the cause was not, by itself, a logically sufficient condition for the effect. It was only the cause together with the will of God which yielded a sufficient condition for the effect. Given just the cause alone, at any time prior to the effect, it was possible for God to choose not to permit the effect. Hence, it was possible for the cause to occur and the effect not to follow. That is to say, the cause was not a sufficient condition for the effect. Nor could theologians like Descartes allow that the cause was necessary for the effect. God, being omnipotent, could have brought about the very same effect by some other, quite different cause.

Take the conclusion of this theological argument, then remove God from the scene, and the result is Hume's theory of causation. Instead of asking us to admit that God could choose not to permit the expected effect to follow a given cause, Hume asks us to admit simply that the effect could fail to follow a given cause. And he is right. (In fact, he asks us to *imagine* the effect failing to follow, and he takes imaginability as a guide to possibility. The shift from a theological to a psychological argument is not an unqualified improvement in the strength of the argument. Yet the conclusion is a compelling one, however doubtful the route which brought us there.) Hume is right in stressing that the effect could fail to follow – and this is not only logically possible, but empirically possible as well. Causes are not sufficient conditions. And the same negative verdict applies to the claim that causes are always necessary conditions. Sometimes the effect would have or could have occurred even if the cause had been absent.

The denial that a cause is a sufficient condition for its effect leads Hume to look elsewhere for his account of causation. Hume's attention is drawn in two directions: one "outwards", the other "in-

3 Hume (1739/1960).

wards". He refers us to the external facts about *regularities* in nature; and he refers us to the internal *expectations* that arise in us after exposure to such regularities.

In one sense, Hume denies that a cause is a sufficient condition for its effect. He denies the sufficiency of the cause in the sense which involves genuine *modality*. That is to say, he denies the sufficiency of the cause if 'sufficiency' is taken under a *realist* construal. Yet he does not deny the sufficiency of the cause if 'sufficiency' is taken, as we might say, more subjectively. He does not deny that the cause is a sufficient condition for its effect *in the sense that* such effects do always follow and we would be surprised if any given one of them did not. For the Humean this is how causal correlations are to be distinguished from "mere" regularities, not by further information about how the correlation relates to things in those parts of nature which are being described, but by information about how the correlation relates to people, their opinions, their purposes, habits, expectations, and so forth.

It is unnecessary here to offer a critique of Hume's account (and Humean accounts generally).[4] We do not accept a nonrealist theory of laws. But we do accept the lesson Hume taught us, that causes are not sufficient for their effects (if sufficiency is underpinned by more than just regularity and subjective expectation). Indeed, we go much further: we also deny the prior assumption that a cause is always necessary or sufficient for its effect, even in the minimal sense that it is an instance of a regularity which has some nontrivial status. Hume did go too far in his rejection of necessity in laws of nature, but he did not go far enough in his rejection of the necessitarian account of causation.

6.2 NEITHER NECESSARY NOR SUFFICIENT CONDITIONS

We shall argue that a cause is neither a necessary nor a sufficient condition for its effect (setting aside subjectivist and antirealist senses of 'necessary' and 'sufficient'). One event may cause another, and yet fail to be a necessary condition for that other event because there is a back-up system which would have brought about the same effect

4 For a survey of and attack on Humean theories of laws (and, indirectly, cause), see Armstrong (1983).

if the actually operative system had failed. Consider, for instance, the food that nourishes you. Eating the particular slice of bread that you did eat will cause a variety of effects; but eating that specific slice was not a necessary condition for the production of those effects. If you had not eaten that slice, you could have eaten another.

You might suspect that eating a different slice would have had slightly different effects. Yet there is no guarantee of this. It is quite conceivable that a wide variety of food intakes could have produced exactly the same outcome. Living things involve a variety of homeostatic systems which aim to preserve a constant state despite the variability of causal impacts from the environment. In general, admittedly, living things do fail to maintain absolute constancy – or so it is natural for us to speculate. However, this omnipresence of imperfection, if it exists, is a contingent factor. There is nothing intrinsic to causation itself which entails that homeostatic systems must always be imprecise and imperfect. Indeed, the quantization of small-scale phenomena in physics would suggest that, at least for some small-scale events, different causes could have precisely the same effect.

Lewis allows that a cause may not be a necessary condition for its effect. Yet he does explain causation, indirectly, in terms of necessary connections. Lewis defends a theory which analyses causation in terms of *chains* of necessary conditions. One event is a cause of another, he says, provided that there is a chain of distinct (nonoverlapping) events, beginning with the former and ending with the latter, in which each of the events in the chain is a necessary condition for the one which follows. And for one event to be a necessary condition for the following event is for a specific counterfactual to hold, namely, that if the former had not occurred, the latter would not have occurred either. In indeterministic cases, Lewis weakens this construal of necessary connections, replacing it by a probabilistic one. We shall leave this aside, however, for the moment.[5]

While Lewis uses counterfactuals in his analysis, Mackie achieves much the same kind of analysis using strict conditionals. Suppose C is a conjunction of conditions which, together with the presence or absence of the cause c, determines the presence or absence of the effect e. Then the Lewis counterfactuals

5 Lewis (1973b), reprinted in Sosa (1975) and with additional notes in Lewis (1986d). Lewis (1979) discusses indeterministic cases.

$$c \text{ does occur } \Box\!\!\rightarrow e \text{ does occur},$$
$$c \text{ does not occur } \Box\!\!\rightarrow e \text{ does not occur}$$

are replaced by the strict conditionals

$$\Box(C \text{ holds and } c \text{ does occur } \supset e \text{ does occur}),$$
$$\Box(C \text{ holds and } c \text{ does not occur } \supset e \text{ does not occur}).$$[6]

Hence, we derive a view summed up by Mackie's mnemonic that a cause is an INUS condition (an *i*nsufficient but *n*ecessary part of an *u*nnecessary but *s*ufficient condition) for its effect.[7]

Lewis and Mackie agree that a cause is connected by a chain of necessary conditions to its effect; they differ only over how the individual links in this chain are to be analysed. Lewis appeals to counterfactuals, whose semantics draws upon a similarity relation which is sensitive to indefinitely many features of the actual world. Mackie appeals to strict conditionals, which include complex antecedents that in practice we can seldom state explicitly. In two different ways, then, Lewis and Mackie provide a background against which a cause is defined in terms of necessary conditions.

These theories of causation allow space for a number of different back-up systems and homeostatic mechanisms. A cause may have a variety of remote effects in virtue of a chain of intermediate causes and effects. It may then happen that such a cause is not a necessary condition for its remote effects. If the cause had not occurred, its chain would not have begun, and yet some other causal chain might have brought about the same final event. In such a case, the cause fails to be a necessary condition in the counterfactual sense. It is not true that, if the cause had not occurred, the effect would not have occurred. Yet Lewis does count it as a cause, because it contributes to a causal chain, a chain of necessary conditions. Mackie, too, can count it as a cause, because it is a necessary part of some sufficient condition – in this case, the sufficient condition which embraces the whole chain of necessary conditions described by Lewis plus the absence of back-up chains.

Yet the theories of Lewis and Mackie do have some drawbacks. These drawbacks pertain not just to the alleged necessity of causes for effects, but also to any alleged sufficiency of causes for effects.

6 For pioneering work on counterfactuals and strict conditionals, see Goodman (1955).

7 Mackie (1965), reprinted in Sosa (1975).

Hence, we will not treat necessary conditions and sufficient conditions separately, but will class them together as subspecies under the same general theory that causation is to be analysed in terms of a modal connection between cause and effect. It is that general theory that we will reject.

Aside from more or less detailed worries, there is a very general objection to theories which explain causation in terms of necessary or sufficient conditions for the effect. Such theories are too closely bound up with the assumption of some sort of *determinism* in nature.

By determinism, we do not mean simply the doctrine that every event has a cause. Even if we grant that every event has a cause, it does not follow that every event is "determined" by the cause, unless the cause is taken to be a sufficient condition for the effect. Yet consider what we are committed to if we take every event to have a cause and we take causes to be sufficient conditions for their effects. It follows that, without sufficient conditions, there will be no causes. That is to say, insofar as there is indeterminism in nature, to that extent we would have to abandon causation.

Yet despite the pervasiveness of indeterminacy in the subatomic realm, we have not been forced to abandon causal talk. In a variety of cases, it is quite clear that we have caused some event to occur, even though it was possible that the effect should have failed to occur. For instance, suppose an electron strikes a copper target, causing an electron to be dislodged from an inner shell of a copper atom. This in turn produces the emission of an X-ray photon, but the photon produced can be of various frequencies. The frequency of the emitted photon depends on which of the other shells of the atom donates the electron that fills the gap in the inner shell. The existence of more than one possible outcome does not affect the appropriateness of our saying that the bombarding electron *causes* the outcome which *does* eventuate.

Defenders of the idea that causes are conditions may modify their theory to make room for causation under indeterminism. Instead of saying a cause is a sufficient condition for an effect, they could say that the cause makes the effect "very probable". This would be the simplest replacement we could make for the notion of sufficient conditions. And yet a variety of examples put pressure on us to bypass that theory in favour of a less direct account. For instance, a person having syphilis can develop paresis, and that case of paresis is caused by the syphilis; and yet the probability of developing paresis, even among those with syphilis, is extremely small. We need a somewhat less direct account of the link between causation and probabilities. The most plausible candidate is the theory that a cause makes the effect *more* probable than it would have been in the absence of that cause.

This theory neatly fuses the idea of cause as a sufficient condition with that of cause as a necessary condition. In the *presence* of the cause, the effect is *more* probable (a weakened notion of a sufficient condition); in the *absence* of the cause, the effect is *less* probable (a weakened form of a necessary condition). This probabilistic account of causation gives extremely plausible accounts of a wide range of cases. Nevertheless, we argue that it is on the wrong track.

Causation is a *local* feature of a cause–effect pair. What makes one thing a cause of another is entirely a matter of the nature of the cause, of the effect, and of what transaction occurs between them. Causation is, roughly speaking, a two-place relation, not an indefinitely-many-place relation. We can leave the causal relation unaltered even if we vary the context in which it occurs. For instance, we may leave the causal relation unaffected even if we institute a back-up system which would have come into play had the cause failed to occur. Cases of this sort are familiar in the literature. Imagine Gorbachev pressing the button that launches the rocket *Glaznost* on its journey to Mars. The causal relation between his pressing the button and the launch occurring is exactly the same, whether or not there is a fail-safe mechanism which would override the intended causal path were Gorbachev to bungle. The relation between cause and effect is, in general, independent of the presence of back-up systems waiting in the wings.

Causation is, we claim, a local matter of the actual, physical transaction between cause and effect. Necessary and sufficient conditions, in contrast, are much more "global" concerns. By instituting a back-up system, what was a necessary condition ceases to be a necessary condition. Yet the intrinsic character of the causal process is not altered.

It is important to clarify our claim here. It has to be acknowledged that there are such things as causal laws, and of course causal laws are, in our terms, global: the truth of a causal law depends on the character of a world as a whole, and not just on one of its constituents. But the truth of causal laws supervenes, we claim, on the existence of a pattern of causal transactions in the world. The law is (or entails) a *generalization over* causal transactions. (In fact, we take a law to be something stronger than a generalization, since we believe laws involve some kind of modality, but that is beside the point here.) The transactions do not count as causal because they are subsumed under laws. The connection works the other way around: the causal laws hold because of the presence of local causal connections. Necessary and sufficient conditions generally are underpinned by causal laws, and hence they, too, are global and depend on the character of a world as a whole. Causal connectedness, in contrast, is local. The causal relation between two events does not depend on the overall pattern of events in the world around.

Causal laws and necessary and sufficient conditions, then, are global, whereas causation is local. That is one reason why we resist analyses of causation in terms of necessity or sufficiency of conditions. And the same reason extends to probabilistic analyses of causation. Whether an event boosts or depresses the probability of another event will be a global matter, not a local one. But a causal process counts as causal entirely because of intrinsic, local, and not global, features.

Consider a more or less macroscopic example, modelled on a variety of subatomic, quantum-mechanical illustrations in the literature. Consider an act of sexual congress between a male with low fertility and a female. This act may cause conception and pregnancy to follow. Yet this act may not be a sufficient condition for these effects to follow. It may be largely a matter of luck that conception occurs. Nor need it be a necessary condition for pregnancy that intercourse should occur with that male. Other males might be willing to take his place. In fact, by coupling with that male, the female may have depressed, rather than raised, the chances of conception. If the female had not coupled with the male, we may suppose that some much more fertile male would swiftly have taken his place, in which case conception would have been almost certain.

In such a case, indeterminacy of outcome prevents the cause from counting as a sufficient condition even in the sense of raising the probability of the effect; and the presence of another eager male prevents the cause from counting as a necessary condition even in the sense that the effect would have been less likely in the absence of the cause.

It could be objected that the effect which results from the nearly infertile male, which we have just called 'pregnancy', was in fact a specific pregnancy. Without the nearly infertile male, the probability of *that* pregnancy would have been zero. If a different male had been involved, then a *different* effect would have resulted. Yet this objection is misplaced. It is a contingent fact that a different pregnancy is brought about by different males. Consider parthenogenesis with frogs' eggs, initiated by pinpricking by a technician in a laboratory. The infertile male could be like a technician with a hand tremor, while the more fertile male could be like a practised, steady-handed technician waiting in the wings. (The ovist preformation theory of generation of the eighteenth century gave the male a role very like that of such a technician.)

We are assuming here that it cannot be ruled out, across the board, that one event might have had different causes. We are assuming that it is not among an event's essential properties that it has precisely the causes it does have. There is a view on the market which dissents from our assumptions here.[8] Davidson has argued that it is a criterion of identity for events that difference in causes entails numerical distinctness of events. We assume this is mistaken, but will not dispute it here. However, even if the Davidsonian view were adopted on event identity, this would not rescue modal theories of causation under anything like their intended interpretation. Modal theories are not intended to be theories about how nonqualitative identities of effects depend on identities of causes. They are intended to be theories which assert that qualitative facts about effects depend on qualitative facts about causes.

It could be objected that, even though the nearly infertile male does not boost the probability of pregnancy compared with the more fertile male, nevertheless he does boost the probability compared with that with no male at all. And provided that probability is boosted with respect to any such comparison class, it may be urged, the probability-boost analysis of causation is vindicated. Yet it is only a contingent matter that coupling with males boosts the chances of pregnancy for a female. Indeed, in the case of some species there is a probability of spontaneous parthenogenesis. The action of the male (or technician) could conceivably block any possibility of spontaneous parthenogenesis and bestow a probability of

8 Davidson (1980).

generation which is *lower* than it would have been in the absence of interference.

It could be argued that such examples are really cases of indirect causation and that upon inspection each case will be found to contain a sequence of causal links, each of which does fit the probabilistic theory of causation. Yet such an appeal to mediating steps of necessary or sufficient conditions (or of boostings of probability) is highly speculative. The burden of proof is squarely on the defenders of such theories to establish that there are always such mediating steps. It is not easy to be confident that mediating steps of the right sort always exist. The pervasive indeterminacy in fundamental physics suggests that, on the contrary, all the mediating steps involved in a process like conception may be ones which raise the very same problems over again, ones in which the causes fail to be necessary or sufficient or even to boost the probability of their effects. Furthermore, even if there *are* mediating steps, this seems not to be a consequence of the nature of causation itself or to be part of what we *mean* when we say that the sexual act caused the pregnancy.

Nevertheless, a debate conducted entirely at the level of speculative counterexamples is likely to be theoretically unproductive. Defenders of the modal theory, Lewis, for instance, have considered examples like the ones described and have found them inconclusive. Thus, we do not rest our case entirely on counterexamples. Our case against the probabilistic theory is also based on a prior theoretical consideration, and we see the counterexamples not just as puzzles for the probabilistic theory, requiring ad hoc fine-tuning, but as symptoms of deeper theoretical concerns.

In part, we rest our case, as we have said, on a construal of causation as a localized, intrinsic transaction between two events. In part also, we rest our case on the role of causation in a wider explanatory context. We take causation to be part of the basic furniture of nature, and as such it functions as an input into the explanation of modalities. The best accounts of modalities make appeal to the framework of possible worlds. It is important to follow this with an account of what such possible worlds contain. We support theories which take causation to be part of what there is in any given possible world. Thus, causation enters into the explanation of modalities and, in particular, into the explanation of necessary and sufficient conditions and also of probabilities. Hence, modal or probabilistic theories, even if they could be adjusted until they become extensionally correct, would nevertheless proceed in the wrong direction from an explanatory point of view. Causation is an input for theories of modality and probability, not an output.

Before abandoning the probabilistic theory altogether, however, it is important to acknowledge one very appealing feature which it does possess. From time to time it has been suggested that causation should be taken as primitive and used as explanatory input for various other theories. Yet there is something profoundly unsatisfying about taking a notion like causation as primitive. Causation does cry out for analysis. We need to know more about the intrinsic nature of the causal relation. It is a strong point in favour of the probabilistic theory of causation that it does say a good deal more about the causal relation. It links the causal relation quite tightly to a network of other theoretical notions. We object that the probabilistic theory links causation *too* tightly to a variety of modal notions, and it gets the priorities the wrong way around. We must admit, however, that there are often quite important links between causation and modality, and it is incumbent upon us to provide an account of these links.

We believe that explanations flow from causation to modality. Sometimes a cause does boost the probability of its effect, as the theory alleges. We argue, however, that in such cases it is *because* of the causal relation that the cause boosts the probability of the effect; it is not because it boosts the probability that it counts as a cause. Nevertheless, we must explain why causes do often boost the probabilities of their effects. And in order to provide such an explanation, we must say more about the nature of the causal relation.

Hence, it is not satisfactory for us simply to reject modal theories of causation. We cannot rest comfortably with a theory which takes causation as a primitive. We must provide an analysis of causation without making any appeal to modal presuppositions.

6.3 CAUSAL TALK

We take it as a plausible metaphysical theory that causation is a relation between events, and we opt for a metaphysical realism which interprets the causal relation not just as a predicate, or a set of ordered pairs, but as a universal. We must construe events widely to include not only changes, but also unchanges or states of affairs. For instance, the unchanging state of pillars may be the cause of a temple's remaining intact over the centuries. But provided that we construe events broadly, it is an attractive metaphysical theory, that causation is a relation between events.

Objections to such a view sometimes arise, however, from examinations of the nature of causal talk. To prepare the ground for our own theory, we must take a stand on the nature of causal talk. To this end, we endorse a semantic theory which derives from Davidson[9] (another echo from the Davidsonic boom).

Davidson's semantics for singular causal statements straightforwardly relates our favoured metaphysical theory of event causation to the most basic form of causal talk: singular causal statements. It neatly explains the truth conditions for singular causal statements which would be problematic if causation were explicated in some other way, as, for instance, in terms of necessary or sufficient conditions. So, for instance,

The short circuit caused the fire

can be true provided that the right relation exists between the two events, the short circuit and the fire breaking out, while it is clear that short circuits are neither necessary nor sufficient conditions for fires. The truth of the singular causal statement depends only on the existence of an appropriate relation between the particular events. It may be true, of course, that we can subsume the events in question under some true causal generalization, but often it will be necessary to redescribe the events before we can do so. Although not all short circuits cause fires, it may be that all short circuits *of the same kind* as this one do cause fires if we are specific enough about the kind in question. Davidson requires, in fact, that the causal relation between events hold just in case there is some way of describing the events so that they can be subsumed under some general causal law. We have already argued against this view, on the grounds that causation is really a more local matter than this would allow. The causal occurrence may be a miracle. When events do fall under a causal law, we argue that these events do not count as causal pairs because they fall under the law. Rather, the law is a causal law because it generalizes over pairs which are causal. Thus, we do not accept Davidson's view that the nature of the causal relation between events derives from the existence of an underlying law. We do agree with Davidson, however,

9 Davidson (1980).

to this extent. The truth conditions of singular causal statements require the existence of a relation between events (not between events-under-a-description), and even when the two events *can* be subsumed under a law, they must often be redescribed before this can be done.

Davidson's semantics do not apply straightforwardly to all causal statements. Some require rewriting; for instance,

The stone caused the window to break

must be taken as being elliptical for something like

The stone striking the window caused the breaking of the window.

Others must be construed as consisting of the conjunction of a Davidson-type causal statement and some other causal information, for instance,

Becker's easy defeat of Lendl surprised the commentators

should (arguably) be construed as something like

Becker's defeat of Lendl caused the surprise of the commentators, and if the defeat had not been easy, it would not have caused the commentators to be surprised.[10]

But allowing for some such rewriting, Davidson's semantics for singular causal statements seems a most attractive theory, and its success in turn reinforces the plausible metaphysical theory that causation is primarily a relation between events.

It is nevertheless arguable that not all causal talk can be adequately expressed using a semantics which has the causal relation holding only between particular concrete events. General causal statements or causal laws seem to require a causal relation either between kinds of events or between properties of events. And causal statements of a counterfactual kind, such as

If Lendl's defeat had not been easy, it would not have caused the surprise of the commentators,

and statements like

Taking antidote caused the delay in Protheroe's death

do appear to require causal relations pertaining to properties of events (such as the easiness of a defeat or a delay in a death). The

10 Jackson and Pargetter (1989).

semantic message is that causal talk may involve construing causation as sometimes relating particular events and sometimes relating universals, and maybe even other categories of causal relation will be required. Some metaphysical response will have to be made to this semantic message. The most obvious response is to allow a variety of causal relations in nature. Sometimes causation relates individual events and sometimes it relates universals (properties) or property instances.

Now as a metaphysical theory, this raises an important problem. Why should the various relations be classed together? Why are they all *causal?* We argued earlier that it would not be acceptable to take the causal relation as primitive. It would be even more unsatisfactory to have a host of primitives or to have analyses for the various causal relations which left them unrelated. This, then, suggests what is required. We must suppose that all the causal relations required by semantics will supervene on some basic causal relation. We will then have to say enough about the nature of this basic causal relation so that it will provide a range of explanations of connections, not only between causation and necessity or probability, but also among the various other causal relations invoked in our causal talk.

In our analysis of causation, we shall focus on basic causal relations, and we shall say little about the many other causal relations which supervene on these and are reflected in causal talk. It is important to recognize that there is a bridgeable but problematic swamp lying between the metaphysics and the semantics of causation. And in offering a metaphysics of causation, we are not pretending to solve all the semantic problems. Hence, it will be no objection to our proposals to cite one or another causal idiom which we have failed to explain.

We do claim, however, that the correct semantics for basic causal statements *cannot* be given in terms of probability or necessary and sufficient conditions. Rather, basic causal statements are made true by the existence of a causal relation whose nature is largely unknown to us. As far as semantics is concerned, this causal relation is primitive. Native speakers cannot pick it out by description – certainly not as 'the relation which holds if and only if such and such probabilities hold'. The best they can do is to recognize causal relations fairly reliably when they are confronted by them. Our task now is to develop a theory about what we *are* confronted with in such cases – not to analyse the meanings of causal statements: our task is metaphysical, not semantic.

6.4 THE HUME WORLD AND CAUSATION

Consider the possible worlds in which some things cause some other things. Call these *causal worlds*. How many causal worlds are there? Some might argue that all worlds are causal worlds. One reason for believing this might draw on reflections concerning individuality. On one theory, an individual is really just a "bundle of properties". And the question arises what "string" ties the bundle together. It might be argued that it is *causal* "string" that ties properties into bundles: a bundle of coinstantiated properties is really just a collection of causally interdependent properties – wherever one of them goes, the others follow.

Another reason for saying all worlds are causal worlds might draw on the theory of time. It might be thought that no world could be entirely timeless. If one also adopts, on, say, Kantian grounds, a causal theory of time, it then follows that no world would entirely lack causation.

However, we do not accept the bundle theory of individuality. We are tempted by the causal theory of time's arrow, of the asymmetry between past and future; but we are not persuaded by a causal theory of time itself. Hence, we are not persuaded that all worlds are causal worlds. On the contrary, as we argued in Chapter 5, we tentatively assume the existence of Hume worlds. We assume that there is a world which is roughly as Hume believed the actual world to be, a world in which there are first-order correlations among events, but no correlations can be distinguished as causal (except in a subjectivist sense, which refers to something like the robustness of our expectation that they will persist). There are also Heimson worlds, which match all the regularities that are laws in the actual world. A Heimson world corresponding to the actual world will be accessible from the actual world. A Hume world may not be accessible, as it matches the actual world only at the first-order level. But for present purposes it is the Hume world and not the Heimson world that we need. It does not matter whether it is accessible from the actual world; all that matters is that it exists. That is, what we require is only the logical possibility of a world which matches ours in first-order regularities but not in causes or laws.

Hence, some worlds do contain causation, while others do not. What is common to all causal worlds but lacking in the Hume world? The difference between causal and noncausal worlds cannot be merely a modal one. All modal differences between such worlds must, we maintain, rest on differences between their contents. Our world and the Hume world could not be completely identical in what they contain and yet different in what modal claims are true of them. There must be something which is present in causal worlds and absent in noncausal worlds.

For this reason, the Hume world should not be defined as the world which is *exactly* like ours except that laws in our world are mere correlations in the Hume world. The difference in status of regularities must rest on some other difference between the worlds. So we define the Hume world to be a world which matches ours at the first-order level of properties and relations. Such a world will match our world in the individuals it contains and in all their first-order properties and relations, yet differ from our world with respect to some of the relations among properties, and relations among relations. If there is to be a Hume world at all, it must be a world which differs from ours only with respect to higher-order universals. Some higher-order universals will supervene on the first-order ones, but there is no reason to think that all of them will.

We can then spell out in more detail the plausible metaphysical theory that causation is to be understood as a relation between events. Causation is a second-order relation between events. It is not supervenient on intrinsic first-order properties of events, but is rather a second-order relation, one which involves not only the individual events and their properties, but also relations among their properties – second-degree relations. Such second-degree relations have to be ones which hold contingently if we allow the existence of a Hume world. Cause and effect could both occur, with the same first-order properties, whether or not those properties stand in the relevant second-degree relations. The relevant second-degree relations are external relations, not internal ones.

Events may have the same first-order properties and yet differ in second-order properties. That explains how it is possible for the Hume world to match ours at the first-order level yet differ in causal connections. However, a lack of match between one *world* and another is not the only way in which a match at the first-order level may fail to accompany a match in causal connections. We noted earlier that causation is local. Hence, even in a *single* possible world, two pairs of events may match at the first-order level yet not match in their causal relations. If the events are in the same world, they will match not only in first-order properties, but also in the properties and relations of these properties. (This is a consequence of our treating them as universals.) Thus, a match at first-order level does entail a match at the second-*degree* level, for all event pairs in the same world (as Michael Tooley has stressed to us in discussion). Nevertheless, such event pairs may fail to match causally – that is what the local nature of causation entails. Hence, causation must be a second-order relation which fails to supervene on both first-order and second-degree properties and relations. Causation relates not just universals, but structures involving both universals *and particulars*.

Thus, causation is a relation holding between structures which involve both particulars and higher-order universals. The causal relation itself, therefore, must be a higher-order relation. The question then arises as to what higher-order properties and relations could constitute the causal relation. To get a grip on this, we draw on another plausible idea about what distinguishes the Hume world from ours.

Consider gravitation and a theological construal which Newton toyed with and Berkeley championed. It was hard to believe that the presence of a planet a vast distance from the sun would exert a force on the sun, apparently without contact or any medium through which contact forces could be transmitted. So the idea arose that the motions of each heavenly body were directly caused by God. The correlations among their motions according to mathematical patterns was then to be explained as part of God's plan. The sun was caused to drift ever so slightly towards, say, Jupiter, but it was God who caused it to do so. It was not the mere presence of the distant Jupiter itself which caused the sun's motion.

Berkeley's world is not a world without causation, but an animistic world where all causation is volitional. Hume stripped volitional causation from the Berkeleyan picture, and the result was a world without causation. The first step towards such a world, however, was the removal of the force of gravity. The Hume world is, first and foremost, a world without *forces*.

The Hume world differs from ours only in higher-order properties and relations. We also maintain that the Hume world differs from ours by the lack of any forces in it. These two ideas can be seen as sides of the same coin: at the most fundamental level causation is the action of forces; and forces are higher-order relations.

6.5 FORCES AND CAUSES

We have elsewhere (with Brian Ellis)[11] considered the nature of forces. We were led to construe forces as relations and as a species of the causal relation. We shall here argue a stronger thesis, namely, that *all* causes supervene on forces; and all forces are higher-order relations between structures involving individual events and their properties.

Forces have been characterized as mediating between causes and effects. Yet suppose a cause *C* and an effect *E* were the *same* general kind of entity as the force *F*. Then in the sequence, *C*–*F*–*E*, we have *F* mediating between *C* and *E*. Yet if *C* requires something to mediate between it and *E*, why should not *F*, too, need something to mediate between it and *E*? A regress threatens (as Hume urged). To reply that forces are by definition "immediate causes" while other entities are not does not solve the problem, because something must be supposed to be the immediate cause *of the force*. So the introduction of *F* as an immediate cause has not really filled any explanatory gap – or rather, it has filled one, but only at the cost of opening another.

One initially tempting way of responding to the regress is by construing forces as *dispositional properties:* either as the dispositions of objects to undergo certain changes or the dispositions of fields to produce certain effects. Dispositions do not themselves take part in the causal sequence of events. The causing is done by some property complex which is the basis of the disposition and in virtue of which the disposition is possessed. Hence, by interposing a disposition between cause and effect we do not create any *causal* gap between this disposition and the events that it mediates between. If we construe forces as dispositions of some sort, therefore, we thereby block any risk of a vicious regress.

Yet dispositions are not the right ontological category for forces. The complete causal story can be given without mentioning any disposition, since the causing is done by the physical basis of any

11 Bigelow, Ellis, and Pargetter (1988).

disposition involved. This stands in sharp contrast to the way forces, unlike dispositions, are essentially involved in causal stories. Dispositions supervene but do not participate, whereas forces do participate. In addition, dispositions differ from forces in that dispositions can be present when they are not acting (the glass is fragile while it sits safely on the shelf), but forces cannot. Forces are necessarily active, that is, they cannot exist without acting, so they are not like dispositions but, rather, like *relations,* which are instantiated ("active") only if they *relate* cause to effect.

With Ellis we argued that forces are best seen as constitutive of causal *relations.* A force is not an event *C* or *E*, but is a relationship holding *between* *C* and *E*. The search for the fundamental properties in nature, like the search for fundamental particles, has provided explanatory power as a result of revealing *what is common* among widely differing phenomena. And the same thing holds true of the search for the fundamental forces in nature. A parsimonious yet sufficiently rich account of the fundamental species of causal interaction in nature should unify our account of physical interactions generally, for it should yield an account of what is common among various laws with quite distinct formal expression, yet which govern the same fundamental kinds of causal interaction. And conversely, we will have an account of the essential differences between interactions governed by laws which may have the same formal expression (say, inverse-square laws) but which involve different fundamental forces.

We have modified our position slightly since the article with Ellis. We no longer identify an *instance* of the causal relation with a *single force,* for reasons that will emerge later. We shall argue, rather, for the view that a number of distinct fundamental forces combine as *constituents* of basic causes. But all the arguments supporting the theory developed with Ellis carry over to the modified theory. Both forces and basic causes must be higher-order relations among events, for if they were first-order relations, a Hume world would be an impossibility. At the fundamental level a force relates a property complex of a state of or change in a field, on the one hand, and a property complex of a state of or change in a particle, on the other. Each property complex has a number of first-order properties and relations as constituents. There will be a Hume world in which these first-order constituents are all there is. But in our world, there is also an external relation, a force, relating these two property complexes. Yet this external relation is separable from the first-order terms it relates. Even in our world, because of the local nature of the causal relation, there may be *other instances* of this property complex of a change in a field, and this property complex of a change in a particle, which do *not* stand in the external, causal relation to each other.

We emphasize that causation is a relation which involves the *aggregate* of forces, not the *resultant* force acting on something. We are realists about component forces.[12] For example, a zero resultant force may be due to a number of balancing real nonzero component forces which then causally explain the nonchange; *or* the zero resultant force may be due to the total absence of any forces (in which case there is no causal explanation of the nonchange, except in the minimal sense that the explanation refers to the fact that no causing is going on). Obviously the component forces and their resultant cannot *all* be real or we would have overdetermination or double causation. There is a principled solution. Sometimes when we speak of component forces and their resultant, it is the components that are real and the resultant that is not; sometimes it is the other way around. Whether it is the components or the resultant that are real, in any given situation, is determined by the physical features of the situation in question, as we shall now explain.

When we have a number of real component forces, the resultant force is a fictional force. It is the force that would have related some change in a field to the effect if a single force had been involved and if the effect, the change of state of the particle, had been the same. Of course, the cause, the change of state of the field, would have to be different in order for only one force to be involved. The reality of components in this kind of physical situation is forced upon us by the kinds of considerations which are illustrated in the following thought experiment. Consider three protons, isolated from outside interference, one at the midpoint of the line between the other two. The predicted outward motion of the ones on the ends will involve forces both between them and between each of them and the one in the middle. Yet the principle of action and reaction of forces, so fundamental to physical theory, entails that the middle proton is subject to two balancing component forces which jointly result in its nonchange.

12 See the discussion in Bigelow, Ellis, and Pargetter (1988). For arguments against the reality of component forces see Cartwright (1980, 1983).

Contrast the case of the three protons with the following kind of case. Sometimes we are concerned with the effect of a particular real force, say on a particle which can move in only one direction distinct from the direction of action of the force. We standardly resolve that force into two fictitious orthogonal resolutes, one in the direction of the force, which if taken as component forces would have when aggregated resulted in the same effect, that is, in the same change of state of the particle. The fictional nature of the component forces in this kind of situation is again forced upon us. The choice of direction is arbitrary, as is the choice of direction of the other orthogonal resolute (in three-dimensional cases) and as is the requirement that the resolved components be orthogonal. All cannot be real, because we would then have gross overdetermination. Hence, in this situation it is the resultant force which is real, not the components.

Thus, which forces are real will be determined by the physical nature of the situation in question. In some situations a single, resultant force is real; but in other situations the resultant force is a useful fiction, and it is a plurality of component forces which are real.

We therefore urge the identification of basic causation with the action of structured aggregates of fundamental forces. The *causal* relation is a higher-order relation which holds between events just when those events are linked by aggregates of forces. We anticipate two kinds of example, which could be thought to tell against us.

There are cases of causation in the realm of quantum theory for which normally no mention is made of forces. A photon absorption may cause an excited energy state for an electron, yet it is not usual to say that the photon exerted a force on that electron. We concede that forces are not *talked* about in at least some of these cases. But we believe that what happens is close enough to the actions of forces, as we conceive of them, for us to describe the quantum-mechanical cases as an extension of a theory framed in terms of "force".[13] There is a relation between some property complex of a field and some property complex of a particle. Perhaps the relation should be somewhat different from the ones discussed in theories prior to quantum mechanics, but it is close enough to inherit the same theoretical role in the analysis of causation. At the quantum level, the distinction between fields and particles becomes very

13 Attention is given by Heathcote (1989) to the formal details involved in the shift from classical forces to quantum electrodynamics.

hazy. It is probably better not to think of a field as one thing and a particle as something quite different. Interactions are not so much changes in fields causing changes in particles as changes in one kind of field causing changes in another kind of field. The term 'force' perhaps harks back too much to corpuscular theories. Yet we claim that quantum interactions are the natural heirs to the old notion of 'forces' which we have used to analyse causation.

The other kind of case that could be thought to tell against us is that in which a field is not interacting with any particle but a field is nevertheless being produced. If a field emanates from a particle, surely there is a causal story to tell about this emanation. And, it may be objected, surely that cannot be a story about forces. A charged particle, for instance, surely does not exert a *force* on its own electrical field, yet plausibly it does *cause* that electrical field.

In response to this objection, we are doubtful whether the emission of a field *is* a basic instance of causation. The objection presupposes a clear demarcation between a particle and the fields around it (or within it), and this separation is out of line with current views in physics. Thus, it seems more appropriate to see a unity between a particle and its accompanying fields – making the fields part of the "essence" of the particle. Of course, this view is tentative and would have to be revised if physical theory were to develop in particular ways. Given the haziness of the distinction between particles and fields, it may in fact turn out that another response to the objection will be open to us. Instead of denying that the particle "causes" its fields, we might admit that this *is* an instance of causation. But we might also maintain that, despite initial appearances, it *does* involve the action of forces, or the quantum-mechanical heir to the notion of force. It is far from an absurd hypothesis that the field is really quantized and consists of fundamental "particle-waves" which are being propelled outwards by a "force". For these reasons we do not regard the apparently causal relation between a particle and its field as a persuasive objection to our reduction of all basic causal relations to forces.

6.6 EXPLAINING SOME DETAILS AND DETAILING SOME EXPLANATIONS

If we are to provide an adequate account of causation, we will have to be convincing on four matters which at least initially seem problematic for our theory. How plausible is it that macroscopic forces, and thus causes, in the world about us, and about which so much of our causal talk is concerned, supervene on the fundamental forces, and hence on basic causes, in physics? How can we justify selecting forces rather than some other ingredient of physical theory for the analysis of causation? Do we have the appropriate explanatory link between causation and the various modal notions, including probability? What sort of higher-order universals are forces and basic causes? These are the matters that we now address.

Macroscopic causes and fundamental forces

Consider the gap between the fundamental forces appealed to in our theory and the macroscopic causes which we encounter in daily life and which are the subject of so much of our complex and varied causal talk. Earlier we argued that all causal relations supervene on the basic causal relations among events. Having now construed this causal relation as the aggregate of the appropriate fundamental forces, we must look again at our supervenience claim to see how plausible it is.

We suggest that all macroscopic causes supervene on the fundamental forces within the macroscopic objects involved. Our current physical theories tell us that all these are reducible to the interaction of a few basic kinds of fields and a few basic kinds of particles. And although the fields and particles may vary, we take it that the forces are instances of the same fundamental relation. This claim that all forces derive ultimately from a single underlying relation is supported by its explanatory power. It explains why there is a single class of laws of composition and laws of action for forces, regardless of the kinds of fields and particles involved. For instance, we calculate a resultant force, given two component forces, by way of one and the same method, the *parallelogram of forces,* whether one of the components is an electrical force and the other a gravitational one, whether both are magnetic forces, or whether they are other kinds of forces.

Not only macroscopic forces but also a variety of other causal matters will supervene on fundamental forces. The basic causal relations, which are aggregates of fundamental forces, relate salient participants in any complex causal interaction, but the causal interaction as a whole may be much more complex than just a relation between fields and particles.

Any network of basic causal relations, and thus of aggregates of fundamental forces, will automatically give rise to a variety of derived relations among chunks of that causal network. Consider causal chains, as employed by Lewis in defining causation. For us, each link in the chain is to be analysed as involving a basic causing, that is, an aggregate of forces rather than counterfactual dependences or probability boostings. The first and last links will in general not be related by the basic causal relation; but there will be a derivative causal relation between them arising out of the basic relations between each of the links. In any causal interaction there will be many such chains running in parallel, with various interconnections. Given two complex aggregates of events, one aggregate earlier than the other, there can be a complex network of causal chains connecting them, and thus there will be derivative causal relations holding between them. Hence, given a single basic causal relation, there will be no shortage of distinct, derivative causal relations. Causal talk may relate, semantically, to any of these derived relations. We should not expect the semantics of causal talk always to relate in a straightforward way to the basic causal relation or to aggregates of forces. We can draw on any of the many derived relations when giving the semantics for singular causal statements, general causal statements, causal laws, causal explanations, causal counterfactuals, causal adverbial modification, and so on. We will not attempt to formulate a full taxonomy of derived causal relations or to employ them in giving a semantics for causal talk.

What is important for our purposes is only to make clear that the metaphysics we offer for causation does not pretend to be, by itself, an adequate account of all causal relations and causal talk. However, the claim that they all supervene on a single basic causal relation does explain, we claim, just what makes all these aspects of causal talk count as causal. The common thread is furnished by the presence of networks of underlying forces.

Causation and physical theory

Let us grant, then, that we can bridge the gap between the fundamental higher-order relations in our theory and the many complex interactions in everyday life. How do we justify the choice of forces as the right thing to select from physical theory for the analysis of causation? Another theory which has the same strategy as ours, that of reducing causation to fundamental occurrences described in physical theory, is that of David Fair.[14] But Fair defines causation in terms of flow of energy rather than in terms of forces. Clearly the two accounts are not equivalent. By Newton's third law, there are two distinct forces, forces associated with changes in states of different particles, for each single instance of energy flow. Thus, for us, there are two forces, hence two instances of causation. For Fair there is one instance of energy flow, hence one instance of causation. His theory is distinct from ours. Yet like ours, it grounds causation firmly in physical theory. What principled reason is there for selecting forces rather than energy flow or perhaps some other fundamental occurrences in physical theory?

We have argued that there are good reasons for explicating forces as higher-order relations. They are thus the right kind of universal to be constitutive of the kind of relation we have argued that causation must be. We think that other candidates from physical theory will not amount to relations of the right, higher-order kind. Forces are the "right stuff"; other things are not. In particular, Fair's candidate, energy flow, is not the right kind of thing, the right kind of higher-order relation, to constitute causation.

Fair's theory requires us to identify "flow" of energy, and this requires us to identify packets of energy across time. The energy present in the effect is the same as, is numerically identical with, the energy lost by the cause. Since there are instances of causation not involving energy transfer, but involving only momentum transfer, in fact Fair needs a disjunctive account. Fair must say that causation is the transfer of either energy or momentum. Hence, he must be able to determine that the momentum possessed by the effect is the very same momentum that was lost by the cause. Identification of either momentum or energy across time raises doubts about Fair's account. In addition, the disjunctiveness of Fair's account makes it difficult for causation to serve as a truly unifying element of physical theory. Also, in cases where both energy and momentum transfer take place, with which is causation to be identified? As they are distinct, the causal interaction cannot be identified with both.

Another worry concerning Fair's account takes us back to the

14 Fair (1979). For discussion, see Earman (1976), Aronson(1982), and Kline (1985).

Hume world. Energy and momentum flows supervene on first-order individuals, properties, and relations. Hence, on Fair's account there could not be a Hume world matching a causal world in all first-order properties. We argued in Chapter 5 that it is desirable to allow for a Hume world, and a theory of causation which allows for it must make causation a higher-order relation. Our theory does this, while Fair's does not.

We cannot prove that there is not some aspect of physical theory more suitable than forces to be constitutive of causation. Our earlier discussion of the Hume world argued for an essential link between forces and causation, and it has not been shown that such a link exists in the case of any other of the fundamental occurrences within physical theory. If causation is to be identified with any such fundamental occurrence, the choice of forces has many natural advantages.

Whither necessities?

We have argued that a cause need not be a necessary or a sufficient condition for its effect. Nevertheless, there are connections between causes and necessity. Given a cause, its effect may not be a necessary consequence in the strict sense which excludes all possibility of overriding. Yet there is a sense in which, given a cause, its effect follows as more than a mere coincidence.

There are two ways in which causation gives rise to modalities. These arise out of an ambiguity in the notion of "the cause" of some event. We will distinguish between "thick" and "thin" causes (on analogy with Armstrong's thick and thin particulars).[15] The thin cause is just the complex of particulars, properties, and relations which stand in the causal relation to the effect in question. The thick cause is the thin cause together with its external relational properties, *including* all the force relationships it stands in. Obviously there will be a necessary relationship between thick causes and their effects. Given a thin cause *together* with all the (component) forces associated with it, the effect of those forces must necessarily follow. This is a consequence of the fact that a relation (at least a relation like this one) cannot be instantiated unless both its relata exist. This explains the fact that forces are necessarily active: a force cannot exist unless it is acting on something, because it is a

15 Armstrong (1978).

relation which holds only when it is instantiated by a pair of appropriate entities.

It is a relatively trivial matter that thick causes are necessarily followed by their effects. Yet this relatively trivial matter does help to explain why we think of causes as necessitating their effects. We often think of a cause as the *remnant* of a causal interaction, which remains after the effect is taken away. Given the *whole* of a causal interaction *except* for the effect, the effect must follow.

The second way in which causation gives rise to modalities is one which rests on thin rather than thick causes. We have argued that the presence of a thin cause is not *always* a necessary or sufficient condition for its effect. Thus, causation cannot be defined in terms of necessary and sufficient conditions. Nevertheless, in many cases causation does give rise to necessary or sufficient conditions. In the right context, a thin cause may indeed be a necessary and a sufficient condition for its effect. In situations in which there is no significant degree of indeterminacy, and there are no fail-safe back-up arrangements, then the nearest or most similar worlds with the cause are worlds with the effect, and those without the cause are ones without the effect. In virtue of this, we have a qualified version of Lewis's counterfactual analysis of causation: if *this* cause were to occur, the effect would follow; and if *this* cause had not occurred, the effect would not have followed. These counterfactuals will hold in any particular case in which there is no significant indeterminacy or back-up. And these qualifications are very often satisfied in the actual world. This explains why most causes are appropriately described as necessary or sufficient for their effects.

The reason the nearest or most similar worlds with a thin cause will also be worlds with its effect is as follows. Nearness or similarity of worlds is determined by the individuals, properties, and relations they share. The more they share, the closer they are. This applies not only to first-order properties and relations, but to higher-order ones as well – including the causal relation. In fact, it is plausible that sharing of higher-order properties and relations should be given greater weight than first-order ones when determining the nearness, or similarity, or degree of accessibility for a world. Hence, the closest worlds with the thin cause will generally be ones in which it stands in the same relationships. Given the thin cause and its relationships (the thick cause, in fact), the effect must be present too.

This explains why causes are usually necessary or sufficient for their effects. There will be an analogous explanation for the fact that the presence of a cause usually boosts the probability of its effect. Given that a thin cause is causally related to its effect, there will in general be a higher proportion of appropriately nearby worlds in which it is accompanied by its effect. And this accounts for the plausibility of probabilistic theories of causation.

Causes as structural universals

Physical theory tells us that fundamental forces have a variety of magnitudes and directions. They are vectors. Basic causes are constructed from aggregates of vectors. They are, we believe, structural universals. Here we can apply the framework established in Chapter 2, concerning quantities, vectors, and structural universals.

First, we recall our discussion, in Chapter 2, of the quantity *mass*. There is a different property for each specific mass, and objects may differ in mass in that they have different specific mass properties. However, all massy objects have something in common: they have mass. The best account of this sameness is to postulate that these specific mass properties stand in higher-degree relations to one another. These relations will be essential, internal ones, rather than external ones – in contrast to causal relations. That is, specific mass properties could not be the properties they are unless they stood in certain definite relations to one another. The relations of proportion that hold between specific mass properties are essential, or intrinsic, to the identities of those properties. Standing in such relationships to other mass properties is what all mass properties have in common. Being different specific mass properties constitutes the way in which they differ.

Some first-order properties and relations, such as two determinate coplanar velocities, stand in *two* distinct relations of proportion to each other (one from each of two families). One of these relations reflects difference in magnitude, and the other reflects difference in direction. That is what makes velocities vectors.

Forces, too, are vectors, but vectors of a slightly more complex kind than velocities since forces are themselves second-order relations. Fundamental forces may have different magnitudes and directions. Each fundamental force has some specific magnitude and direction, and in virtue of this, it stands in a cluster of higher-degree internal relations to other fundamental forces. The existence of this family of higher-degree relations will ensure that all fundamental forces have something in common – they are of a single kind. Yet they also differ from one another, because they instantiate different members of the same family of higher-degree relations. The quantitative relations between forces are intrinsic ones. The fact that one force is twice the intensity of another, or orthogonal to another, is not a contingent matter. Whether two things are related by a force is contingent, but whether two forces stand in certain quantitative relations is not. They would not be the forces they are if they were not related as they are.

Second, we recall now our discussion of structural universals. Consider methane. Methane molecules consist of a carbon atom and four hydrogen atoms bonded in a particular configuration. Methane molecules instantiate the universal "methane". So methane is intrinsically related to three other universals: hydrogen, carbon, and bonded. Methane is what is called a structural universal. But methane is not merely the mereological sum of the other three universals. Our account of structural universals is that they are higher-order relational properties, which will then stand in higher-degree internal relations to their constitutive universals.

Now the constitutive properties of a structural universal can be specific determinate quantities. In fact, they can also be vectors. So there is nothing problematic about having a structural universal the constituents of which are specific fundamental forces of certain magnitudes and directions. And there will be a host of internal relations between such a structural universal and those specific forces involved. In fact, the internal relations involving these structural universals will also have to relate them to certain individuals. These structural universals will have to incorporate an element of particularity. If causation is to have the *local* character we have claimed for it, then it cannot be a relation between entirely nonlocal universals. So the structural universals which constitute causal relations must be ones that include relational properties embedding particulars as well as universals.

If a basic cause is conceived as a structural universal involving the aggregate of the associated fundamental forces, this structural universal will be a conjunctive higher-order relation among the particular events: roughly, it will be the relation of standing in relation F_1 *and* standing in F_2 *and* . . . , where F_1, F_2, and so on are the constitutive forces. Hence, by identifying basic causes as structural universals which involve aggregates of forces and by identifying forces with vectors, we endow causal relations with very rich essential natures. Thus, we are far removed from the sort of theory we objected to earlier, which takes the causal relation as primitive, unexplained, and mysterious. Our theory does not treat causation as a primitive. Causation is explained in terms of vectors and structural universals. Both of these are important for many purposes in science at large. They are not cooked up just to solve the problem of causation.

The world contains not only causes and effects, but also causal relations holding between causes and effects. Because causal relations enter into the structure of the world, their presence has various modal and probabilistic consequences. Causation and necessary and sufficient conditions do often go hand in hand. Causation, however, is a robust ingredient within the world itself, whereas modalities and probabilities supervene on the nature of the world as a whole and on the resulting relations between one possible world and others. Some modalities, therefore, are essentially causal; but causation is not essentially modal.[16]

16 Much of this chapter is based on Bigelow and Pargetter (1990). We are especially indebted to comments from David Armstrong and Michael Tooley on material in this chapter.

7

Explanation

7.1 HOW AND WHY

One thing which many people seek, and science is especially effective at finding, is knowledge of what things there are in the world, how they are related to one another, what things have happened, and when. For a long time it was not certain how long there had been human inhabitants on earth. Aristotle thought there had been humans on earth for millions of years – in fact, for an indefinite, even an infinite, expanse of time stretching into the past. The Hebrew, Christian, and Islamic traditions taught, in contrast, that there have been humans on earth for only a few thousand years. Geologists a few hundred years ago established that the mountains and oceans of the earth are hundreds of millions of years old, whereas human life has existed on earth for a much shorter time. Science has discovered how old the earth is and how young humanity is.

Enquiries of this sort are of the greatest importance and interest, and constitute an essential and substantial component of science. Yet there are also elements of science which are somewhat different. The acquisition of knowledge of what happens, where and when, is sometimes characterized as being concerned with *description* rather than *explanation*. Science, however, is concerned with both description and explanation. Yet it is misleading to draw a sharp line between description and explanation. Description and explanation are not mutually exclusive categories. It is better to distinguish among kinds of explanation. We can explain *how* things are, or we can explain *why*.

Consider an example. Copernicus offered an explanation of *how* the planets move. He also wondered *why* they appear to move as they do when viewed from the earth. Yet he did not adequately explain why they move in the way they do. A century later, Kepler gave a much more accurate description of how the planets move – namely, in ellipses. He also attempted to explain why they move in that manner (by appeal to the rotation of the sun and to the magnetic field of the earth); yet the explanation he offered was in many ways unsatisfactory and was indeed mistaken. Newton's explanation in terms of universal gravitation attraction was a much more successful explanation of why the planets move as they do. Explaining how is very important; but even very successful explanations of that sort leave us hungry for more. We want to know why.

All explanations proceed by offering information about what there is in the world. But, we shall argue, explaining why something is so is, in fact, giving a very particular kind of information about what there is in the world. To explain *why* is to give information about the *causal relations* in nature.

Information about causal relations is information about just one of Aristotle's four categories of "causes".[1] For Aristotle a cause of a thing is (roughly) something *without which the thing would not be*. An object, then, has its matter as one of its causes, since without its matter the thing would not exist. This is called its *material cause*. The structure, properties, and relations within an object, too, count as one of its causes, since if the same matter were altered radically in form, the object would cease to exist and another distinct object would have been created with the same matter. So the form of an object is also a cause – it is the *formal cause*.

In giving these sorts of causes, we are explaining what there is in the world or explaining *how* things are. We are not, however, explaining why things are as they are. To explain why, we maintain, is to give information about causes in a narrower sense, causes in the sense of things which stand in the causal relation to the thing to be explained. And the causal relation, we have argued, is a specific sort of physical relation in the world – in fact, a structural universal the constituents of which are forces ultimately supervening on the fundamental forces of nature. This is what Aristotle calls an *efficient cause*. To explain why something is so, we have to give information about efficient causation.

Aristotle also lists a fourth kind of cause: *final cause*. The final

1 See, e.g., Aristotle, 'On the Four Causes', in Ross (1928).

cause of something is the purpose or goal for the sake of which the cause exists; for instance, the final cause of the court jester is (perhaps) to amuse the king, or the final cause of the eye is to give us sight. And this seems to provide us with an answer to some kind of why-question.

Science has been impressive at supplying us with material, formal, and efficient causes of things, at telling us what and wherefrom, but it has been less impressive at supplying final causes, at telling us what for. Efficient causes and final causes both tell us why, but in different ways. We shall attend first to efficient causation, and later in the chapter we shall discuss final causation.

It may be wondered what is held in common between how-explanations and why-explanations, and what is common between the efficient-cause and the final-cause subspecies of why-explanations. Why are all these things called by the same name, 'explanation', despite their differences?

All explanations are aimed at giving more information about something of concern to us. But there are constraints on what sorts of information are appropriate. Any information at all may, at a stretch, be said to explain more about *what* is the case. Yet sometimes such information is completely uninteresting. There is an inescapable pragmatic dimension to explanations in general; explanations give information which is intended to increase *understanding*. And what counts as an increase in understanding depends on what a person knew before hearing the explanation and on what more it was useful to know. Explanations provide "relevant" information. We lay down no strict rules of relevance.[2]

The kinds of information which an explanation can supply will depend on the kind of thing which is being explained. If we are asking for an explanation of a particular event, we could be given a specification of its cause; or we could be told more indefinite information about its cause; or we could be told about the purpose for which the event occurred; or we could simply be told more about exactly what did happen; and so on. In contrast, if we ask for an explanation, not of a particular event, but of a *kind* of event, we are faced with a slightly different range of options. To illustrate, compare a request for an explanation of why a particular person stole a

2 Most writers on explanation have stressed the pragmatic element: see, e.g., Hempel (1965), Putnam (1975), and Lewis (1986e).

camera with a request for an explanation of why people steal. The range of answers which may be appropriate to the first is not the same as the range appropriate to the second.

Why-explanations are most obviously appropriate when we are considering explanations of particular events. Why-explanations, we have argued, offer information about causes; and causation is a relation between particular events which supervenes on the fundamental *causings* in the world – the fundamental forces. Note that both efficient causation and final causation are relevant to a why-explanation. This is because, as we shall argue, explanation by final causes does involve reference to the same causal relation that is appealed to in explanations by efficient causation. The role of the causal relation is different in the two cases, but it is the same causal relation that is involved, as we shall see when we discuss functional explanation.

Yet although particular events are the most basic target for why-explanations, it is also possible to seek why-explanations for kinds of events or for laws governing kinds of events. We may ask, for instance, why people steal, or we may ask why unsupported objects fall. Explanations of such things will, we claim, count as why-explanations only if they supply information about causes. Since kinds and laws are not themselves events, they are not the appropriate category of entity to stand in causal relations to things. They cannot be acted on by a force. Hence, if we are to supply causal information to explain them, we cannot do so by giving their causes or their effects; we must, rather, give causal information concerning the particular events which fall under the kinds or laws which we hope to explain.

The standard pattern for this is to find *causal laws* which subsume whatever it was we aimed to explain. For instance, to explain why unsupported objects fall we appeal to the force of gravity. This means that we appeal to laws which explicitly govern the operation of forces in nature, and then we show that the descriptive generalization 'Unsupported objects fall' is entailed by those laws governing forces. We have supplied information about the generalization by showing that it is entailed by certain other generalizations. And we have supplied *causal* information about the generalization, because the other generalizations which entail it are concerned with causal relations.

7.2 LAWS AND EXPLANATION

The theory we accept for causation will constrain the range of theories we can hold about explanation. To give the causes of an occurrence is obviously to contribute to an explanation of that occurrence. There may be other sorts of explanation as well, but clearly the giving of causes can contribute to *one* sort of explanation. We should therefore take stock of what conclusions we have reached concerning causation and what consequences they have for our account of explanation.

We have argued that causes are neither necessary nor sufficient conditions for their effects. Nor need a cause even boost the probability of its effect. Thus, the existence of a causal relation does not by itself guarantee the existence of laws, even statistical laws, linking cause to effect. There can be cases of singular causation. For instance, there could be miracles: events which constitute exceptions to the laws of nature. There could be magic. There could be indeterminacy. Yet even miraculous, magical, and indeterministic events may nevertheless be caused.

Hence, when we explain an occurrence by giving its causes, there is no guarantee that this will furnish us with any laws of nature which link the occurrence to its cause. There is no guarantee that we will be able to show that the occurrence followed from laws of nature, given the cause. We are committed to a degree of independence between explanation, on the one hand, and laws of nature, on the other, and to the possibility that *some* explanations do *not* involve the subsumption of events under laws.

Causal explanations subsume events under the causal relation. If that causal relation can be subsumed under a causal law, we will be able to follow our causal explanation with a nomological explanation. If the laws involved are statistical laws, we will be able to furnish a statistical explanation. Yet there need be no such further explanations. And even if there are such further explanations, this does not reduce causal explanations to nomological explanations. Causal explanations tell us *why* something happened. Nomological (including statistical) explanations do not explain why, except insofar as they entail information about causes. Nomological explanations explain *how* things happen; only causal or teleological explanations tell us *why*.

Our views on causation thus lead us to deny the widely held theory that all explanation is subsumption under laws. Yet we allow that *some* explanations do involve subsumption under laws. It would be extremely implausible to rule out all explanations which proceed by way of subsumption under laws. The nomological model of explanation has been the orthodoxy from Aristotle to Hempel, and it has been reinforced again and again by examples from the history of science. Scientists have been astonishingly successful in subsuming phenomena under laws, and lower-level laws under deeper laws. It cannot be denied that this has been done. And it must be allowed that it is often explanatory, in some sense. Hence, *some* explanations do proceed by subsumption under laws.

The basic idea, that explanation is derivation from laws, is very old; but it has received its fullest articulation in recent times by Hempel.[3] We shall give an outline. The basic pattern underlying a Hempelian explanation is that something to be explained, an outcome, say '*O*', is derived by a sound inference from a (conjunction of) the laws of nature, '*L*', together with the so-called initial conditions, '*C*':[4]

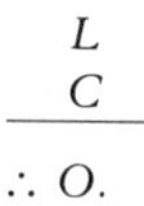

$$\frac{\begin{array}{c}L\\C\end{array}}{\therefore O.}$$

In some cases, no initial conditions will be required, and explanation may succeed in deriving *O* from laws alone. So *C* may be vacuous in some cases. But in other cases, laws alone will not do the job. If we want to explain why, say, Halley's comet returns about every seventy-six years, we will use the laws of mechanics, but we will also have recourse to matters of particular fact. For instance, we will have to feed in details on the mass of the comet, its position and velocity at some given moment, the paths of the planets, and so forth. These facts are clearly not laws of nature – they are particular, not general, and they not only could have been other-

3 Hempel and Oppenheim (1948), Hempel (1965). The idea that explanation relates to derivation from laws also stems from Mill (1843/1950), Jevons (1877), Ducasse (1925), Feigl (1945), Popper (1945), and Hospers (1946).

4 Strictly '*L*' and '*C*' are thus sets of sentences, all of which are to be seen as premisses of the argument, and '*O*', for uniformity, could also be taken as a set of sentences each of which is "explained" as it is entailed by the premisses. However, for simplicity we shall speak of *L*, *C*, and *O* as complex, conjunctive sentences, rather than sets of sentences.

wise, but have been otherwise, and will be otherwise again. They are what we call *initial conditions*. We have abbreviated them as '*C*', but remember that this stands in for a long conjunction of particular facts – just as '*L*' may stand in for a long conjunction of distinct laws of nature.

Hempel distinguishes statistical from nonstatistical explanation. We shall deal with the latter first. For nonstatistical explanations, Hempel requires that when *L* and *C* explain *O*, then *L* and *C* must *logically entail O* – the inference to *O* must be a *valid* one, that is, there must be no possible world where the laws and initial conditions hold and the outcome fails to occur. If there is any possibility of *L* and *C* holding and *O* failing, Hempel says that what we have is not an explanation but (at best) an *explanation-sketch*. We have something that could be turned into an explanation if various unstated assumptions were articulated and added, and if various promissory notes were redeemed. This can be illustrated by a simple case: why did the bowl break? Because it was taken straight from the freezer and placed in a hot oven. The outcome to be explained, *O* (the bowl breaking), is not logically entailed by the information following the word 'because'. There are possible worlds in which the bowl is taken from the freezer to the oven and yet does not break. To obtain a strict entailment, more information would have to be supplied concerning, for instance, the materials from which the bowl is made; the laws giving rates of expansion when heated; the variations, for different parts of the bowl, in the ratio of surface area to volume; the presence of microfractures; and so on. It is hard to imagine anyone ever filling in all the required laws and initial conditions. The best we can hope for are explanation-sketches. Nevertheless, Hempel suggests, the way to understand the sorts of things which are actually offered as explanations is by seeing them as sketches of an often unattainable ideal.

In the ideal case, then, Hempel requires that the outcome to be explained, *O*, be logically entailed by laws *L* and initial conditions *C*. Yet this does not fully express the intent behind the idea that explanation is "derivation from laws". It is clearly part of the intent that the derivation should make genuine *use* of the laws. It is not enough for the derivation to mention laws of nature in passing, while in fact deriving the outcome without any essential reliance on those laws at all. Derivations of that sort will not be genuine explanations but *bogus* explanations. Confidence tricksters, whether medical quacks, magicians, mediums, economic advisors, or whatever, often have in their bag of tricks a variety of ways of dressing up plausible "explanations", which refer to reputable laws of nature along the way but which turn out under scrutiny to be mere vacuous circularities under a screen of window-dressing.

To rule out trickery, Hempel adds requirements that the inference

$$\frac{\begin{array}{c} L \\ C \end{array}}{\therefore O,}$$

if it is to be a genuine explanation, must be one in which O follows *validly* from L and C, both L and C are true (and hence consistent with one another), and O would not have followed from C alone in the absence of the laws L.

There is further fine-tuning to be done to Hempel's account of nonstatistical explanation. There is also much to be done by way of examination of problem cases: cases which do seem to be satisfying explanations even though they do not fit Hempel's bill and cases which do satisfy Hempel's account but strike most people as terrible explanations. Problem cases of these sorts tend to stimulate a process of successive modifications – what Lewis calls the "one patch per puncture" method.

However, we will not pursue the fine detail further here, except that we must note one very major puncture, which calls for a very large patch. Science, these days, makes heavy use of statistics. As a result, a great many explanations involve statistics in the laws L. And this sometimes results in its being impossible to meet Hempel's strict requirement of logical entailment of the outcome by the laws and initial conditions. It is simply impossible to find laws and initial conditions which will guarantee each outcome. The best we can manage in many cases is to find laws and initial conditions which make an outcome very probable. Sometimes, indeed, we cannot even make the outcome very probable – the best we can do is to find laws and initial conditions which make it much *more* probable than it would have been had the laws or initial conditions failed to hold. And with some cases even this requirement seems beyond reach.

For these reasons, Hempel supplements his account of nonstatistical explanation with a separate account of statistical explanation.[5]

5 Hempel (1965).

He does not construe statistical explanation as just one subclass of ordinary explanations comprising those which happen to feature some statistics. Rather, he construes them as derivations of the thing to be explained, which proceed by inferences that are *not valid.* Hempel claims that a statistical explanation of an outcome O is an inference of the form

$$\frac{\begin{array}{c} L \\ C \end{array}}{\therefore O,}$$

in which L and C are laws and initial conditions, as before, but in which the conclusion is *probable given the premisses*. There are several variants. We could require that the conditional probability of O on L and C be high. Or we could require that the conditional probability of O on L and C be higher than the prior probability of O. Or we could conjoin these conditions. Or we could accept the weak requirement that L and C make O probable to at least some degree. And there are further possibilities we could explore.[6]

For our purposes, however, the Hempelian account of statistical explanation does not significantly diverge from the framework Hempel set up to account for nonstatistical explanation. The fundamental shift occurs in bringing to light a special class of things to be explained. Instead of setting out to explain why an event occurred, we may set out to explain why it was probable for such an event to occur. Or we may set out to explain why it was probable to such-and-such a degree (a higher degree than was expected). We may now apply the ordinary Hempel account of nonstatistical explanation to such a case. We may look for appropriate sorts of valid entailments of the conclusion *that the event had such-and-such probability.* What we obtain are ordinary Hempelian explanations of statistical facts. They are not really statistical explanations of nonstatistical facts, but rather ordinary explanations of statistical facts. They may still be called "statistical" explanations, for two reasons: first, because the laws they appeal to generally include statistical laws; and second, because what is explained is only a probability.

This Hempelian scheme can be used, however, to stretch the use of the term 'explanation' a little. When we have an explanation,

6 An example of such a variant is found in Salmon (1982). See also Forge (1986).

$$\frac{\begin{matrix} L \\ C \end{matrix}}{\therefore \text{ Probably } O,}$$

of the *probability* of an outcome, there is also a sense in which we have explained the outcome O itself. By explaining 'Probably O', we have in a sense explained O. Hence, Hempel suggests, to explain an outcome O is to derive, by valid inference from laws and initial conditions, *either* that the outcome holds *or* that the outcome is probable.

This is the fundamental idea behind the Hempelian theory of statistical explanation. Hempel himself did not construe the probabilistic qualification as attaching to the thing to be explained, 'Probably O'. He claimed, rather, that the probabilistic qualification should be attached to the link between premisses and conclusion by way of a weakening of the strict requirement of logical validity. This localization of the statistical element in the passage from premisses to conclusion rather than in the conclusion itself is in our view not worth quarrelling over. It may affect subtle details. In logic, it does sometimes make a difference whether we construe something (modus ponens, say) as an axiom or as a rule of inference; and similar issues arise in inductive logic. Yet we are working at a level of generality at which such worries need not be addressed. Consequently, we shall continue to construe Hempel's theory in its nonstatistical form: to explain is to derive by valid inference from laws and initial conditions. We regard statistical explanation as only a special case of this.

One important feature of Hempel's theory should be noted: the open-endedness of the explanatory process. To explain is to derive from laws and initial conditions. Yet those laws and initial conditions, too, may be explained. To explain these initial conditions, we shall call on prior initial conditions; and repeating this process will lead us back in time to earlier and earlier conditions. In a similar way, laws themselves stand in need of explanation. They, too, can often be validly inferred from other laws and initial conditions.

Consider an example. Kepler's laws of planetary motion (that planets move approximately in ellipses around the sun) cry out for explanation: why are they so? Newton explained them in a very Hempelian manner, by showing that deeper laws (the laws of gravity, etc.), together with initial conditions (concerning the solar and planetary masses, distances, etc.), logically entail Kepler's laws. Kepler's laws hold because, among other things, gravity decreases as the square of the distance between two bodies. They are explained by deduction from deeper laws. Those deeper laws, in turn, may invite curiosity: why are *they* so? They, too, may be explained by yet deeper laws, and this process is open-ended. It is always possible to ask for more explanations, although at any moment in history we can presumably give only finitely many of them. This open-endedness of Hempel's theory, of course, exactly matches the open-endedness of science as it is currently conceived. This is a strength of Hempel's theory.

The Hempelian account of explanation is still suffering growing pains.[7] Yet we do not argue that its overall drift is wrong. It may be in error over details, but not over the broad characterization of some of the most central kinds of explanation to be found in science. Our claim is only that Hempelian explanation is not the only sort of explanation to be found in science. Some explanations do proceed by subsumption under laws. Yet not all do.

Some explanation is causal explanation. Some explanation is nomological. Yet even when both kinds of explanation can be given, they are explanations of different sorts. Roughly speaking, causal explanations tell us *why,* whereas nomological explanations tell us *how.* Nomological explanations may tell us why, as well as how, but only insofar as they involve specifically causal laws, laws governing the instantiation of the causal relation. Given our assimilation of the causal relation to the operation of *forces,* it follows that nomological explanations tell us why only insofar as they are laws governing the operation of *forces* or causal interactions involving those forces. Nomological explanations which give no information about forces will explain how but not why.

Laws of nature fall into two broad categories. On the one hand, there are laws governing forces – causal laws. On the other hand, there are laws which do not explicitly govern the operation of forces – noncausal laws. Noncausal laws do not describe any specific causal relations in the world, and yet it is possible for such laws to give indirect information about causes. Any descriptive information about the world will place constraints on causal relations; it will place "boundary conditions" on the causal laws. In some cases,

7 A very useful discussion of Hempel's account is provided by Railton (1978).

very strict constraints will be placed on causal relations, and so a great deal of information about causes will be derivable from the noncausal information. Thus, even noncausal laws may carry significant information about causes.

Consider again the example of Kepler's laws governing planetary motion. Although these laws do not tell us anything directly about the forces that drive the planets, they do tell us something about which laws could *not* be correct descriptions of the planetary forces. Given the elliptical Keplerian orbits, one can no longer believe the theory, widely held from Aristotle to Galileo, that heavenly bodies move naturally in circles without the application of any forces. Newton thought that Kepler's laws also ruled out Descartes's theory, that the planets are swept along their paths by a fluid, which circulates in a vortex around the sun. Even Kepler's own speculations about a kind of rotating solar wind interacting with the earth's magnetic field could not (under inspection) be squared with his own descriptive laws. Thus, Kepler's laws of planetary motion do enable us to *rule out* several theories about the forces which act on the planets. Kepler's laws are noncausal, in that they do not describe the forces which drive the planets. Nevertheless, they are not entirely devoid of implications concerning the planetary forces.

In general, noncausal laws will always have implications for an adequate theory of forces, but the extent of those implications will come in degrees. Some noncausal laws will be rich in causal implications; others will have only the sketchiest of causal implications. There will be a spectrum of laws, the explicitly causal ones shading off through less and less causally informative ones.

It must be noted, however, that in some cases the causal information lying behind a law of nature may sometimes be negative causal information: the information that there are no forces operative. Consider, for instance, the comparison between the Newtonian law of gravitation and the Cartesian law of inertia (Newton's first law). By means of the law of gravitational attraction, Newton succeeded in explaining *why* the planets move in ellipses (approximately). The gravitational law is an explicitly causal law, since it governs forces and it generates why-explanations. The law of inertia also gives causal information. It states that an object in motion will continue in the same direction, unless acted on by a force. That is, the law of inertia gives information about the behaviour of objects in the absence of forces. In one sense, this is noncausal information; yet in another sense, it is information, albeit negative information, about causes. The explanations generated by the law of inertia will thus be noncausal explanations, yet they will carry some fairly specific causal information. The law of inertia is a *conservation law* – a law of conservation of motion. It is worth noting that the remarks we make about inertia will carry over to other conservation laws in science. Conservation laws in general explain how but not why, except to the extent that they entail indirect information about causes.

Suppose that we observe an object moving in a straight line with a constant speed, and we ask for an explanation of why it is moving in that way. One possible explanation would be by reference to frictional forces balancing the forces of acceleration which the object is being subjected to. This will, in fact, most commonly be the correct explanation of any such observed motion below the earth's atmosphere. Yet another possibility would be an explanation by way of subsuming the object's motion under the law of inertia. We explain that there are no forces acting on the object, and so the object moves with constant speed in a straight line. The law of inertia, by itself, explains *how* things move in the absence of forces. It does not explain *why* things move in that way in the absence of forces. And yet this noncausal law can enter into explanations of *why* particular things move as they do. Such explanations do count as why-explanations because they do give causal information, even though that causal information is of a negative sort, to the effect that *there are no* forces acting on the object.

Compare the laws of motion with some of the laws from optics, some of which give even less causal information than is provided by the law of inertia. Take *Snell's law,* for instance. This law governs the behaviour of a ray of light when it strikes a surface dividing two distinct transparent media, say air and glass. Suppose the surface is horizontal. Then Snell's law tells us that the height of a point on the ray *above* the surface will always be a fixed multiple of the distance of a point on the ray *below* the surface, provided that the two points are an equal distance from the point of impact (see Figure 7.1).

Snell's law describes the behaviour of light. It explains *how* light behaves. It does not, however, explain *why* light behaves as it does. We may explain the path of a given ray of light by subsuming it under Snell's law, yet this gives very little information about the

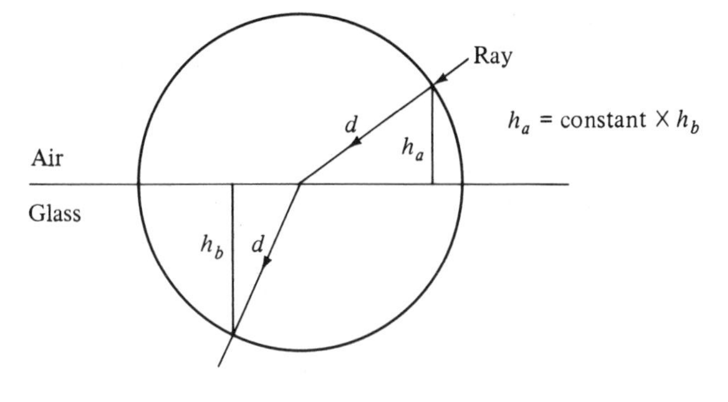

Figure 7.1

causes, that is, the forces (if any) which are acting to generate that path. Hence, subsumption under Snell's law gives a very low degree of explanation of why a ray follows the path it does. It tells us only that the forces operating are *the same as usual* – there is nothing unusual or queer about this case; light generally behaves like that. This is causal information of a sort, but it is extremely unspecific: the information that the causes are such that the path of the ray of light is such that h_a = constant × h_b.

Now compare Snell's law with another optical law, *the law of least time*. This law states that, given two points on a ray of light, the ray takes that path between those points which ensures that the light takes the minimum possible time to get from one point to the other. Suppose, then, that light travels at different speeds in air and in glass. Take one point along a ray as it travels through air towards glass and another point along that ray after it has entered glass. What is the path of the ray connecting those points? Given a constant speed in air and a constant but slower speed in glass, and given the principle of least time, it follows that the path of the ray will accord exactly with Snell's law.

The derivation of Snell's law from the law of least time succeeds in subsuming Snell's law of refraction under a more general law. The law of least time explains more than just refraction. Given a constant velocity of light through a constant medium, the law of least time explains the straight-line path of rays of light. And it also explains the law of reflection: that the angle at which a ray strikes a reflective surface will equal the angle at which it is reflected.

Does the law of least time "explain" Snell's law? In one sense, it does. In Aristotelian terminology, it tells us much more about the formal cause of refraction. And yet it does *not* yet tell us *why* light behaves in that way.

It may appear that the principle of least time explains the path of rays of light by appeal to final causes: the principle of least time seems to tell us that the light takes the path it does "in order that" the time taken should be as little as possible. Yet (almost) no one took this seriously as a genuine explanation by "final causes".

It is interesting to ask why the principle of least time was not taken seriously as a why-explanation by appeal to goals or purposes. It is *not* because it failed to reveal a pattern in phenomena. This confirms our contention that why-explanations are not simply explanations by subsumption under a pattern. In order for the principle of least time to generate a why-explanation by way of final causes, it would have to be backed by a complex pattern of causal relations of a specific sort. As we shall argue in Section 7.4, the required pattern of causal relations would have to involve causal relations which increase propensities for survival and replication for a given system. It is obvious that causal relations of this sort are not involved in the optical phenomena under discussion. Hence, it is obvious that the principle of least time is not to be interpreted as a why-explanation by way of final causes. We are left with an intriguing question, whether final causes are ever appropriate in physics – a question which has been raised again recently over the so-called *anthropic principle*.[8] At this stage, we will side-step the question whether we have been wrong to banish Aristotle's final causes from physics. We merely comment on the one special case: the optical principle of least time.

Neither Snell's law nor the more general law of least time really explains *why* light behaves as it does. But they do furnish some constraints on any adequate why-explanations. Hence, it would be misleading to characterize them as entirely descriptive or as contributing nothing at all to an explanation of why light behaves as it does. Why-explanation comes in degrees. And since the law of least time subsumes Snell's law, the law of least time will provide greater constraints on the causal explanations which are compatible with it. Hence, it carries greater implicit causal information. And that en-

8 The anthropic principle is discussed in Lesley (1982) and Barrow and Tipler (1985).

sures that it is *more* explanatory than Snell's law. It does not explain why, but it takes us closer to a correct why-explanation.

Explanation in science often takes the form of subsumption under ever more general laws. This is partly because science is not concerned merely with explaining why things are as they are; it is also concerned with explaining what there is in the world and how things behave. Yet subsumption under laws is not all there is to explanation. Sometimes subsumption under laws gives very little indication of why things happen as they do. And laws explain why things happen as they do only insofar as they carry information (of an indirect *or* direct sort) about causal relations, that is, about forces. Nevertheless, subsumption under laws does nearly always carry a nonnegligible amount of information about causal relations. Hence, laws are explanatory in both senses: they explain *how* the world is, and they also tell us (to a greater or lesser extent) *why* things happen as they do; for minimally they tell us that the causes are such that they produce the patterns which the laws describe.

Statistical laws frequently describe how certain things are and frequently give little direct information about the causal relations involved. Subsumption under a statistical law relates the event to the chance of it occurring. We see that the likely thing came about, or perhaps the unlikely thing happened. In either case, however, we do know more about the event and indirectly more about its underlying causes. This insight also shows why Hempel was demanding too much in requiring that the laws and conditions provide an explanation only if they made the event highly probable. To know that an event is an unlikely law-governed occurrence and the unlikely thing happened is as explanatory as to know that an event is a likely law-governed occurrence and the likely thing happened. Subsumption under statistical laws, therefore, just like subsumption under other noncausal laws, does explain *how,* but does not explain *why,* or at least, it explains why to only a very limited extent.

7.3 CAUSING AND TRACKING

Sometimes we know what caused an event, even though we do not know what laws (if any) subsume this particular causal interaction. So, since to explain why is to provide information about causes, we can sometimes explain by identifying particular causes, even though we cannot explain by citing laws.

Sometimes the contrary situation arises. Sometimes we do not know what causes an event, but we do know what laws the event is governed by. The case of falling bodies provides a clear illustration of this sort: it was known long before Solomon that (roughly) all unsupported objects fall, even though it was not known why this was so, since the force of gravity had not yet been hypothesized.

Yet there is a third class of cases which often arise – ones in which we know neither the details of the causal processes nor the laws under which an event falls. Cases of this sort arise unavoidably when we are dealing with complex processes, particularly with living organisms. Explanation in the biological sciences, therefore, and still more in the psychological realm, rests heavily on ways of giving general information about laws and causes without identifying those laws or causes in detail. Scientists could explain the death of animals as being due to suffocation, for instance, before they understood the role of oxygen in respiration, the role of the lungs in the circulatory system, and so on. They could similarly explain particular psychological phenomena fairly well without detailed knowledge of the laws and causal mechanisms involved. Hume could explain some instances of association of ideas quite adequately:[9] it is quite true in some cases, for instance, that a song reminds us of a person simply because that song was playing when we first met her. Yet the causal details and the laws of association were even less well understood by Hume than they are by contemporary psychologists. Hence, it is crucial that we find ways of conveying general information about laws and causes, even when detailed understanding of them eludes us. One way in which such general information may be conveyed is by functional explanations, as we shall see in Section 7.4. Another way of conveying such general information, however, is by the use of counterfactuals, and we shall explore this avenue a little further here.

In many cases, we know very little about laws or causal mechanisms, yet we do know enough to be able to make informed choices about which action to perform. Such choices rely, as decision theory says, on our knowledge of what the outcomes would be for the different actions we could perform. We may know that, if we were to, say, push the joystick to the left, the icon on the monitor would move left; if we were to pull it towards us, the icon would

9 Hume (1960/1739).

move down the screen; and so forth. We know how the movement of the icon depends on (counterfactually)[10] the movements of the joystick. There is a *functional dependence* of icon on joystick; the movement of one "tracks" the movement of the other.

The general form of such counterfactual dependence can be captured by a bank of counterfactuals:

$$\begin{array}{c} p_1 \mathbin{\Box\!\!\rightarrow} q_1 \\ p_2 \mathbin{\Box\!\!\rightarrow} q_2 \\ \vdots \\ p_i \mathbin{\Box\!\!\rightarrow} q_i \\ \vdots \end{array}$$

In the case of the joystick, there are four possible states, $p_1, \ldots, p_4$ (left, right, away from you, towards you), and four corresponding movements of the icon, $q_1, \ldots, q_4$ (left, right, up, down). If the joystick were in any state p_i, the icon would have motion q_i, for any choice of i between 1 and 4.

Note also, however, that counterfactual dependence need not require finiteness of the range of alternatives. It is in principle possible, for instance, for there to be an accelerator pedal of infinite sensitivity, such that the rate of acceleration of a vehicle will counterfactually depend on the positions of the accelerator. Then there may be continuum-many possible positions of the accelerator. For each real number x between, say, 0 and 1, there will be the proposition p_x that the accelerator is depressed a distance of x metres. Corresponding to this will be the proposition q_x that the vehicle accelerates at $f(x)$ m/s^2, where f is some function mapping real numbers onto real numbers. This may then generate a bank of continuum-many counterfactuals of the form

$$p_x \mathbin{\Box\!\!\rightarrow} q_{f(x)}.$$

Many of the laws of nature are, in fact, equations which, together with initial conditions, entail banks of counterfactuals of this sort. Statistical laws may be seen as arising from a weakening of the counterfactual requirement that $q_{f(x)}$ *would* be the case if p_x were true,

10 See Lewis (1973a).

to a requirement that $q_{f(x)}$ *probably would* be the case if p_x were the case. We may even specify how probable $q_{f(x)}$ would be, thus obtaining a bank of statistical counterfactuals of the form

$$p_x \Box\!\!\rightarrow (P(q_{f(x)}) = r).$$

Alternatively, our laws and initial conditions may also entail a bank of probabilistic counterfactuals:

$$p_x \textcircled{r}\!\!\rightarrow q_{f(x)}.$$

Consider, then, the manner in which such counterfactuals are related to laws of nature. It may be that such counterfactuals do hold in a specific situation, without there being any obvious way to generalize from this situation to other slightly different situations. It may be, for instance, that the joystick (or accelerator) is in good working order right at the moment, so we are confident that the relevant motions do counterfactually depend on what we do with the instrument. Yet we would not dream of asserting that it is a law, or even a true generalization, that *whenever* we move the instrument thus and so, the relevant motions depend on the position of the instrument. Only when the instrument is in good working order is this so. Yet we may have no independent way of determining whether the instrument is working, except by seeing whether the relevant motion does depend on the position of the instrument. We may presume that there are laws of nature which underpin the bank of counterfactuals; yet we may be in no position to sketch the form of any such laws. By establishing the bank of counterfactuals which hold in a given case, we do impose severe constraints on the laws of nature. Thus, the counterfactuals do yield indirect information concerning laws. Yet they do not require knowledge of the detailed form those laws will take.

Counterfactual dependences, therefore, provide indirect information about laws. And laws provide either direct or indirect information about causes. Hence, counterfactual dependences do provide causal information; and thus, we argue, they do help to explain *why* something is so. In fact, as we noted in Chapter 5, there can be cases of causation *not* governed by any nontrivial laws but which nonetheless could sustain counterfactual dependence. Thus, banks of counterfactuals will provide causal information whether or not this is mediated by underlying laws.

When we have a bank of counterfactual dependences, we can

often say in more detail exactly what sort of a why-explanation we have. Suppose p_3 is true and q_3 is true, and we have the fourfold bank of counterfactuals

$$p_1 \Box\!\!\rightarrow q_1,$$
$$p_2 \Box\!\!\rightarrow q_2,$$
$$p_3 \Box\!\!\rightarrow q_3,$$
$$p_4 \Box\!\!\rightarrow q_4.$$

For instance, suppose we are dealing with the motion of an icon on a monitor and its dependence on the position of a joystick. We may then say that q_3 is true *because* p_3 is true – the icon is moving up the screen *because* the joystick is pushed away. Yet we may say more. We may say that p_3 explains why q_3 is true, rather than q_1 or q_2 or q_4. This is not, however, an explanation of why q_3 is true *rather than* some entirely different possibility. Why is it that the icon is moving up the screen rather than, say, swelling in size? Why is it moving up the screen rather than blinking on and off? Why is it moving up the screen rather than whistling *Dixie?* Who knows? Such alternative possibilities were not in our mind when we asked why the icon was moving up the screen. Once such alternative possibilities are raised, we must look for a different bank of counterfactuals.[11]

Thus, explanations are often pragmatically tied to a range of alternative possibilities which are salient for the person who seeks to understand something. This is illustrated neatly by a well-known story about a priest who asked a bank robber why he robbed banks and received the reply, 'Because that's where the money is'. It is not to be denied that explanation is, in some measure, linked with understanding and that what provides understanding depends on subjective, psychological factors. And yet explanation also has an objective aspect. Not just any method of "removing puzzlement" will count as an explanation. The use of counterfactual dependences yields a neat way of balancing the subjective and the objective aspects of explanation. What is pragmatically determined is which range of alternative possibilities are salient for a person at a time. What is objective is whether a given bank of counterfactuals, which involve those possibilities, are true or false.

Counterfactual dependences, therefore, can be used to account for the specific, context-sensitive manner in which we often ex-

11 The relativized or contrastive nature of explanation is recognized, for instance, in Dretske (1973), Peacocke (1979), van Fraassen (1980), and Sober (1986).

plain why something is so. Sensitivity to context is allowed for; ignorance of the laws of nature is also allowed for. And yet information of a general nature is still conveyed concerning underlying causes, and hence on our account there is genuine explanation of *why* something has occurred. The why-explanation, however, is relativized to a range of salient possibilities. A why-explanation which explicitly identifies causes will not be so relativized. The relativity of why-explanations arises when the causal information which is supplied is very indirect. Then the kind of indirect information which is called for will depend heavily on the purposes at hand.

One way in which counterfactual dependences provide causal information is by mediation through laws of nature, as we have said. The counterfactuals constrain the laws, and these in turn constrain the causes. However, counterfactuals also provide causal information more directly, without requiring the mediation of laws.

Counterfactual dependence and Lewis causation

The relationship between counterfactual dependences and causes is, in fact, very close and also very tricky. Our understanding of the notion of counterfactual dependence was, in fact, derived initially from Lewis's seminal paper 'Causation'.[12]

Lewis defends a counterfactual analysis of causation. Yet it is interesting that he presents this analysis as a special case of counterfactual dependence. Lewis begins with counterfactual dependence, as we have defined it. That is, one sequence of propositions $\langle q_1, q_2, \ldots, q_i, \ldots \rangle$ counterfactually depends on another sequence $\langle p_1, p_2, \ldots, p_i, \ldots \rangle$, just when

$$\begin{array}{c} p_1 \;\Box\!\!\rightarrow\; q_1 \\ p_2 \;\Box\!\!\rightarrow\; q_2 \\ \vdots \\ p_i \;\Box\!\!\rightarrow\; q_i \\ \vdots \end{array}$$

12 Lewis (1973b).

One special case of this arises when we consider two sequences of the form

$$\langle e \text{ occurs}, e \text{ does not occur} \rangle,$$
$$\langle c \text{ occurs}, c \text{ does not occur} \rangle,$$

where c and e are distinct (nonoverlapping) events. The former sequence counterfactually depends on the latter just when

$$c \text{ occurs} \; \Box\!\!\rightarrow \; e \text{ occurs},$$
$$c \text{ does not occur} \; \Box\!\!\rightarrow \; e \text{ does not occur}.$$

When two such counterfactuals hold, Lewis claims that c is a *direct* cause of e.

Lewis then notes that the relation of being a direct cause is not transitive. It may be that c is a direct cause of d, and d is a direct cause of e, and yet c is not a direct cause of e. This is because counterfactual conditionals are not transitive. Suppose c, d, and e all occur. Then both the counterfactuals

$$c \text{ occurs} \; \Box\!\!\rightarrow \; d \text{ occurs},$$
$$d \text{ occurs} \; \Box\!\!\rightarrow \; e \text{ occurs}$$

are vacuously true under Lewis's semantics, and they are probably also true under our favoured semantics too (even though we require only what we have called weak centring, see Sections 3.2 and 3.3). These provide half of each of the counterfactual conditionals required for c to count as a direct cause of d, and d as a direct cause of e. The other required counterfactuals are

$$c \text{ does not occur} \; \Box\!\!\rightarrow \; d \text{ does not occur},$$
$$d \text{ does not occur} \; \Box\!\!\rightarrow \; e \text{ does not occur}.$$

If these two hold, then c is a direct cause of d, and d of e. Yet since counterfactuals are not transitive, there is no assurance that

$$c \text{ does not occur} \; \Box\!\!\rightarrow \; e \text{ does not occur}.$$

Hence, there is no assurance that c is a direct cause of e.

Lewis, however, is not content with the intransitive relation of direct causation. He wishes to define, in addition, a notion of indirect causation. When something is either a direct or an indirect cause of an event, we may say it is simply a "cause" of that event. In order to ensure transitivity for the relation 'being a cause of', Lewis allows that, whenever c is a direct cause of d and d is a direct cause of *e*, then *c* is a cause of *e*. And in general, if *c* is either a direct or an indirect cause of *d* and *d* of *e*, then *c* is a cause of *e*.

What Lewis has defined is, we acknowledge, a very useful special case of counterfactual dependence, together with the transitive extension of this counterfactual dependence. We have argued earlier that the ontologically basic causal relation is to be identified with the operation of one or more forces. Hence, the relation Lewis has defined is distinct from the relation we have identified as causation. However, both his relations and ours are real enough, and it is fruitless to quibble over which is to receive the honorific title of 'causation'. So we could grant Lewis the relation 'is a cause of', defined as above.

The question then arises as to what relationship there is between Lewis's kind of causation and counterfactual dependence more generally. Counterfactual dependence may hold between one sequence of propositions and another, without any corresponding causal relations. One reason for this is that a counterfactual dependence yields a Lewis causal relation only if it holds between *distinct events* (or, more properly, between the propositions which assert that those events occur). We can say that *c* is a cause of *e* only when the occurrence of *e* rather than not-*e* counterfactually depends on the occurrence of *c* rather than not-*c*, *and c* and *e* are both events, *and c* and *e* are nonoverlapping events (i.e., nothing is a part of both).

Counterfactual dependence is broader than Lewis-style causation for three reasons. The first reason is that counterfactual dependence may hold among propositions which do not concern the occurrence of events. For instance, consider counterfactuals of the form 'If there had been more rats, there would have been more cats; if there had been twice as many rats, there would have been twice as many cats . . .'. These constitute counterfactual dependence of the cat population on the rat population. The propositions involved, however, do not report the occurrence of events. There are, of course, events which are closely related to cat and rat populations: births and deaths, for instance. But the counterfactual dependence of populations is only indirectly linked to counterfactual dependence among events.

The first reason that counterfactual dependence is broader than causation, then, is that it need not be directly concerned with events. The second reason is that, even if the counterfactual dependence does concern events, it need not concern distinct, nonoverlapping events. The occurrence of compound events, for instance, will often counterfactually depend on the occurrence of their components; yet it is inappropriate to say that an event is caused by one of its own components. Consider a compound event like walking across a field. If you had not walked across the last quarter, you would not have walked across the field: so walking across the field counterfactually depends on walking across the last quarter. Yet walking across the last quarter is not a cause of walking across the field. (This example is a variant of an example raised against Lewis by Jaegwon Kim. His case was: "If I had not written double 'r', I would not have written 'Larry' ".)[13]

The third reason that counterfactual dependence is broader than Lewis-style causation is as follows. There is a difference between saying that *which* of these events occurs depends on which of those events occurs and saying that *whether* one of these events occurs depends on whether one of those events occurs. It may be that counterfactuals of this form are true:

$$c_1 \text{ occurs } \Box\!\!\rightarrow e_1 \text{ occurs,}$$
$$c_2 \text{ occurs } \Box\!\!\rightarrow e_2 \text{ occurs.}$$

And yet the following counterfactuals need not be true:

$$c_1 \text{ does not occur } \Box\!\!\rightarrow e_1 \text{ does not occur,}$$
$$c_2 \text{ does not occur } \Box\!\!\rightarrow e_2 \text{ does not occur.}$$

If the latter are not true, there is no Lewis-style causation between the *c*-events and the *e*-events, even though there is counterfactual dependence of the *e*'s on the *c*'s. Suppose c_1 occurs and e_1 occurs. Then we may explain why e_1 occurred rather than one of the other *e*'s: the explanation is that c_1 occurred rather than one of the other *c*'s. Yet this is not to explain why e_1 occurred, that is, why e_1 occurred rather than not.

Consider an illustration. A person in a sensory-deprivation experiment may be hearing a particular tone. It may be that which of several tones he experiences depends on which of several sounds is reaching his ears. Yet it may be that, if no sound reaches his ears, he will nevertheless experience one or another of these tones spontaneously generated in his brain. So we may have the counterfactuals

13 Kim (1973).

Sound *a* reaches the ears $\Box\!\!\rightarrow$ Tone *a* is experienced,
Sound *b* reaches the ears $\Box\!\!\rightarrow$ Tone *b* is experienced.

.
.
.

Yet we may not have

Sound *a* does not reach the ears $\Box\!\!\rightarrow$ Tone *a* is not experienced.

Rather, it may be that, even if sound *a* had not reached the ears, tone *a* might nevertheless have been experienced. We may have

Sound *a* does not reach the ears $\Diamond\!\!\rightarrow$ Tone *a* is nevertheless experienced,
Sound *b* does not reach the ears $\Diamond\!\!\rightarrow$ Tone *b* is nevertheless experienced.

.
.
.

It is interesting that in this case it is plausible to say that the sound reaching the ears is a cause of the experience. And this is so, despite the fact that Lewis's official requirements for causation deem this not to be a case of causation. This indicates that it is advisable to rely on the more general notion of counterfactual dependence, rather than to focus too heavily on the special case on which Lewis rests his official analysis of causation.

Counterfactual dependences, then, furnish one very rich and flexible means of conveying indirect causal information when we are ignorant of the details of both the laws and the causal mechanisms underlying the phenomena. Counterfactual dependences are ubiquitous in science. Laws expressed by equations are often convenient summaries of counterfactual dependences. For instance, Boyle's law, relating the pressure of a gas to its temperature and volume, encapsulates indefinitely many counterfactual dependences of pressure on temperature and volume: 'if the temperature were this and the volume that, then the pressure would be as follows . . .'.

In biological phenomena in particular, very rich and complex patterns of counterfactual dependence may be found. And their role becomes still more prominent when we turn to the phenomena of perception and action in advanced organisms, notably humans. In general, there is a very rich pattern of counterfactual dependence of our perceptual experiences upon the input to our sensory organs. In the process of reasoning, there is counterfactual dependence of the conclusions we reach upon the premisses we start with. And in the process of action, there is counterfactual dependence of our actions upon our beliefs and desires.

The currently fashionable family of doctrines known as functionalism[14] would lie well with the idea that the key to the mind lies in the nature of such counterfactual dependences. Perhaps there is more to the mind than this. Yet it is scarcely credible that counterfactual dependence would not play a prominent role, in one guise or another, in the understanding of the nature of the human mind.

Lewis's causal explanations

Lewis views Hempelian explanations as we do – as roughly correct in details and right-headed in principle, and yet not exhausting the full range of explanations employed in science. Lewis argues that, in addition to Hempelian explanation by subsumption under laws, there is also such a thing as *causal explanation* of particular events, and that sort of explanation need not be Hempelian.[15]

The insight that some explanations come from identifying causes also arises from some probabilistic accounts of explanation.[16] Take someone who, as we believe incorrectly, holds the view that a cause is anything that increases the probability of an effect. Then it is not surprising to find such a person attracted to the view that explanations make that which they explain more probable. It means that their explanations will coincide with their causes. Since we have already argued that probabilistic accounts of causation are misguided, we shall look at causal explanations in another way, a way which we develop by considering Lewis's account. However, it should be noted that, as pointed out at the end of Chapter 6, there is a strong connection between our accounts of causation and probability, and not surprisingly there is a carryover of this connection to our accounts of explanation and probability.

Lewis's account of causal explanation proceeds in the following five stages.[17] First, he uses laws of nature as an input into a theory of

14 For a general review of functionalist accounts of the mental, see Dennett (1978) and Hofstadter and Dennett (1982).

15 Lewis (1986e).

16 Such probabilistic theories are discussed in Suppes (1970), Ellis (1970), Mellor (1976), Cartwright (1979), and Skyrms (1980).

17 Lewis (1986e).

counterfactuals. Second, he uses counterfactuals to define a relation between events, namely, counterfactual dependence, as already explained. Third, Lewis uses counterfactual dependence to define causation by two principles: (1) counterfactual dependence is causation, and (2) a cause of a cause is a cause. For Lewis, causation is transitive.

Fourth, Lewis constructs what he calls the *causal history* of a given event. This is a tree structure which traces back the causes of the causes of the causes, and so on. Given some event *c* in the remote causal history of an event *e*, it may not be true that *e* counterfactually depends on *c*. It may be that *e* would still have come about, though in some other way, even if *c* had not occurred. Yet *c* will still be a cause of *e*, and so part of the causal history of *e*. The causal history of *e* will be the set of all events which stand in chains of counterfactual dependence which lead up to event *e*. Or rather, the causal history will be the set of all such events paired with the ordering relation of counterfactual dependence.

Fifth, Lewis claims that anything which supplies some information about the causal history of an event serves thereby to give a causal explanation of the event. Complete information about the causal history (if this were achievable) would furnish a complete causal explanation of the event. Partial information yields partial explanations. And partial information may be partial in different ways. One kind of partial information, for instance, gives complete information about the causal tree at a specific time, but no information about earlier or later stages of the tree. This sort of cross-sectional information, incidentally, is exactly what is needed for a Hempelian explanation. But other sorts of information could also be given. For instance, we could be given just one of the chains of cause–effect pairs tracing back indefinitely into the past, like a patrilineage that entirely ignores the matrilineal and the mixed patrilineal–matrilineal lines. Or we could be given merely holistic, gestalt information about the tree – say, that it is complicated, that it is simple, that it has very little input from Madagascar, and so forth. The possibilities are endless.

Lewis recognizes that some information may be newsworthy while other bits of information may be silly, or irrelevant, or too obvious to be worth noting, or objectionable in many other possible ways. All this he consigns to pragmatics. Any information about a causal history counts as *an* explanation, in the abstract; it just may not be the sort of explanation anyone wanted. Some explanations may be more pragmatically appropriate, in a given context, than others, although all are nevertheless causal explanations *sub specie aeternitatis*.

Our theory has much in common with Lewis's, but there are points of difference. Some differences flow from the underlying differences in our theories of counterfactuals and of causation. Lewis uses laws to explain counterfactuals, while we do not but rather use degrees of accessibility for both. Lewis uses counterfactuals to explain causation, while we do not. These differences will have some consequences for our accounts of causal explanation. More crucial still is the role which forces play in our conception of causation. As a consequence of our theory of causation, we do not accept Lewis's principle of the transitivity of causation. We do not accept that the cause of a cause of *e* is a cause of *e*. Hence, for us, information about the causes of an event will not in general trace back indefinitely into the past. We take a causal explanation of an event to be information about the causes of that event. And information about the *causal history* of the event need not be information about *the causes* of the event. Tracing back the causes of causes of causes, and so on, leads back to Adam and Eve. For instance, for Lewis, the fact that Adam took the apple from Eve gives a causal explanation of the victory of Nelson at Trafalgar – a pragmatically inappropriate one, but a causal explanation all the same. It is an advantage of our theory that it puts less strain on a theory of pragmatics. We do not grant that Adam and Eve's activities give any sort of explanation of Nelson's victory, since we do not grant that there is a causal relation between Adam and Eve and Nelson.

In summary, then, our theory differs from Lewis's in that he allows any information about causal history to furnish a causal explanation, whereas we allow only information about the *causes* of an event, where causation is ultimately to be analysed in terms of the action of forces. Nevertheless, there are similarities between Lewis's theory and ours. Our views on Hempelian and causal explanations are very similar to his. And like Lewis, we view causal explanation as information about causation. And information about causation can be more or less informative – so there are degrees of explanation, depending on the degree of information one gives about causes. Such information can also be more or less pragmatically appropriate; so although we want to reduce the pragmatic element below that required by Lewis, we do agree, in principle, on the need for pragmatic distinctions between appropriate and inappropriate, or interesting and uninteresting causal information. We also agree on the fundamental point that not all explanation is subsumption under laws. Some explanation is just provision of information about causes.

Explanation can be viewed as an attempt to find our location in logical space. One such method of localization is to be told about the worlds accessible from our world. This is the sort of information we get from information about the laws of our world. Another sort of information is to be told other particular matters of fact about our world – but this by itself does little to locate us in logical space, for it plays a low-level role in determining the degree of accessibility of other worlds. Yet another way of locating us is to give us causal information, for this tells us highly relevant information concerning the accessibility of other worlds.

This picture shows what binds all explanations together as explanations, but also shows the difference between explanations of how and of why. Yet it also makes it clear that no explanation can be devoid of some why-information, and conversely that no why-explanation can be devoid of how-information.

7.4 FUNCTIONAL EXPLANATION

In describing the *function* of some biological character, we describe some presently existing item by reference to some future event or state of affairs. So the function of teeth at one time is to pulp food at appropriate future times. This seems to present an exact parallel to the function of human artefacts; for instance, the function of the nutcracker at one time is to break open nuts at appropriate future times. We explain *why* we have teeth by pointing to their function, and we explain *why* we keep nutcrackers by pointing to their function. We also explain why more nutcrackers will be produced in the future by explaining what useful function they perform. These are what are called *functional explanations.* They seem to be answers to why-questions, but they are different from the explanations which cite efficient causes.

In the case of biological functions and in other cases where human intentions are not obviously causally active, explanation by reference to function has been difficult to assimilate into our scientific view of the world. There are several reasons, but we shall here concentrate on one which arises directly from the fact that, in describing a present structure in terms of its function, we mention a future outcome of some sort, the "final" cause of that structure. The future outcome may be, in many cases, nonexistent. A structure may never be called upon to perform its function. The function of a bee's sting, for instance, is relatively clear; yet most bees never use their stings. Likewise, teeth may never pulp food, just as nutcrackers may never crack nuts.

Thus, when we describe the function of something in the present, we make reference to a future event or effect which, in some cases, will never occur. Hence, prima facie, we cannot really be describing any genuine, current property of the character.

Even when a character does perform its supposed function, the future events that result from it cannot play any significant "scientific" role in *explaining* the nature and existence of the character. The character has come into existence, and has the properties that it does have, as a result of prior causes. It would still exist, with just the current properties it does have, even if it had not been followed by the events that constitute the exercise of its alleged function. Hence, its existence and properties do not depend on the exercising of its function. So it is hard to see what explanatory role its functions could have. Crudely put, backward causation can be ruled out: structures always have prior causes, so reference to future events is explanatorily redundant. Hence, functions are explanatorily redundant.

Of course, there is nothing inappropriate about describing a character *and mentioning* its future effects. But describing a character as having a function is not just mentioning that it has certain effects. Not every effect counts as part of its function. And some functions are present even when there are no relevant effects to be mentioned – as with some bees and their stings. Future effects are not unmentionable; but they are explanatorily redundant in characterizing the existence and current properties of a character. Hence, what role can functions have in a purely scientific description of the world. How can they be "placed" within the framework of current science?

Three main theories attempt to construe functions in a way that allows them to fit smoothly into the scientific, causal order. We believe that each is nearly right or partly right. Yet they are all unsatisfactory in one crucial respect; they do not restore to functions any significant explanatory power. In particular, they deny to functions any causal efficacy. So, for instance, they will not permit us to explain the evolution of a character by saying that it evolved *because* it serves a specific function.

We shall offer an account of biological function and of functions generally which, although it shares much with the most promising extant theories, is crucially different from them in that it bestows greater explanatory role upon functions.[18] Our theory will be a realist one. We defend realism not only about modality and causation, but even about functions and purposes in nature. But before explaining our theory, we shall briefly consider the three most prominent prior contenders.

The usual theories of functions

Three responses arise naturally in the face of the tension between functions and the scientific standpoint. The first is eliminativist. It is assumed that functions, if there were any, would have to be important, *currently existing,* causally active, and explanatory properties of a character or structure. It is also assumed that functions, if there were any, would essentially involve reference to future, possibly nonexistent events. Yet something involving essential reference to future, possibly nonexistent events could not possibly characterize currently existing, explanatory properties. It is thus concluded that there really are no functions in nature.

To add functions to the scientific biological picture, on this view, is like adding final causes to physics. Most think final causes have no place in the scientific account of the physical universe, and if the psychological pressures are resisted, we find we can do without them and final causes just fade away. To the eliminativist, the same will be true of functions; as the biological sciences develop, any need for function talk will vanish, and the psychological naturalness of such talk will fade away with time and practice.

A variant on this eliminativist view adds an account of why the attributions of function seem to serve a useful purpose in everyday and scientific discourse. The eliminativist does not believe in functions as genuine, currently existing properties of a character. But the eliminativist does believe in future effects of a character. And

18 Our theory is a cousin of theories which are sometimes called "goal theories" and which have been advocated, for instance, by Wimsatt (1972) and Boorse (1976, 1977). Some important differences will be noted later.

nothing stops us from mentioning whichever future effects we take an interest in. Consequently, an eliminativist can interpret talk of functions as being merely the specification of effects one happens to be interested in. Which effects are deemed to relate to functions of a character will depend not on the nature of the character itself, but on our interests. The function of kidneys is different for the anatomist than it is for the chef. Insofar as function talk makes sense, it does *not* describe the current nature of a character (there are no functions in nature); rather it relates a current character to a future outcome, in an interest-dependent, extrinsic manner.[19]

The best answer to an eliminativist theory is to come up with an adequate analysis of functions within the scientific view. This is what we shall attempt to do shortly. But a motive for seeking such a noneliminativist account can be cited; and this motive will also provide a less than conclusive, but nevertheless weighty argument against eliminativism. In the biological sciences, functions are attributed to characters or structures, and these attributions are intended to play an explanatory role which cannot be squared with the eliminativist's account of function talk. For instance, it is assumed that biological structures *would* have had the functions they do have even if we had not been here to take an interest in them at all. And some of the effects of structures that we take an interest in have nothing to do with their function. And some functions are of no interest to us at all. Furthermore, biology usually treats 'function' as a central, explanatory concept. None of this rests easily with an eliminativist theory. A powerful motive for resisting eliminativism, then, is that an adequate analysis of functions, if we can find one, will enable us to take much biological science at face value: we will be relieved of the necessity of undertaking a radical reformation of the biological sciences. Eliminativists' vision of functions "fading away" is as yet just a pipe dream; and their explanation of the apparent usefulness of function talk fails to explain away more than a fragment of the functions which functions serve in biological science.

It is not the biological sciences alone which could be cited here. Psychology could be mentioned too. And even physics has facets that raise problems for an eliminativist view. Suppose someone were to suggest that the function of water is to refract light and the

19 A theory of this sort is given in Cummins (1975). See also Prior (1985a).

function of mists to create rainbows. Presumably this is simply false. Yet it describes something in terms of future effects in which we take an interest, which is exactly what an eliminativist takes to be the business of function talk. So eliminativists have no good explanation of why the physicist takes such attributions of function to be plainly false. They cannot explain the manifest difference between 'The function of mists is to make rainbows' and 'The function of teeth is to pulp food'.[20]

There is a response to the tension between the scientific view and functions which rejects any role for future events in the characterizing of functions. The future effects of a character do not themselves play an explanatory role in characterizing that character. Yet sometimes there exists, prior to the character, a *plan,* a representation of that character *and* of its future effects. Such a representation of future effects may exist, whether or not those effects ever come to pass. And this representation exists prior to the character and so contributes to the causal processes that bring that character into being by the usual, forward-looking, causal processes that rest comfortably within our overall scientific image of the world. On this view, we can account for a function not by direct reference to any future event, but rather by reference to a past representation of a future event. Theories of this sort have frequently been called *goal theories;* but they are best construed as a subcategory within the class of goal theories. They are goal theories in which the identification of a goal depends on the content of prior representations.[21]

This account of functions fits best with attribution of functions to artefacts. The idea of breaking open nuts seems to have played a causal role in the production of the nutcracker. It does not fit so neatly into the biological sciences. Of course, it used to provide a persuasive argument (the teleological argument) for the existence of a Creator: there *are* functions in nature; functions require prior representations; yet the creatures themselves (even when creatures are involved) have no such foresight or were not around at the right time; hence, the prior representations must have been lodged in

20 This problem is treated at length by Nagel (1961, pp. 401–28). Nagel manages to blunt the force of such objections, but only by augmenting his initial theory, which differs only superficially from eliminativism, producing a theory similar to those of Boorse and Wimsatt.

21 Woodfield (1976) argues for a view that takes the primary case of functions to rest on a prior plan and all other cases of (unplanned) "functions" to be mere metaphorical extensions of the primary cases.

some awfully impressive being; and so on. Nowadays, however, in the clear and noncontroversial cases of functions in nature, it is taken that they can be accounted for from the standpoint of a theory of evolution by way of *natural* selection – and in such a theory there can be no room for any analysis of biological functions which rests on prior representations. Even if God foresaw the functions of biological structures, that is a matter outside biology; functions, however, are a biological and not a theological matter.

It is worth noting that, even though the representational theory seems to rest comfortably with attributions of functions to artefacts, nevertheless some artefacts prove more problematic than might first appear. Many artefacts evolve by a process very like natural selection. Variations often occur by chance and result in improved performance. The artisan may not understand fully why one tool performs better than others. Yet because it performs well, it may be copied, as exactly as possible. The reproduction of such tools may occur for generations. The features of the tool which make it successful and which lead it to be selected for reproduction are features that have specific functions. But they were not created with those functions in mind. They may have been produced with an overall function in mind (say, hitting nails); but the toolmaker may not have in mind any functions for the components and features of the tool which contribute to the overall function. For instance, the toolmaker may copy a shape that has the function of giving balance to the tool – but he need not foresee, or plan, or represent any such function. He knows only that tools like this work well for banging nails, or sawing wood, or whatever the overall function might be. Consequently, even with artefacts, structures can serve specific functions even though there exists no prior representation of that function.[22]

There is a further reason for feeling uneasy about the representational theory. The theory analyses the *apparent* forward directedness of functions by an indirect, two-step route. The forward directedness of functions is analysed as comprising a *backward* step to a representation, which in turn has a *forward* directedness towards a possibly nonexistent future state. Thus, the seeming forward directedness of functions is reduced to another sort of forward di-

22 Millikan (1984) makes several intriguing points about artefacts, their reproduction, selection, survival, and so forth. She advances a sophisticated version of the etiological theory that is discussed later in this section.

rectedness: that of representations – plans, beliefs, intentions, and so on. And this is worrying. The worry is not just that these are "mentalistic" and just as problematic as functions – and just as hard to assimilate into the scientific picture of the world. Rather, an even greater worry is that of vicious circularity. Many find it plausible that the notion of representation will turn out to be analysable in terms that at least include functional terms.[23] And functional terms presuppose functions. Hence, the future directedness of representations may turn out to presuppose the future directedness of functions. This threatens to do more than just restrict the scope of representational theories; it undermines such theories even in their home territory, as applied to artefacts.

In addition to the eliminativist and the representational theories, there is a third response to the tension between the scientific view of the world and the concept of function. This response again involves rejecting any role for future events in the characterization of functions. Like representational theories, the theories arising from this response look to the prior causes of the thing which has the function, and for this reason, theories of this sort are called *etiological*. Etiological theories, like representational theories, define function in terms of origin. Yet etiological theories also eschew reference to prior representations of future effects, as well as reference to the future effects themselves.[24]

The difference between representational and etiological theories recalls the distinction Darwin drew between artificial selection and natural selection. When animal breeders select, they represent to themselves the characters they wish to develop. Natural selection has closely analogous results, but it operates in the absence of (or at least without any need for) any conscious or unconscious representations of future effects. The etiological theory of functions explains biological functions by reference to the process of natural selection. Roughly, a character has a certain function when it has evolved by natural selection *because* it has had the effects that constitute the exercise of that function.

Clearly, there is room here for an overarching, disjunctive theory which unites the representational with the etiological. Such a unifying theory would say that a character has a certain function when it

23 See note 14.

24 The etiological theory is widely held. See Wright (1973, 1976). We have also been influenced by unpublished work of Karen Neander on this theory.

has been selected because that character has had the relevant effects. In the case of artefacts, the selection typically involves conscious representations. In the case of (Darwinian) sexual selection, representations may enter the picture, but reproduction, heredity, and evolution also play a part; and in the case of natural selection, representations drop out altogether. But on the etiological theory, a character has a *biological* function only if that character has been selected for by the process of *natural* selection because it has had the effects that constitute the exercise of that function.

We take this etiological theory of biological function as the main alternative to the account of biological function we shall develop. The big plus for the etiological theory is that it makes biological functions genuinely explanatory, and explanatory in a way which is relatively comfortable with the modern biological sciences. But we shall argue that this explanatory power is still not quite right: the etiological theory offers explanations that are too backward looking.

But before we turn to this matter, we should note another worry with the etiological theory, a worry that extends even to the overarching disjunctive theory of which it is a part. This worry is that there is too great a dependence of functions on contingent matters – matters which, had they been (or even if they are) otherwise, would rob the theory of any viability.

The etiological theory characterizes biological functions in terms of evolution by natural selection. Most theorists take evolutionary theory to be true, but contingently so. What if the theory of evolution by natural selection were found to be (or had been) false? Clearly then, on the etiological theory of biological functions, as we have specified it, there would be no biological functions. Whether there are biological functions at all, on the overarching, disjunctive theory, will depend on what replaces the theory of evolution by natural selection. Suppose it is creationism. Then the representational theory would apply; for we have the representations in the mind of the Creator that would have the appropriate causal role in the development of biological structures, so the representational theory would become a general theory of functions. Biological structures would then be granted functions – though not "biological" functions, according to the etiological theory. Biological structures would have what we might call theological functions.

For creationism there would, of course, be an enormous epistemological problem of discovering what the functions of biological structures were; for this would depend on discovering what the Creator had in mind. So we would have great difficulty discovering whether the function of the heart is to produce the sound of a heartbeat, in line with the Creator's idea of a beating rhythm in nature, with the circulation of the blood as a nonfunctional effect; or whether the reverse is true. It would be much like an anthropologist discovering the nature of ancient artefacts without any access to the intentions of the earlier cultures. Holy scriptures might fill the epistemological gap in some cases, but not in all.

But suppose creationism is not the alternative. Consider the possible world identical to this one in all matters of laws and particular matters of fact, except that it came into existence by neither of the processes our culture has postulated: neither by planned creation nor by evolution via natural selection. Then, according to the overarching "representational or etiological" theory, there would be no functions. There would be no biological functions on the etiological theory and no causally active representations as required by the representational theory, and hence no functions at all. This we take to be counterintuitive. You would wear it if you had to: but do you?

We have the intuition that the concept of biological function and views about what functions biological characters have are not thus contingent upon the acceptance of the theory of creation or evolution by natural selection. We are also inclined to think that the representational account of the function of artefacts gives too much importance to the representations or ideas of the original planner, even in cases when there is one. As we indicated earlier, we believe a satisfactory account of functions in general, and biological functions in particular, must be more forward looking.

Fitness, function, and looking forward

It emerges from our discussions that the tension between functions and modern, causal science has generated, fundamentally, two stances on the nature of functions. The first is the eliminativist stance. This has the merit of giving full weight to the forward-looking character of functions, by specifying them in terms of future and perhaps nonexistent effects; and also to the explanatory importance of functions. It is mistaken only in its despair of reconciling these two strands. The second stance is backward looking. This embraces theories which look back only to prior representations, those which look back only to a history of natural selection, and those which look back to a history of either one sort or the other.

We shall argue for a forward-looking theory. Functions can be characterized by reference to possibly nonexistent future events. Furthermore, they should be characterized that way, because only then will they play the explanatory role they should play, for instance in biology. The way to construe functions in a forward-looking manner, we suggest, is to construe them as analogous to dispositions in an important way. The shift we recommend, in the analysis of functions, has a precedent: the analysis of the evolutionary concept of fitness.[25]

One wrongheaded but common objection to the Darwinian theory of evolution is that its central principle – roughly, 'the survival of the fittest' – is an empty tautology which cannot possibly bear the explanatory weight Darwin demands of it.[26] This objection assumes that fitness can be judged only retrospectively: that it is only after we have seen which creatures have survived that we can judge which were the fittest. Moreover, it assumes that the fact that certain creatures have survived, whereas others have not, is what constitutes their being the fittest.

The etiological theory of biological functions rests on the same sort of misconception as that which underlies the vacuity objection to Darwin. On this theory, we can judge only retrospectively that a character has a certain function, when its having had the relevant effect has contributed to survival. And indeed, it is not only our judgements about the function that must be retrospective. The property of possessing a function is itself, according to the etiological theory, a retrospective property. It is a property which is *constituted* by a history of contributions to survival.

Consequently, the notion of function is emptied of much explanatory potential. Under the etiological theory it is not possible to explain why a character has persisted by saying that the character has persisted because it serves a given function. To attempt to use function in that explanatory role would be *really* to fall into the sort

25 See Pargetter (1987).

26 See, e.g., Waddington (1957), Cannon (1958), Smart (1968), and Barker (1969). Manier (1969), Ruse (1971), and Pargetter (1987) are among those who have replied to this view.

of circularity often alleged (falsely) against the explanatory use of fitness in Darwinism.

This comparison with fitness serves another purpose. It has displayed why functions would lack explanatory power on the etiological theory, but it also shows how to analyse functions so as not to lose this explanatory power. Fitness is not defined retrospectively, in terms of actual survival. It is, roughly, a dispositional property of an individual (or species) in an environment which bestows on that individual (or species) a certain survival potential or reproductive advantage. This is a subjunctive property: it specifies what would happen or what would be likely to happen in the right circumstances, just as fragility is specified in terms of breaking or being likely to break in the right circumstances. And such a subjunctive property supervenes on the morphological characters of the individual (or species).[27] Hence, there is no circularity involved in casting fitness in an explanatory role in the Darwinian theory of evolution. In the right circumstances fitness explains actual survival or reproductive advantage, just as in the right circumstances fragility explains actual breaking. In each case the explanation works by indicating that the individual has certain causally active properties that in such circumstances will bring about the phenomena to be explained.

What holds here of fitness holds, too, of biological functions. Fitness, in fact, will be *constituted* by the possession of a suitable collection of characters with appropriate functions. Fitness and functions are two sides of a single coin. Thus, the etiological theory is mistaken in defining functions purely retrospectively, in terms of actual survival. Hence, there need be no circularity in appealing to the functions a character serves in order to explain the survival of the character. Fitness is forward looking. Functions should be forward looking in the same way and hence are explanatory in the same way.

A propensity theory

Here is one way to derive a forward-looking theory of functions. Let us begin with the etiological theory. Consider a case in which

27 For more on dispositions and their supervenience on categorical bases, see Prior, Pargetter, and Jackson (1982), Pargetter and Prior (1982), Prior (1985b), and Pargetter (1987).

some character regularly has a specific effect and has been developed and sustained by natural selection because it has had that effect. In such a case the etiological theory deems that it is (now) a function of the character to produce that effect.

Let us look more closely, then, at the past process that has "conferred" a function according to the etiological theory. The character in question must have had the relevant effect on a sufficient number of occasions – and in most cases, this will have been not on randomly chosen occasions but on *appropriate* occasions, in a sense needing further clarification. (For instance, sweating will have had the effect of cooling an animal – and it will have had this effect on occasions when the animal was hot, not when it was cold.)

The history that confers a function, according to the etiological theory, will thus display a certain pattern. The effect that will eventually be deemed a function must have been occurring in appropriate contexts; that is to say, it must have been occurring in contexts in which it contributes to survival, at least in a statistically significant proportion of cases.

Further, this contribution to survival will not, in realistic cases, have been due to sheer accident. One can imagine individual incidents in which a character contributes to survival by mere coincidence. Laws of probability dictate that a long run of such sheer accidents is conceivable but it will be very rare. Nearly all cases in which a character has a long run of survival-enhancing effects will be cases in which the character confers a standing propensity upon the creature, a propensity that increases its objective chances of survival over a significant range of likely environments.

If we imagine (or find in the vast biological record) a case in which a character is sustained by a chance sequence of accidents rather than by a standing propensity, it would not be appropriate to describe that character as having a function. This can happen when a character is linked with another character that does bestow a propensity and where variations in the character have not occurred to allow selection against the inoperative character. It can also happen by sheer chance, by a long-run sequence of flukes. Such a sequence is very improbable; but biology offers a stunningly large sample. It is very probable that many improbable events will have occurred in a sample that large. Consequently, what confers the status of a function is not the sheer fact of survival due to a character, but rather survival due to the propensities the character bestows upon the creature.

The etiological theory describes a character *now* as serving a function when it *did* confer propensities that improved the chances of survival. We suggest that it is appropriate, in such a case, to say that the character *has been serving that function all along*. Even before it had contributed (in an appropriate way) to survival, it had conferred a survival-enhancing propensity on the creature. And to confer such a propensity, we suggest, is what constitutes a function. Something has a (biological) function just when it confers a survival-enhancing propensity on a creature that possesses it.

A number of features of this propensity theory of biological functions should be made explicit. First, like the corresponding account of fitness, this account of functions must be relativized to an environment. A creature may have a high degree of fitness in a specific climate, but a low degree of fitness in another climate. Likewise, a character may confer propensities which are survival enhancing in the creature's usual habitat but which would be lethal elsewhere. When we speak of the function of a character, therefore, we mean that the character generates propensities that are survival enhancing in the creature's natural habitat. There may be room for disagreement about what counts as a creature's natural habitat; but this sort of variable parameter is a common feature of many useful scientific concepts.

Ambiguities will arise especially when there is a sudden change in the environment. At first, we will refer the creature's natural habitat back to the previous environment. Later we will come to treat the new environment as the creature's new natural habitat. The threshold at which we make such a transference will be vague. The notion of natural habitat is, for this reason, especially ambivalent as applied to domestic animals.

In its most obvious use, the term 'habitat' applies to the physical surroundings of a whole organism. But we can also extend its usage and apply the term to the surroundings of an organ within an organism – or to the surroundings of a cell within an organ. In each case, the natural habitat of the item in question will be a functioning, healthy, interconnected system of organs or parts of the type usual for the species in question. When some of the organs malfunction, other organs will go on performing their natural functions, even though in the new situation this may actually lower the chance of survival. Why do these harmful operations still count as exercises of functions? Because they *would* enhance survival if the other organs were performing as they do in healthy individuals.

Consequently, functions can be ascribed to components of an organism in a descending hierarchy of complexity. We can select a subsystem of the organism, and we can ascribe a function to it when it enhances the chances of survival in the creature's natural habitat. Within this subsystem, there may be a subsubsystem. And this may be said to serve a function if it contributes to the functioning of the system that contains it – provided that all other systems are functioning "normally" (i.e., provided that it is lodged in its own natural habitat). And so on.

Similar hierarchies may occur in the opposite direction: a microscopic organism may have a function in a pond, which has a function in a forest, which perhaps has a function in the biosphere, and so on. Functions in large-scale systems are possible in principle, on our theory. In practice, large-scale systems may fail to meet the required conditions for possession of functions; but such functions are not ruled out by definition alone.

Functions, then, are construed by the propensity theory as *relational* properties. The next thing to note is that functions are truly dispositional in nature. They are specified subjunctively: they *would* give a survival-enhancing propensity to a creature in the creature's natural habitat.[28] This is true even if the creature does not survive or is never in its natural habitat. Likewise, fragility gives to an object a propensity to break, in an appropriate manner, in the right circumstances – yet, of course, some fragile objects never break. And fitness gives a propensity to an individual or species to survive, in an appropriate manner, in a specified environment, and in a struggle for existence, even if there is no struggle for existence in its lifetime or if the individual or species fails to survive.

28 It is this central role we give to propensities that distinguishes our theory from others, like those of Boorse and Wimsatt, which fall back on the notion of statistically normal activities within a class of organisms. On their theories, a character has a function for a creature when it *does* help others "of its kind" to survive in a sufficiently high proportion of cases. On our view, frequencies and statistically normal outcomes are important evidence for the requisite propensities. But there are many well-known and important ways in which frequencies may fail to match propensities.

Of course, when functions do lead to survival, just as when dispositions are manifested, the cause will be the morphological form of the creature and the relationship between this form and the environment. Functions supervene on this in the same way that dispositions supervene on their categorical bases. But the functions will be explanatory of survival, just as dispositions are explanatory of their manifestations; for they will explain survival by pointing to the existence of a character or structure in virtue of which the creature has a propensity to survive. They will explain survival because they give relevant causal information, and as we have already argued, explanation of *why* things are so is simply the provision of causal information.

Functional properties, then, are things which confer survival-enhancing propensities. This theory entails that there should be some way to spell out the notion of a survival-enhancing propensity in formal terms, employing the rigours of the probability calculus. Clearly, there will be a spectrum of theories of this general form. These theories will vary in the way they explicate the notion of enhancement. They may construe this as the raising of the probability of survival above a certain threshold. Or they may construe enhancement of survival-probability as increasing it significantly above what otherwise it would have been. And so on. We are not however, attempting to find and defend the correct theory of propensities. We are only arguing that propensity theories offer the most promising account of biological functions.

The question then arises whether the scope of the propensity theory is limited to biological functions or whether it can be extended, in some sense, to artefacts. Obviously, like the etiological theory, the propensity theory could be part of an overarching, disjunctive theory which analyses biological functions in terms of propensities for survival and the function of artefacts in terms of prior representations of their creators. Yet we noted earlier some problems for a backward-looking theory, even for artefacts.

We propose a general, overarching theory, one that is forward looking for both biological and artefactual functions, a theory that concentrates on the *propensity for selection*. A character or structure has a certain function when it has a propensity for selection in virtue of its having the relevant effects. Biological functions rest on a propensity for survival in the natural habitat, under conditions of natural selection. In the case of artefacts, we have a selection process often involving representations in the minds of the users. But the representations are those at the time of selection – now, so to speak. They need not be blueprints that antedate the first appearance of the function. The representations play a role in qualifying the mode of selection; but they do not disturb the structural fact that the relevant functions are grounded in propensities for selection. Thus, we argue, there is a sense in which all functions have a commonness of kind, whether they be of biological characters or of artefacts.

Comparisons

On most biological examples, the etiological theory and our propensity theory will yield identical verdicts. There are just two crucial sorts of case on which they part company. One sort of case is that of the *first* appearance of a character that bestows propensities conducive to survival. On our theory, the character already has a function, and by bad luck it might not survive, but with luck it may survive, and it may survive *because* it has a function. On the etiological theory, in contrast, the character does not yet have a function. If it survives, it does not do so because it has a function; but, after time, if it has contributed to survival, the character will have a function.

We think our theory gives a more intuitively comfortable description of such cases, at least in most instances. But there are variants on this theme for which our theory gives less comfortable results. Suppose a structure exists already and serves no purpose at all. Suppose, then, that the environment changes, and as a result, the structure confers a propensity that is conducive to survival. Our theory tells us that we should say that the structure now has a function. Overall, this seems right, but there are cases in which it seems counterintuitive. Consider, for instance, the case of heartbeats – that is, the *sound* emitted when the heart beats. In this century, the heartbeat has been used widely to diagnose various ailments; so it has come to be conducive to survival. The propensity theory thus, it seems, deems the heartbeat to have the function of alerting doctors. That sounds wrong. The etiological theory says the heartbeat has no such function because it did not evolve for that reason. That sounds plausible.

And yet, we suggest, the reason we are reluctant to grant a function to the heartbeat is not that it lacks an evolutionary past of the required kind. Other characters may lack an evolutionary past, yet intuition will not resist the attribution of a function. Rather, our reluctance to credit the heartbeat with a function stems from the fact that the sound of the heartbeat is an automatic, unavoidable by-product of the pumping action of the heart. And that pumping action serves other purposes. Although the heartbeat does contribute (in some countries, recently) to the survival of the individual, it does not contribute to the survival of the character itself. The character, heartbeat, is present in everyone, whether or not doctors take any notice of it. Although it "contributes" to survival, it is a redundant sort of contribution if it could not fail to be present whether or not it was making any contribution. Heartbeat does *not* contribute, in the relevant sense, to survival. It does not confer a propensity for survival which is higher in those who have a heartbeat than in those who do not, since there *are* none who do not.

So much for cases of *new* survival-enhancing characters. There is a dual for these cases: that of characters that were, but are no longer, survival enhancing. These cases, like the former ones, discriminate between the etiological and propensity theories. If a character is no longer survival enhancing (in the natural habitat), the propensity theory deems it to have no function. The etiological theory, in contrast, deems its function to be whatever it was that it used to be and was evolved for.

In general, we think the propensity theory gives the better verdict in such cases. Under some formulations, our judgement may be swayed in favour of the etiological theory. We may be inclined to say that the function of a character is to do such-and-such, but unfortunately this is harmful to the creature these days. Yet surely the crucial fact is, really, that the function *was* to do such-and-such. It serves no pressing purpose to insist that its function still is to do that. Especially not once we have passed over the threshold at which we redefine the creature's natural habitat. If a character is no longer survival enhancing because of a sudden and recent change in environment, we may refer to its "natural" habitat as the habitat before the change. Consequently, our propensity theory will continue to tie functions to what would have been survival enhancing in the past habitat. In such cases, there will be no conflict between the judgements of our theory and those of the etiological theory.

The test of examples and counterexamples is important. Yet in this case, in the analysis of functions, there is a risk that it will decay into the dull thud of conflicting intuitions. For this reason, we stress the importance of theoretical grounds for preferring the propensity theory. A propensity can play an explanatory causal role, whereas the fact that something has a certain historical origin does not, by itself, play much of an explanatory, causal role. Consequently, the propensity theory has a theoretical advantage, and this gives us a motive for seeking to explain away (or even overrule) apparent counterintuitions. In a similar way, Darwinian evolutionary theory provides strong theoretical motives for analysing fitness in a certain way. Our intuitions – our unreflective impulses to make judgements – have a role, but not an overriding one.

Paul Griffiths has urged a theoretically motivated objection to the propensity theory, one which does not rest on mere intuitions. Functions, we urge, are grounded in forward-looking propensities, just as fitness is. We have used the parallel with fitness as an argument in favour of our propensity theory of functions. Griffiths turns this argument back on itself. He has objected that, precisely because fitness is forward looking, functions should be backward looking. There is no need to use the term 'functions' for forward-looking properties, since biological science already has the perfectly good term 'fitness', tailor-made for the description of forward-looking propensities.

Our reply is that several distinct forward-looking explanatory roles need to be performed, and functions and fitness can play distinct forward-looking roles without any undue duplication. Fitness is a property of an organism. Functions specify the properties which collectively confer fitness on an organism. If certain morphological characters confer fitness on an individual, we may ask why this is so. And the answer may be that they confer fitness because they have the functions they do. One attribution of fitness breaks down into attributions of many functions. Thus, functions are in one respect less, but in another respect more, informative than fitness. They do not tell us the resultant degree of fitness of which they are the several components, but they do each tell us not only *which* features contribute to fitness, but also *why* these features contribute to fitness. Identifying functions will tell us which of the effects of a feature are the ones which contribute to fitness.

Thus, although both fitness and functions are grounded in propensities, they play significantly distinct explanatory roles. Just as distinct component forces give rise to a single resultant force, so too distinct functions give rise (or at least contribute) to a single overall degree of fitness.

Artefacts have functional features, but no biological fitness. A propensity theory for artefactual functions, therefore, suffers no threat of being made redundant by artefactual *fitness* in the biological sense. Yet Griffiths could mount an analogous challenge by asking what need there is for propensity functions when we could easily use instead the propensity concept of degrees of *utility*. Our reply to that challenge would be the same as our reply concerning biological functions. Utility is a resultant which depends on the several distinguishable functions of an artefact. It is a merit of our theory that it provides a unifying concept of function which applies both to organisms and to artefacts and, furthermore, that this concept serves to knit together the kindred concepts of *fitness* and *utility*.

The upshot of this propensity theory of functions, for both artefacts and organisms, is that the identification of functions does provide fairly specific causal information. We have argued that causal information is precisely what is needed to explain not just *how* things are, but *why* they are as they are. Hence, we argue, functional explanations do give us legitimate, not merely metaphorical, information about why things are as they are. Aristotle's final causes have a place in science, both in culture and in nature. Scientific realism can extend beyond realism about the entities of physics and encompassing realism about pockets of purpose in this cosmos of ours.[29]

7.5 REALISM AND EXPLANATION

Various sorts of realism are supported by what are known as arguments to the best explanation. We ask why certain things are so. We then note that, if there were things of some other sort, related thus and so, that would explain why the things which puzzled us are as they are. If the explanation is good enough, and we can find none better, it is reasonable to grant some probability to the hypothesis that things really are as our explanation supposes them to be.

For instance, we may ask why objects expand on heating. We then note that, if the process of heating an object is in fact a matter of forcing fire particles into it, that would explain why an object expands on heating. Alternatively, we may note that, if the process of heating an object is, in fact, a matter of speeding up the motions of the particles of which it is composed, then that, too, would explain why an object expands on heating. After much consider-

29 This discussion of functions is based on Bigelow and Pargetter (1987a).

ation, we might settle on a judgement that one of these explanations is better than the other. Then it is reasonable to believe that things are as the better explanation supposes them to be: middle-sized dry-goods are indeed composed of moving corpuscles, and the hotter they are, the faster their corpuscles are moving.

Many kinds of realism rest on inferences to the best explanation. However, not every kind of realism must be grounded in that way. In order to mount an inference to the best explanation, we must begin with something which is to be explained. Unless we are realists about whatever we set out to explain, it is hard to see how we could be realists about whatever we use to explain it.

The most natural realist picture is a foundationalist one. That is, realism naturally falls into a hierarchical structure, resting on a foundation of realism about some basic class of entities which do not explain anything which is more basic than they are, but which furnish the materials that all other posits are introduced to explain.

One objection to this hierarchical kind of realism is directed not at its foundationalist structure, but only at the standard choice of foundations. Traditionally, it has been held that the raw materials which all explanations address are the raw materials of *experience*. We begin with "appearances", and we mount an inference to the best explanation of these appearances, and thereby arrive at a theory of the "reality" which underlies appearances.[30] This perspective can represent all forms of realism, without exception, as resting on inferences to the best explanations. At the foundation of all these explanations, we have the raw material which, ultimately, they all serve to explain: namely, appearances. Yet it is not appropriate to call such a view realism *about appearances*. The term 'realism' is normally restricted to a contrastive use, in which it refers only to the status of things other than appearances.

We are dubious about this traditional realist perspective. In the first place, it assumes that in some sense we view the world "from the outside", so that we ourselves and the appearances which are present to us are in some important sense not *within* reality itself. This is, we take it, a mistake.

It is also doubtful whether all the things we are realists about are introduced in order to explain appearances. There is considerable appeal in the doctrine of *direct realism,* according to which we per-

30 For a classic discussion of this see Russell (1912).

ceive material objects "directly"; that is, *without inferring* their existence from anything more basic.[31] Even if we grant that we can, on reflection, distinguish an object from its appearances, it is doubtful whether the former is simply the best explanation for the latter. The epistemic ordering of objects and appearances is, in fact, quite a tricky matter. There does seem to be some sense in which it is only when we begin introspecting and attending to our own inner states that we begin to form judgements about appearances. The extroverted perceiver is the more basic case. It is only *after* we have had considerable practice in perceiving objects that we can turn our eye inwards to inspect the nature of this process. Extroverted realism about objects must precede introverted reflections upon appearances. Thus, basic realism about physical objects around us does not rest primarily upon any actual inference to the best explanation of something more basic.

Nevertheless, once we have become direct realists about material objects, we may then notice the appearances of these objects. And we may then notice also that the existence of these objects does provide the best explanation of these appearances.

This illustrates a very important way in which the hierarchical image of realism is blurred in practice. Even if we begin with one kind of realism in the absence of a second kind of realism and then mount an argument to the best explanation, nevertheless once we have arrived by this route at the second kind of realism, these two kinds of realism fuse into one and the same reality. The hierarchical order in which we arrived at them does not remain in the things themselves, but was merely an extrinsic feature of their relation to us as epistemic agents in the world with them. And so it is entirely appropriate to search for ways in which the things we noticed first, our initial sphere of realism, may provide the best explanations of anything else we are now realists about. And so things which may have grounded our inferences to something else may now be used as the best explanation of features of those other things. Our initial sphere of "uninferred realism" then acquires support by way of inference to the best explanation of other things.

The resulting image of realism, when we allow this kind of feedback from inferred entities to the initially uninferred, is a form

31 Congenial, traditional, sympathetic presentations of direct realism can be found in Armstrong (1961). See also the critical expositions and discussions in Hirst (1959) and Jackson (1977b).

of *holism*.[32] In this guise, we support holism; and as we argued earlier, this sort of epistemic holism, properly understood, is no threat to realism.

If realism rests on inference to the best explanation, we must face the question of what kinds of explanation may be appealed to in support of realism, for explanations can take several forms. Explanations, we have argued, provide information about causes; but such information may be more or less direct, and besides, the "causes" which it informs us about may be described, in Aristotelian language, as "material", "formal", "efficient", or "final" causes.

The most convincing arguments for realism are those which appeal to efficient causation. So it seems, if we are to judge by history and the progressive mechanization of the world view of the sciences. And it has been persuasively argued recently (e.g., by Hacking and Cartwright)[33] that the strongest arguments for realism about unobserved entities are arguments to the best causal explanations – in the sense of efficient causation.

If the only persuasive arguments for realism were arguments from effects to causes, that would be bad news for the modal realism which fills this book. Yet we do not accept that arguments to "efficient causes" are the only kinds of explanations which can support realism. The argument to modal realism, as presented in this book, may perhaps be aptly described as an argument to the best "analysis" of science.

Quine's deeply disturbing essay, 'Two Dogmas of Empiricism',[34] has made many philosophers very nervous about using the word 'analysis'. And, indeed, that word has many associations which we would like to tear ourselves away from. By describing their goal as one of "philosophical analysis", philosophers have too often persuaded themselves not only that they need not get out of their armchairs and perform experiments, but even that they need not pay much heed to reports from scientists who *have* performed experiments. If "analysis" is intended to justify ignorance of science, we disown it.

Yet we also resist the extreme view that the only arguments for realism are arguments to efficient causes. There is also a place for

32 Here we are endorsing elements of epistemological holism of the sort expressed by, e.g., Quine (1960) and Pollock (1974).

33 Hacking (1983) and Cartwright (1983).

34 The second essay in Quine (1961).

arguments to the best explanation by material causes and by formal causes. The question of what *constitute* the possibilities and other modal elements in science is a genuine question. So is the question of *how* things must be if there are to be such possibilities. Modal realism is justified as the best explanation of these matters.

Thus, we have argued, modal realism is firmly supported as offering the best account of the nature of science and of the world. Modal realism furnishes an account of the material and formal causes of things, which then supports a fruitful account of efficient and final causes as well. Efficient and final causes cannot profitably be mooted at all, until at least some groundwork has been laid concerning material and formal causes. Some of the what- and how-questions must be tackled before we can make any headway with why-questions.

The metaphysics which underpins our modal realism, a kind of Platonic realism about universals, can thus be defended by a kind of inference to the best explanation. This inference is still more strongly reinforced by the role which this same metaphysical realism can play in the interpretation of mathematics. It is appropriate that we should return to mathematics here, since mathematics not only is central to science, but also is the primary source of the modalities in science. It is a great merit of our metaphysics that it permits a realist account of mathematics, one which thoroughly integrates mathematics into general scientific realism. So it is to mathematics that we turn in the final chapter.

8

Mathematics in science

8.1 MATHEMATICAL REALISM

Many mathematicians think that the things studied in mathematics are human creations. They think that integers, real numbers, imaginary numbers, groups, topological spaces, and the rest have been invented by human beings. Such things would not exist if there were no people. They came into existence as a result of human activity. They even have birthdays.

We agree with the positive claims of such antirealists: we agree that there are indeed human creations which would not exist if there were no people and which have birthdays. These human creations are words, ideas, diagrams, images, concepts, theories, textbooks, academic departments, and so forth. Yet as realists, we insist that these human creations are not the only things worth studying; there are more things than these. In addition to human creations, there are things which are not human creations, things which would still exist even if people did not, things whose birthdays, if they have any, probably coincide with the birth of the universe and certainly precede the origin of human life.

Realism about mathematics rests comfortably and naturally with scientific modal realism. It is not compulsory for all scientific realists to take a realist stand about mathematics along with all the rest of science. No one can sensibly maintain realism about absolutely all the bits and pieces bandied about in science. It must be admitted that *some* parts of science are merely useful fictions and do not accurately represent anything beyond themselves. Hence, there is nothing wrong, in principle, with being a realist about a great deal but being an antirealist about mathematics.[1]

1 Field (1980) has made an impressive attempt to marry realism about space-time, particles and fields, and so forth with fictionalism about all things mathematical.

And yet, although there may be nothing wrong in principle, there is quite a bit wrong in practice with a marriage between scientific realism and mathematical antirealism. Mathematics is not just one small element of science – it is a *huge* element of science. And it is not easy to draw a clear line between the mathematical and the nonmathematical portions of science; mathematics permeates nearly every component of science. If we were to try to leave the mathematics out of science, it is hard to see what would be left.

Consider an illustration. In his attempts to understand how bodies fall, Galileo thought hard about the speed with which they fall.[2] It was noticed that an object which falls farther is moving faster by the time it finishes its fall than one which falls a shorter distance. The question arises, then, *how much* faster an object will move when it falls, say, twice as far as another. If it falls twice the distance, will it fall at twice the speed? Is its speed proportional to the distance covered? (Suppose the two objects have the same shape, size, density, and whatever other characteristics one might suspect to be relevant to their manner of falling.)

Galileo did not have access to any adequate notion of instantaneous speed. For him, the speed over a distance was defined simply as the distance divided by the time. If a body *a* falls twice as far as a body *b*, at twice the (average) speed of *b*, then *a* and *b* must take the same amount of time to complete their fall. And yet consider more closely the body *a* which falls twice the distance in the same time. How fast does it cover the first half of its path? It must cover the first half of its path in less time than it takes for its whole journey. Since object *a* takes as long for its whole journey as object *b*, this means that it must take less time for the first half of its journey than *b* takes for its journey. Hence, *a* travels faster for the first half of its journey than *b* travels on its journey. But the first half of *a*'s journey is the same as *b*'s whole journey, except for the fact that it is followed by further falling. So *b*'s journey differs from the first half of *a*'s in speed and time taken, even though it differs in no other relevant way.

Yet this means that the speed with which a body covers a given distance depends on whether it finishes its fall at that point or whether it will continue to fall further. And this is absurd; or at any

2 Galileo's *Dialogues Concerning Two New Sciences* is the source for his most mature thoughts on motion; for an English translation, see Crew and de Salvio (1954).

rate, it is certainly known to be false. It would require a falling object to have something akin to precognition or be subject to reverse causation. Hence, the (average) speed of a falling object cannot possibly be proportional to the distance it falls.

This result tells us something important about falling bodies. If we are to be realists about anything at all, we should surely be realists about this: an object which falls twice as far does not have twice the average speed. After all, this is equivalent to being a realist about how long an object takes to cover a given distance, and any scientific realist should be a realist about *that*. Indeed, we need not go beyond a commonsensical, prescientific realism to be a realist about whether an object which falls twice as far travels with twice the speed. And yet in being realists about such commonsense matters, we are, on the face of things at least, being realists about certain mathematical relationships among speed, time, and distance.

It might be objected that we should not be realists here, because the objects *a* and *b* discussed here are idealized objects falling through an idealized medium. Yet that is a red herring. We do not, in fact, have to suppose *a* and *b* to be identical in mass, volume, and so forth. We do not have to suppose them to be falling through a perfect vacuum or a perfectly homogeneous medium. We can take *a* and *b* to be ordinary bricks falling through ordinary air from any heights between, say, the height of a dandelion and the height of a church steeple. The reasoning will follow through just as forcefully: such ordinary bricks will not fall twice as far at twice the average speed.

"Average speed" may be alleged to be a mere mathematical abstraction. Yet that, too, is a red herring. It is simply a physical fact that brick *a* falls twice as far as brick *b;* and it is simply a physical fact that *a* does not take the same time to fall as *b* does. That is exactly what it means to say that the average speed of *a* is not twice that of *b*.

Galileo thought speed to be physically real. He thought that the force of impact (e.g., the depth of hollow caused by impact from a falling brick) would be roughly proportional to speed. So speed was, he thought, causally efficacious. It was as real as anything could be. Of course, we now think that what is causally efficacious is not the average speed over the whole path, but rather the instantaneous velocity. Yet the mathematical definition of instantaneous velocity is immeasurably more abstract than Galileo's definition of speed. The abstraction and complexity of its definition, however, do not in any way diminish its physical reality. An object which falls twice as far does not have twice the instantaneous speed; and this explains why it does not cause twice the impact. Instantaneous speed is real and causally efficacious. It was not properly understood until after Galileo; but it was possessed by falling bricks and caused damage to people and property long before it was properly understood.

The fact is that the physical world contains quantities as well as qualities. One object may fall farther than another; and more specifically, it may fall *more than twice as far* or *less than twice as far,* and so forth. One object may also take *longer* than another to fall, or it may take *less* time to fall. And there are relationships between distance of fall and duration of fall. The equations which describe such relationships are obviously human constructions. Yet the relationships themselves are not human constructions.

For instance, Galileo's equation, which states that distance of fall is proportional to the square of the time of fall, is indeed a human creation. Yet it is not a human creation that some pre-Galilean falling brick covered about one length of a specific magnitude in the first unit of time, three lengths of that magnitude in the second unit of time, five lengths of that magnitude in the third unit of time, and so on. The brick *did* fall about three times as far in the second unit of time as in the first: that is as objective a fact about the pre-Galilean brick as any fact about anything. And this entails that *what is asserted* by Galileo's equation is objectively so, and would have been so whether or not anyone asserted it.

In being a realist about such things, it is perverse to resist realism about the mathematics involved. To drum the point home further, consider a little more closely the Galilean law that the distance a body falls is proportional to the square of the time taken. It is illuminating to see how that mathematically abstract law is grounded in real physical properties and relations.

Galileo claimed that, in the first unit of time an object falls, it will cover one length of a specific magnitude; in the second unit of time it will cover, not one or two such lengths, but three lengths of that magnitude; in the third unit of time it will cover five lengths of that magnitude; and so on. In this claim, Galileo was anticipated by medieval theorists who not only hit on the same pattern, but also advanced a very interesting theory about why such a pattern should occur.[3] This helps to illustrate the way in which mathematics contributes to explanations of phenomena in science. The medieval idea was that, in each successive unit of time, a new *increment* is added to its speed. As equal increments are added to the speed, so too equal increments are added to the distance covered in each successive unit of time. That, in outline, is why the distances covered in successive units of time increase by equal increments: 1, 3, 5, 7, . . . ; each of these is two more than the one before. The general rule is that, in the nth unit of time, the nth odd number will be the number of lengths of that magnitude which the falling body traverses. To obtain the total distance travelled, up to and including the nth unit of time, we must therefore sum the first n odd numbers. The sum of the first n odd numbers is n^2. This is a typically mathematical fact, and it seems to be a fact about *numbers,* which in turn seem to be abstract entities. Indeed, it may seem to rest on nothing more than rules for manipulating symbols, which permit us to prove, algebraically, an equation of the form

$$(1 + 3 + 5 + \cdots + (2n - 1)) = n^2.$$

Yet this is a misconception. Numbers may be abstract, in some sense; but physical things can instantiate various number properties. In an important sense, a number is *present* in a collection of individuals which have that number. Consider, for instance, a collection of objects which have the number n^2, a collection arranged in a manner indicated in Figure 8.1, for instance. You can see, in such an array, that there are n rows and n columns (i.e., that there are n^2 objects). You can also see that the total aggregate is the sum of the ∟-shaped "bent rows". (The Greeks called them *gnomons*.)[4] You can see that the numbers in each of these bent rows form the sequence of odd numbers. And each one *has* to contain two more than its predecessor, so this pattern *must* continue. Perception and reflection on what you are perceiving enable you to see that the number of the objects is n^2 and that the number of the objects is the sum of the first n odd numbers.

This relationship between numbers is instantiated within square arrays of objects. But this very same relationship carries over to

3 Dijksterhuis (1961) is a good source for the core history and further references.
4 Heath (1921) is a good source for early Greek mathematics.

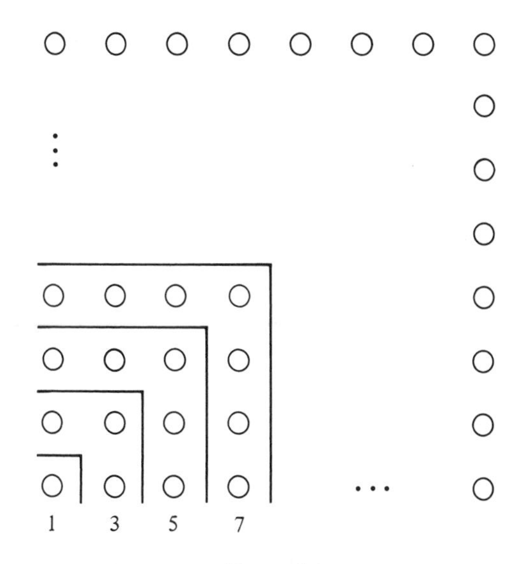

Figure 8.1

Galileo's problem of falling bodies. Instead of dealing with discrete objects, which can be arranged in a square grid, we are dealing with the number of lengths of a given magnitude which a falling body passes through in successive intervals of time. The very same relationship holds, however, in this case as in the previous case. Using this relationship between numbers, we can infer that the total distance fallen in n units of time will be the square of the distance fallen in the first unit of time.

The differences between distances fallen and material objects should not blind us to the similarities. Both may be numbered: both may instantiate the very same numbers. And if they instantiate the same numbers, the same proportions will hold between them. That is why the mathematical relations we perceived when instantiated by arrays of ○'s will also be instantiated (though less perceptibly) by sequences of distances travelled.

The facts about falling bodies, then, are ones that any scientific realist should be a realist about. That takes us a considerable way towards realism about mathematics. After all, Galileo's reasoning concerning velocities was deeply entangled with mathematics. The equations involved were simple: for instance, the equation which (wrongly) states that speed is proportional to the distance travelled.

Yet, simple as it is, this already illustrates most of the philosophical problems posed by mathematics. And it illustrates the case for realism about mathematics. The equation merely provides a convenient, concise description of real relations among real properties of real things.

This leads to a position which may be described as a broadly Platonist one. Yet we hasten to deflect the sort of reflex rejection that Platonism frequently evokes. The elements of Platonism that we endorse do not include the most notorious of Plato's doctrines. We do not suppose there to be some class of entities, the Forms or Ideals, which were apprehended by us only in a former life, which cannot be seen or felt in this world of illusion, and to which mundane objects bear only a remote resemblance. As we explained earlier, in Chapters 1 and 2, we reject those otherworldly Platonic doctrines. They are not required by a realist interpretation of mathematics.

The realism we advocate is in some respects closer to that of Aristotle than to that of Plato. Universals are not to be construed as inhabitants of some otherworldly kingdom of ideal objects. Rather, they are here in the same world that we are in. The speed of a falling object is no more otherworldly than the object itself is. And the proportion between the speeds of the two objects is no more otherworldly than the two objects and their speeds are. This is not to say that we endorse all of Aristotle's views on universals. In particular, we deplore Aristotle's tendency to favour qualitative over quantitative characteristics of things.

We urge that it is possible to wed the clarity and rigour which Plato projected onto a mythical mathematical realm with the this-worldliness of mathematical universals that are as causally efficacious, as observable, and as real as any physical properties or relations.

8.2 NUMBERS AS RELATIONS

Mathematics embraces many things. A full understanding of mathematics would require explanations of the nature of set theory, geometry, topology, and a great many other things. We shall focus here on just one of the leading actors in the never-ending story of mathematics: numbers and, more specifically, real numbers.

We shall argue that real numbers are universals. They are higher-order universals. More particularly, they are relations between relations (or between properties). They are, in fact, the very same higher-order relations, or proportions, which we appealed to in our theory of quantities in Section 2.5. We have argued that what makes a property, say length, count as a quantity is its standing in proportions to a family of other properties. We now argue that these proportions should be identified with real numbers.

On this theory, real numbers are thoroughly physical, as physical as any properties or relations. They are instantiated by physical quantities like length, mass, or velocity, which in turn are instantiated by physical individuals like photons, electrons, elephants, or bricks. Their being instantiated makes a causal difference. Thus, although they are "abstract" entities, in the very basic sense in which all universals are abstract, they are nevertheless not at all abstract in the sense which implies causal impotence, or unobservability, or aloofness from the realm of space and time, or in the sense in which an abstract entity is the product of a mental process of abstraction. Numbers are not denizens of a Platonic heaven, nor are they symbols which refer to nothing outside the mind. We invent symbols, and we perform mental acts of abstraction; but such symbols and abstract ideas refer to properties and relations of physical quantities which we succeed in noticing as a result of a process of abstraction. Abstraction is merely a process of focussing attention upon one or another of the many universals which are instantiated around us. Galileo, for instance, focussed on the speed of a falling object and gave little attention to the endless other properties it might have – who owns it, where it came from, what it smells like, what colour it is, how old it is, and so on. This focussing of attention, however, does not create any new thing which was not there before. It does not generate any mere "object of thought". Rather, it permits us to notice one of the many things that is there before us and that would have been there whether we had noticed it or not.

Of all the many salient mathematical entities we could have discussed, numbers best serve our purposes. This is because numbers seem, on the face of it, to be the most difficult case for a scientific realist. Other parts of mathematics are very easy to construe as being concerned with patterns and structures – that is, with properties and relations. Numbers are among the hardest things to construe in that way. Although numbers do generate patterns and structures, there is a strong tendency to think that the numbers themselves are *objects* which instantiate properties and relations – but are not themselves properties or relations. Numbers do *seem* to be "abstract objects".

We shall argue that this appearance can be explained without undermining the basic conception of numbers as universals. If we can sustain an account of *numbers* as universals, this will win the hardest battle of all for the mathematical realist. If numbers can be identified with universals, it is difficult to believe that any other mathematical subject-matter would prove more resistant than numbers.

Let us see, then, how real numbers might be explained in terms of higher-order universals. In searching for such an explanation, we have no need to start from scratch. There is a formal theory available, which is expressed in a set-theoretical framework but which can be reconstrued to yield a metaphysical theory of numbers as universals. This formal theory may be traced back to Frege and to Whitehead (who, it seems, arrived at the same theory independently) and was further developed by Wiener and Quine. The formal details are ferocious, but the basic idea is very neat.[5]

The theory we shall describe deals with relations between relations. It can be extended, however, to deal also with relations between properties. We may, for instance, wish to compare the *lengths* of two objects, and length is presumably a property of the objects. However, it is also possible to consider the *relationship* between the *ends* of the object rather than the *property* of length for the object as a whole. And more generally, we can move from properties to relations by another technique. Suppose one object has a particular property (say, a specific length), and another has another particular property (say, another specific length). Then there will in general be a relation which holds between the objects in virtue of their properties. For instance, in virtue of their lengths, there will be a relationship of one's being longer than the other and, indeed, one's being just "so much longer" than the other. If we define a relationship between these relations, it follows that there will be a derivative relation between the properties which generate those relations.

The Frege theory rests on one key kind of relationship between relationships. Consider the parenthood relation and the grandparenthood relation.[6] They are, of course, quite distinct relationships:

5 Frege (1893), Whitehead and Russell (1910), Wiener (1912), Quine (1941, 1969b), and Bigelow (1988a).

6 This kinship illustration derives from Quine (1941), reprinted in Quine (1966).

if a is a parent of b, then in most cases a is *not* a grandparent of b. Yet obviously there is a very intimate relation between these two relationships. If two things are linked by the grandparent relation, those same two things will be linked by a chain involving two instances of the parenthood relation:

If a is a grandparent of b, there is a c such that a is a parent of c and c is a parent of b.

It will help if we recall an abbreviation that we introduced in Section 2.6. Given any relation R and any two things x and y, we will write

$$x\ R^n\ y$$

to mean that we can get from x to y by n applications of the relation R – that is, there are $x_1, x_2, \ldots, x_{n-1}$ such that

$$\begin{array}{c} x\ R\ x_1 \\ x_1\ R\ x_2 \\ \cdot \\ \cdot \\ \cdot \\ x_{n-1}\ R\ y. \end{array}$$

Using these abbreviations we can sum up the relation between parenthood and grandparenthood by saying that, whenever x is a grandparent of y, then x parent^2 y. It is important to note that the *ancestral* of a relation has played a crucial role in the development of the theory of natural numbers, and this is just a generalization of the sorts of relations we are discussing here. The number zero is a kind of Adam, which begets all the numbers by repeated applications of the *successor* relation, which is the analogue of the parenthood relation. Just as 'human' may be defined to include all and only the descendants of Adam, so too 'natural number' may be defined to include all and only descendants of zero. The relation of descendanthood is the *ancestral* of successorhood. It is significant that the definition of the ancestral of a relation has been one of the most insuperable barriers for nominalism. In order to define the ancestral, we must take a realist stance over relations; we must speak as though relations exist just as much as the things which they relate.

The Frege–Whitehead theory of real numbers works by extracting more than could be foreseen from the kind of relationships which hold between parenthood, grandparenthood, and ancestorhood. Notice first that parenthood stands in one relationship with grandparenthood, in virtue of the fact that

$$x \text{ grandparent } y \text{ if and only if } x \text{ parent}^2\ y.$$

Parenthood stands in a different relationship with great-grandparenthood, in virtue of the fact that

$$x \text{ great-grandparent } y \text{ if and only if } x \text{ parent}^3\ y.$$

Because grandparenthood and great-grandparenthood are linked, in different ways, to the same basic relation, this automatically generates a relationship between them. We can, for instance, set down the principle

$$\text{If } x \text{ great-grandparent}^2\ y, \text{ then } x \text{ grandparent}^3\ y.$$

Anywhere we can get by two applications of the great-grandparent relation, we can also get by three applications of the grandparent relation.

This relation is an instance of a general pattern. Given two relations R and S, we may have a relationship between them in virtue of which

$$x\ R^n\ y \text{ if and only if } x\ S^m\ y.$$

These relationships may be called *ratios* or *proportions*. When R and S are related as described, then R stands to S in the ratio $m{:}n$.

By modifying the order of the variables x and y in the principle above, we can obtain a subtly different effect:

$$x\ R^n\ y \text{ if and only if } y\ S^m\ x.$$

The ratios obtained by this means are *negative* ones. For instance, the grandchild relation stands in the ratio $-2{:}1$ to the parenthood relation, since

$$x \text{ grandchild } y \text{ if and only if } y \text{ parent}^2\ x.$$

Thus, positive and negative ratios emerge very naturally out of the degrees of ancestorhood for a given relation.

It is useful to add a recursive clause that recognizes a ratio between two relations R and S in virtue of their joint connection to some other relation Q. If there is a ratio between R and Q, and a ratio between S and Q, then there is a derivative relation between R and S. Wiener used a special case of this in his modified version of Whitehead's theory.[7] Wiener said that, when

> the ratio of R to Q is n:1,
> the ratio of S to Q is m:1,

then we can conclude that

> the ratio of R to S is $n:m$.

This enables us to establish a ratio of $n:m$ between R and S even if it is not possible to iterate R or S. For instance, consider your ancestral relation to Eve and your mother's ancestral relation to Eve. On Wiener's definition, there will be a ratio $n:(n + 1)$ between these ancestral relations, for some n, and this will be so even if the future of the human race is too short to iterate either of these ancestral relations. There may be no one who is related to you as you are related to Eve, and no one who is related to your mother as your mother is related to Eve. In that case we cannot say that, anywhere you can get by $(n + 1)$ iterations of the former relation, you can also get by n iterations of the latter. So the ratio between these two ancestral relations does not emerge directly from iterations of those relations themselves. It emerges only recursively, in the manner outlined by Wiener, through their joint connections back to the basic parenthood relation.

In order to obtain the full complexity of the rational number system, we must suppose that the given relation has the right pattern of instances. The parenthood relation, for instance, probably does not have enough instances to generate infinitely many ratios. It is more likely that spatial relations or temporal relations will generate the full complexity of the rational number system. However, if you seek an *a priori* assurance of instantiations for all ratios, you should turn to set theory. In fact, the primary motive for the invention of set theory was the desire to set pure mathematics on a secure, *a priori* foundation. Yet the fact that ratios are instantiated in set theory should not blind us to the fact that they are also instantiated by physical quantities of many different sorts.

7 See Wiener (1912) for the recursive extension of Whitehead's theory.

8.3 RATIOS AND THE CONTINUUM

Ratios are special cases of real numbers. Yet not all real numbers are ratios. The notion of *proportion* is more general than that of ratio, and it is this more general notion which gives rise to the system of real numbers. The notion of proportion applies, for instance, in geometry, and some of the proportions in geometry do not correspond to any ratio. That is what the ancient Pythagoreans discovered.[8]

Consider regular pentagons. There is a proportion between the diagonal and the side of any regular pentagon. This proportion is the same for any such pentagon, no matter what size it is. Hence, in Figure 8.2, the proportion between AB and AC will be the same as that between DE and DF. Notice also that it can be shown that AB is parallel to DF, and hence that the triangle ABC has the same shape as triangle DFC: they are similar triangles. Thus, the proportion of AB to AC will be the same as the proportion of DF to DC.

So by the similarity of pentagons we have

$$AB{:}AC = DE{:}DF.$$

By similarity of triangles we have

$$AB{:}AC = DF{:}DC.$$

Hence,

$$DE{:}DF = DF{:}DC.$$

We can therefore use these lines to construct what is known as a *golden rectangle,* as in Figure 8.3. It can be shown that, in this rectangle, $EC = FD$. This means that, when a square (on base EC) is taken away from a golden rectangle FDC, the remainder is another golden rectangle EDF. These two rectangles differ in size but not in shape. As we have already seen, the proportion of short side to long side for the smaller rectangle, $DE{:}DF$, is the same as the proportion of short side to long side for the larger rectangle, $DF{:}DC$.

Then because the smaller golden rectangle EDF has the same shape as the larger one, it follows that a square may be subtracted from it, leaving a yet smaller golden rectangle GFH, as in Figure 8.4. This golden rectangle, too, can be diminished by a square to yield yet another golden rectangle, and so on ad infinitum.

There is a genuine relation between the long side and the short

8 Heath (1921) is a classic source for these early geometrical discoveries.

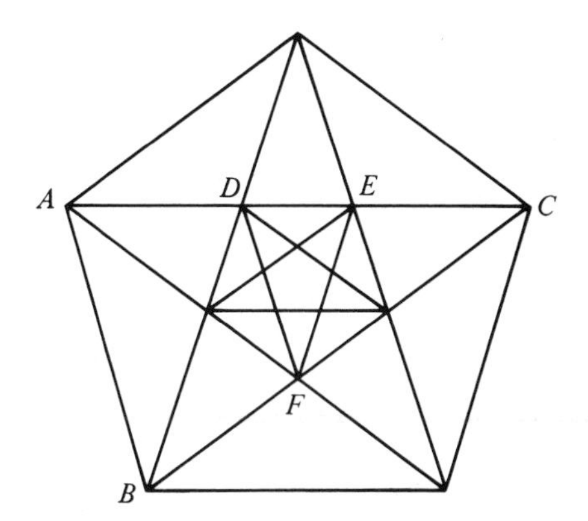

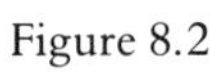
Figure 8.2

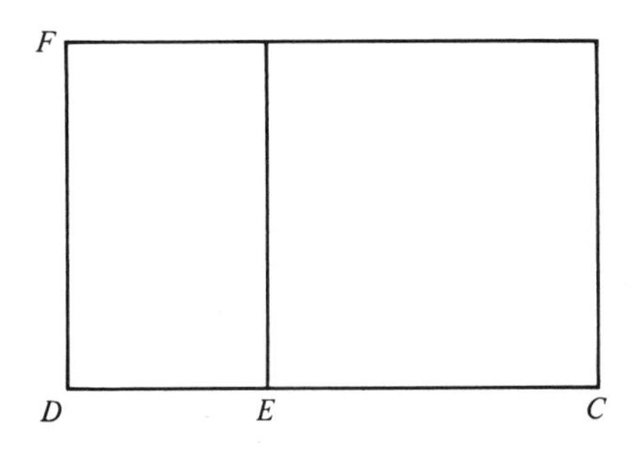

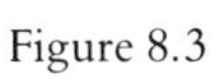
Figure 8.3

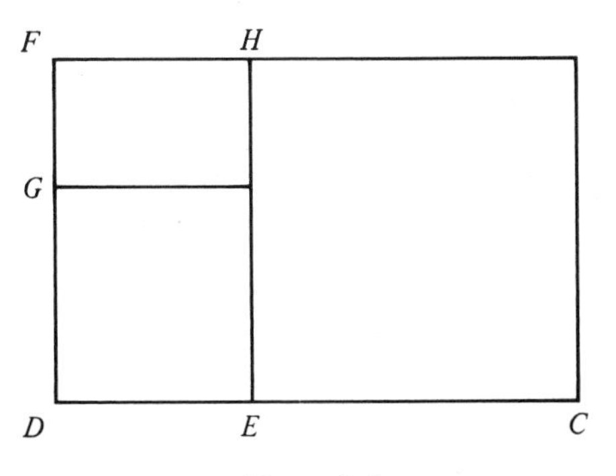

Figure 8.4

side of any golden rectangle; this proportion is real enough. Yet it is *not a ratio,* in the sense we have defined. It can be proved that the proportion of, say, *DF* to *DC* is not a ratio, in the following way. If we lay down lines of length *DF* end to end, over and over, and if we lay down lines of length *DC* end to end, over and over, the multiples of *DF* will *never* measure out the same distance as *any* multiples of *DC*. Four lengths *DF* do not equal five lengths *DC*, seven lengths *DF* do not equal nine lengths *DC*, . . . , n lengths *DF* do not equal m lengths *DC*, no matter what numbers n and m we may choose. If we start from a point x and measure n steps of length *DF* to reach a point y, we can never get from x to y by any number of steps of length *DC*. This shows that, on the Whitehead analysis, there is no *ratio* between *DF* and *DC*.

The same result can be shown for the Wiener modification of the Whitehead theory. There is no small distance which can be repeated, say, m times to yield line *DF* and n times to yield *DC*. These lengths are, as they say, *incommensurable*. The proportion is, in fact, $2:(1 + \sqrt{5})$ and since $\sqrt{5}$ is irrational, this proportion cannot be represented as any ratio $a:b$ for any whole numbers a and b, and so cannot be represented as any finite (or recurring) decimal number.

The incommensurability of *DF* and *DC* can be proved by a *reductio*. Suppose they were commensurable. That is, suppose there were some length d which divided evenly into length *DF* and into length *DC*. Then consider the rectangle *FDC* in Figure 8.4. Length d divides evenly into *DF*, and *DF* equals *EC*. This length divides into *DC* as well as into *EC*. Hence, this length must divide evenly into *DE*. And so this length evenly divides *DE* as well as *DF*, and these are the sides of the second of the three golden rectangles in Figure 8.4. Thus, the very same length which evenly divides the sides of the larger rectangle *FDC* must also divide the sides of the smaller rectangle *EDF*. This argument can be repeated to show that this same length d can also evenly divide the sides of the third golden rectangle in Figure 8.4 *GFH*. And so on ad infinitum.

Hence, if there were a length which evenly divided the sides of the larger golden rectangle *FDC*, that same length would also evenly divide the sides of the smaller and smaller golden rectangles ad infinitum. Therefore, no finite length can evenly divide both sides of a golden rectangle. The sides of a golden rectangle are incommensurable.

Initially, of course, this seems to be bad news for theories such as ours. And yet irrational proportions can be accommodated by a very natural extension of the definition of ratios. The required extension necessitates the introduction of a third relation in addition to the two being compared. When two relations R and S are incommensurable, then whenever

$$x\ R^n\ y,$$

it follows that

$$\text{not: } x\ S^m\ y,$$

whatever the values of n and m. Repetition of n applications of R will never replicate m applications of S. And yet we may find that, although n applications of R and m of S always lead to different destinations, those destinations may stand in a determinate relationship to one another. More specifically, it may be that the two destinations stand in some determinate *order,* under some linear ordering $<$. That is, it may be that, for some given n and m,

$$\text{if } x\ R^n\ y \text{ and } x\ S^m\ z, \text{ then } y < z.$$

(The relation $<$ must match up in a specific way with the natural ordering of the ratios, but we omit the formal details of this here.)

Consider again the short and the long sides of a golden rectangle. The proportion between them is $2:(1 + \sqrt{5})$. Now, $\sqrt{5}$ is more than 2 but less than 3. Hence, the proportion for the golden rectangle is not equal to 2:3, but it is quite close. Hence, three of the shorter lengths will *almost* match two of the longer lengths. Three of the shorter lengths will fall just a little short of two of the longer lengths. With appropriate definitions of R, S, and the ordering relation $<$, this falls into the pattern

$$\text{If } x\ R^3\ y \text{ and } x\ S^2\ z, \text{ then } y < z.$$

This is just one instance of the general pattern

$$\text{If } x\ R^n\ y \text{ and } x\ S^m z, \text{ then } y < z.$$

Where R and S and $<$ are appropriately defined for the proportions of the golden rectangle, there will be an infinite number of instances of this pattern. The golden proportion between R and S is, in fact, uniquely identified by the list of numbers n and m for which the above schema holds.

This can be generalized. Every proportion between relations R

and S can be uniquely characterized by the list of natural numbers n and m for which the schema

$$\text{If } x\ R^n\ y \text{ and } x\ S^m\ z, \text{ then } y < z$$

holds.

The resulting theory of proportions traces back to Eudoxus's contribution to Euclid's *Elements* (particularly book 5, def. 5).[9] The modern construal of this theory of proportions as a theory of real numbers developed at the end of the nineteenth century, with the work of Dedekind and others.[10]

What is required for this theory is that the natural ordering generated by the *ratios* be extended. We must ask what the grounds are for this extension. The clearest justification derives from geometry. Distances between points sometimes stand in simple ratios to one another, but sometimes they do not. The golden rectangle shows this. More specifically, what has been shown is that we must choose among the following two options. The first option is that of denying that there could be any such thing as a golden rectangle. The second option is that of allowing the possibility of a golden rectangle, and hence allowing spatial relations which stand in no ratios to one another but which do obviously stand in other relations. These other relations need a name, so we call them 'proportions'.

As realists, we urge the latter option. It does not matter that there are probably no physical objects whose shape is precisely that of a golden rectangle. All that matters is that there are spatial relations of that sort which are available to objects. Even a subatomic particle probably has no precise location, but is "smeared" like a wave over an indefinite region. Yet that indefinite region includes definite subregions which stand in spatial relations to one another. And that is all that is required for a realist theory of proportions, or real numbers. Besides, spatial relations are not the only relations which exhibit proportions. Nature furnishes us with many quantities, and the very same proportions which hold among, say, lengths also hold among other quantities.

So if nature contained no quantities, would there be no real numbers? Is the existence of real numbers contingent upon the existence

9 The standard English translation of Euclid is Heath (1956).

10 For a translation of Dedekind's classic work on real numbers, see Dedekind (1888/1963); for Cantor's groundbreaking work, see the translation by Jourdain (Cantor, 1915/1955).

of quantities? There are two ways we could go: Plato's or Aristotle's. Aristotle requires any quantity to be instantiated if it is to exist. This makes the subject-matter of mathematics depend for its existence on whether the quantities in question are instantiated. But this, it would seem, would make mathematical truth contingent on empirical matters – on geography, history, and so forth. The Aristotelian option gives too little breathing space, too little autonomy, to pure mathematics. For the Platonist, in contrast, all quantities exist whether or not any contingent beings happen to instantiate them (or to study them). This grants self-government to pure mathematics. Hence, there has been a powerful drive towards Platonism, at least among pure mathematicians. It is that drive which has spawned set theory, as we shall explain in the following section.

8.4 EQUATIONS AND SETS

Set theory is the child of a union between arithmetic and geometry. Descartes was one of the matchmakers behind this union; the Cartesian plane was the bed on which set theory was conceived.[11]

Geometrical objects include such things as points, lines, planes, circles, and triangles. Each point will be a determinate distance from any given line. Such distances will stand in relations of proportion to one another. Some of these relations of proportion will be identical with certain ratios between natural numbers. Other such relations will be "irrational" ones, and these never equal any ratio between natural numbers. Such irrational proportions will, however, be greater than some ratios and less than all others. They can thus be slotted into the ordered sequence of ratios. This is, we have argued, the way the real numbers were identified and described.

Given the real numbers, it is possible to associate each of the basic geometrical objects, points, with real numbers. Each point will be a determinate distance from any given line, and this distance can be described by a number. Choose some fixed distance as a unit. Then the distance between point and line will stand in a specific proportion to the unit distance. That proportion is, we have argued, a real number. If we consider the distance of a point from *two* straight lines, the x axis and the y axis, we can then associate *two* real numbers with the point. Furthermore, we can identify any point on

11 See Descartes's *The Geometry*, translated by Smith and Latham (1954).

a plane by giving a number which measures its distance from the x axis and a number which measures its distance from the y axis.

What set theory permits us to do is to extend this correlation between geometrical and numerical objects. A point corresponds to an ordered pair of real numbers. A line is a set of points; and so a line corresponds to a set of ordered pairs of real numbers. Other geometrical objects can be construed in an analogous manner.

Many such sets can be concisely described by *equations*. Consider a circle. Suppose we have been given an x axis and a y axis (chosen at right angles), lying in the same plane as the circle. Suppose we have been given a unit distance. Then we can report, for each point on the circle, how far it is from the x axis and how far it is from the y axis. We can report these distances by giving the proportions in which they stand to the fixed distance we have been given as a unit. Let x be the distance of the point from the y axis, and y be its distance from the x axis. Then the set of points on the circle is the set of all and only the points such that

$$(x - a)^2 + (y - b)^2 = c^2$$

for fixed numbers a, b, and c. This set of points corresponds to a unique equation; and the equation corresponds to a unique set of points.

Geometry deals with universals which are instantiated, if at all, by spatial things. That means that geometry is vulnerable to empirical discoveries about space. Suppose we find that real, physical space does not instantiate the properties and relations we thought it did. Then it seems to follow that those properties and relations (at least some of them) are (or may very well be) *uninstantiated.*

Aristotelian realism asserts that any properties or relations which are uninstantiated will be *nonexistent*. If nothing *has* them, if they are not *in* anything, then they do not exist. From this perspective, then, geometry is extremely vulnerable to empirical discoveries. If space does not instantiate the universals we thought it did, then our geometry is false, across the board, unless deviously paraphrased in such a way as to remove all existential entailments.

Yet the propositions of Euclidean geometry are not false, even if straightforwardly (existentially) construed and even if physical space does turn out to be non-Euclidean. The truths of pure mathematics are not empirical hypotheses about contingent matters. So says a Platonist.

Initially, we can contrast a Platonist with an Aristotelian by saying that the Platonist asserts that universals may *exist* even if they are *uninstantiated*. Yet that is not quite right. If a universal exists, the Platonist will find (or invent) something which instantiates it. So the Platonist is, in fact, unlikely to believe in uninstantiated universals. The Platonist is distinguished, rather, by the belief that pure mathematics is autonomous and *a priori* and that the universals it studies will be instantiated by *something*, whether or not they are instantiated by the fickle physical phenomena around us.

The Platonist drive is what gives exceptional importance to set theory. Geometry, construed as a theory of physical space, is vulnerable to empirical discoveries; but by correlating geometrical objects with sets of numbers and studying those instead, we can insulate our pure mathematics from contingent threats.

The amazing thing about set theory was that it not only reduced geometry to numbers and sets, but also reduced numbers themselves to sets. This, again, served the purpose of freeing pure mathematics from empirical contamination. Real numbers are, we have argued, proportions which hold between *quantities*. If there were no quantities in nature, therefore, these proportions would be uninstantiated. If there were too few quantities, then even if some proportions were instantiated, others might not be. Number theory would be vulnerable to recalcitrant experience. But by correlating numbers with special sets and studying those instead, we can remove our pure mathematics from all threats of contingency.

In this book, we give qualified endorsement to Platonism. The qualifications will emerge soon; but first we shall acknowledge the reasons our view is more appropriately described as Platonist than Aristotelian. Given modal realism, we can be sure that, for any logically consistent universal, there will be some *possibilia* which instantiate it. Logical consistency ensures instantiation. It does not, however, ensure instantiation in the actual world. In endorsing *actually* uninstantiated universals, our theory is Platonic.

From this modal realist perspective, then, there is no *need* to invoke set theory in order to secure *a priori* instantiations for geometrical properties. It is here that we part company with the kind of Platonic theory described earlier. Geometrical properties can be studied whether or not they are instantiated in the actual world.

Having renounced one of the motives behind set theory, however, we nevertheless endorse set theory – for other reasons. We do not need to believe in sets as insurance against inconvenient discoveries in physics. But nevertheless *there are* sets of numbers corresponding to the various (possible) geometrical objects. In dealing with equations for various curves on a plane, it is quite clear that there does exist a geometrical object, an entity: the curve. This curve contains various points, but it is nevertheless a single thing, albeit a thing with many parts. Each of the points corresponds to a pair of numbers – an ordered pair, in fact. So there is a plurality of pairs of numbers corresponding to the simple geometric object, the curve. This plurality of pairs of numbers constitutes a *set*. This set is a single entity, which corresponds to a geometrical entity, the curve. Just as the curve has parts, namely the points it contains, so too the set has members, one for each of the points on the curve.

Note that one and the same geometrical object corresponds to infinitely many sets of number-pairs. Each different choice of axes and each different choice of a unit of distance generates a different set of number-pairs. All these sets of number-pairs are related to one another in virtue of the fact that they all describe the same geometrical object. It is obviously true that they have something in common. Yet it is the *sets* that have something in common. It is not the number-pairs, belonging to the sets, which have this something in common. In recognizing the universal which they share, we thereby recognize the things which share this universal. Mathematicians do not need a Platonic motive for "inventing" sets as insurance against contingency. Mathematicians discover sets whether they want them or not.

They discover, for instance, that two superficially different equations describe the same curve. Two such equations stand in a similar relation to number-pairs: they have the same extension. They pick out the same set of number-pairs. This is something which seems to be true, and that is a legitimate reason for asserting it. It is not just something which mathematicians posited, ad hoc, solely to secure an *a priori* foundation for geometry.

8.5 SETS AS UNIVERSALS

What, then, are sets? It is our contention that sets are universals. They are universals of the same sort as recurrences, although not all sets are recurrences in the sense of being in several places at once. An individual is something which instantiates universals but is not itself instantiated by anything. Sets instantiate universals; in fact, they instantiate all the structures and patterns studied in mathematics, and that is why they are so important. But sets are also instantiated by things. That is why they are not individuals, but universals.

If they are universals, which universals are they? What things instantiate them? Several theories are worth pursuing. We focus primarily on the most direct theory: that a set is a universal which is instantiated by its members. The membership relation, on this view, is just the instantiation relation restricted in scope to those universals which qualify as sets.[12]

It may be objected that set membership and instantiation differ from one another in certain fundamental respects. A difference between set membership and instantiation more generally is strongly suggested by the axiom of *extensionality*. This axiom states that, whenever two sets are distinct, nonidentical entities, there must be something which is a member of one but not the other.

This fails to hold for universals and instantiation more generally. It is not in general the case that, when two universals are nonidentical, there must be something which is instantiated by one but not by the other. At least, it is fairly clear that distinct universals need not have distinct *actual* instances. It is arguable that the situation is different for a modal realist, who quantifies over all possibilia. It could be argued that, if two universals are distinct, there must be some possible entity which instantiates one but not the other. This would ensure that universals do also satisfy the extensionality axiom. This could, as we have said, be argued. We are not persuaded, however. We wish to keep open the possibility of there being universals which are distinct, even though they have the same actual and possible instances: distinct yet necessarily coextensive universals. This seems to us to be especially likely in the case of some complex universals. For instance, the universal, being trilateral and equilateral, is necessarily coexistensive with being a trilateral and equiangular. Yet arguably, these universals are not numerically identical.

In any case, the extensionality axiom does apply to sets if its scope is limited to actualia, and this is not so for universals more generally; so there remains a clear contrast between sets and other universals.

There is a further crucial difference between sets and other universals. Not only do sets have to differ from one another in their

12 A classic discussion of sets as the modern version of the Forms is found in Quine (1969a). For a good, brisk listing of the basic axioms of set theory, see Halmos (1960).

membership, but they could not have had any other membership than that which they do have. Sets are in this regard different from political parties. A political party could have one extra member, but a set could not. If a set had one extra member, or one fewer member, it would not have been the set that it is. This can be summed up in the formula

$$\forall y(\Box(y \in x) \vee \Box\sim(y \in x)),$$

'Either something is a member of a given set in all possible worlds, or it is a member of that set in no possible worlds – it cannot be a member in some worlds but not others'. This is an instance of the principle of predication, discussed in Section 3.2. As we explained earlier, this principle fails for universals generally; but it does hold for sets. In fact, this can be used as a necessary condition for sethood: a universal is a set only if any given thing instantiates it either in all possible worlds or in none.

Essential properties have very much the same character as sets. A property is essential for a given object when the object cannot exist without having that property. So in every world where it exists, the object has that property. Not only individuals but universals too can have essences. Sets, then, can have essences. Having the membership it does have is an essential property of a set.

This helps to clarify the status of sets within the realist metaphysics we have presented in this book. Sets are universals; and their distinctive feature, extensionality, is in fact a reflection of the distinctive essences of these universals. The essential link between a set and its members, in fact, works in two directions. The set cannot exist without its members, as we have said. But equally, the converse holds, that the members cannot all exist together without instantiating that set. Of course, one member might have existed without the others. And then, arguably, the set would not have existed. So, arguably, each member, taken singly, might have existed without instantiating the set. So belonging to the set is not an essential property of the members of the set, taken singly. Nevertheless, it is an essential property of the members, taken all together. The members cannot *all* exist without all being members of that set. Hence, the set is a kind of *plural essence*.[13] It is of the same

13 For a slightly different defence of this identification of sets with plural essences, see Bigelow (1988a). The view developed here is also discussed in Bigelow (1989).

general character as essential properties, but it is distinctive in that it is essential for a plurality of things rather than for a single thing. It may be thought of as the property of *being one of these things*. A *unit* set is an essential property of its only member, the property of *being this thing*. All other sets (except the empty set) are essential to a plurality of things; but the essentiality of sets is just the same whether they are instantiated by a single thing or a plurality.

The axiom of extensionality serves to distinguish one set from another, among any sets which for one reason or another we believe in already. But it does not by itself entail the existence of any sets at all. Before extensionality can be used to distinguish among sets, we must infer on some other basis the existence of some sets.

How do we infer the existence of some sets in the first place? The famous axiom, or schema, which was supposed to serve this purpose is the *comprehension* or *abstraction* schema. This provides one of the most dramatic examples of the attempt to draw ontological conclusions out of semantic assumptions. The comprehension schema asserts that, given any *description* whatsoever, there is a *set* of things which meet that description. Let us suppose that we are given some description, in the form of an open sentence, with free variable x: $\Psi(x)$. Then comprehension asserts that there will be a set y containing all and only the things which satisfy $\Psi(x)$:

Comprehension

$$\exists y \forall x ((x \in y) \equiv \Psi(x)).$$

This extracts the existence of something, a set, out of the meaningfulness of a description: ontology out of semantics.

Fortunately, or unfortunately, the comprehension schema entails a contradiction. There are those who take seriously the suggestion that this provides a valid argument, with true premisses, establishing that some contradictions are true (e.g., Routley and Priest).[14] We follow the majority, however, in concluding that the comprehension schema is not valid. The rejection of comprehension is, in fact, just one further step along a path we (and others) have been travelling for some considerable time. Decisions about what exists in the world arise, in the first instance, before any introverted reflections about semantics. Our ontological speculations may then be modified in the light of feedback from semantics. But we should

14 See Priest (1979) and Routley (1980).

not expect an ontology to be simply peeled off language. So we should not expect the comprehension schema to be valid.

We recommend that a set be thought of as something all its members have in common: roughly, the property of being one of *these* things. The question is whether, for each description, there is the property, that of being one of *these* things, which is possessed by *all and only* the things which meet the description. The inconsistency of comprehension shows that there are cases in which there is no such property. Consider some very general description, of what we might call an *open-ended* sort. Perhaps there is some property of being one of *these* things which is shared by *many* of the things which meet that description. But there may be other things which meet the description, ones which do not have the property of being one of *these*. In fact, no matter how many things there are which fall under the property of being one of *these* things, there may be other things which lack the property but meet the description. It may be that *every* property of the form 'being one of *these* things' is instantiated by some, but not all, of the things which meet the description.

The most widely used set theory, Zermelo–Fraenkel (ZF), postulates a substitute for comprehension. This ersatz comprehension is sometimes called separation; with a little help from some further axioms, it entails the existence of enough sets to serve the purposes of mathematics. In particular, it furnishes enough sets for the reduction of geometry to number theory.

The general idea runs as follows. Given that there is a property of being one of *these* things, we can separate, among the instances of this property, those instances which in addition meet some specific description. Consider, then, the things which qualify as instances of *both* being one of *these* things *and* meeting some further description. For those things, there will be the corresponding property of being one of *those* things. That is, given the property x,

being one of *these* things,

we can separate the further property y,

being one of *those* things.

In the language of set theory, given a set x, there will be a subset y containing all and only the members of x which satisfy some further description. In symbols:

Separation

$$\forall x(\text{Set } x \supset \exists y \forall z((z \in y) \equiv ((z \in x \wedge \Psi(z)))).$$

Under our construal of sets as plural essences, the separation schema is intrinsically plausible. It is *much* more plausible, in fact, than comprehension.

This is very good news for our metaphysical construal of set theory. From the more common, semantic perspective, comprehension has always been the assumption with the most intrinsic plausibility, the one which seemed self-evident. From that perspective, separation seemed to be an ad hoc stab in the dark. Its plausibility derives primarily from its avoiding the fate of the sadly inconsistent comprehension axiom, while preserving as much of its spirit as possible. The usual reason for settling for separation, rather than the comprehension we secretly hanker after, is simply that it has not (yet) been shown to be inconsistent. This unsatisfactory state of affairs has been largely brought on by a too semantic perspective on set theory. A more metaphysical stance offers a remedy.

The comprehension schema was supposed to furnish us with enough sets to reduce geometry to number theory. Its watered-down replacement, separation, does not by itself entail the existence of any sets at all. It entails only the existence of various subsets, given that we already have reason to believe in some set to begin with. Neither extensionality nor comprehension entails the existence of any sets at all, unless there is some independent way of establishing the existence of some initial sets to begin with.

There are, in fact, several more axioms of this sort in ZF. *Pairing* states that, *if* there are two sets, there is a set containing just these two sets as members. *Union* states that, *if* there is a set of sets, there is a set which is the union of those sets. *Power set* states that, *if* there is a set, there is the set of all its subsets. None of these axioms states that there is any set at all. They tell us what sets would be like and what their existence would entail if there were any.

The axioms of ZF, then, are recursive. Most of them generate new sets from old. There are just two axioms which serve as the "basis" for the recursion:

Empty set asserts that there is a set with no members.
Infinity asserts that there is a set with infinitely many members.

In fact, the infinity axiom is usually formalized in a way which subsumes the empty set axiom, since it asserts the existence of a special infinite set which, as it happens, does include the empty set. That is, the axiom of infinity asserts that there is a set containing all the so-called von Neumann natural numbers, namely the set

$$\omega = \{\phi, \{\phi\}, \{\phi, \{\phi\}\}, \ldots\}.$$

According to our mathematical realism, the sets in the sequence ω are not identical with the natural numbers. They are important because they instantiate the natural numbers – for every natural number, there is a member of ω with that many members. That is why the axiom of infinity is so important.

This is the axiom with real ontological punch. It is the one which ensures that there are enough sets to instantiate all the rich patterns studied in pure mathematics. Given instantiations for the natural numbers, we ensure instantiations for all the ratios between natural numbers. Relations between ratios then capture the structure of the real number system. Allowing sets of real numbers, sets of such sets, and so on, we are then able to reduce geometry to number theory. This enables us to construct the number patterns studied in the infinitesimal calculus. We need the calculus to study velocity, acceleration, and so forth. Set theory ensures that we can study these things in pure mathematics, without looking at anything beyond set theory.

It is the axiom of infinity which ensures that there are enough sets to instantiate all the patterns we need to study in physics. It would be nice for pure mathematicians if the axiom of infinity were true. *Is* it true? Is it plausible, under our metaphysical theory of sets?

Part of the axiom flows inevitably from our construal of sets as plural essences. Given an initial segment of the sequence of sets in ω, the existence of the next member in the sequence is assured. Suppose that there is a property of being *these* things, and there is one extra thing not included under the former property. Then it is exceedingly plausible that there will be the property of being *those* things, which applies to all the previous things *plus* the one extra thing. This is plausible, granted that there are any such properties in the first place. Furthermore, if we are seriously realist about such properties, such a property can itself count as the "one extra thing". Thus, suppose there is the property of being these things. Then there are not only these things but also that property as well. Hence, there is the property of being *those* things, which embraces all of these *plus* the *property* of being *these* things.

This ensures that, given an initial segment of ω, the next member of the sequence is assured. But infinity requires more than this. Not only must there be an assurance that each member of ω exists: we also need an assurance that the whole set ω itself exists. On our metaphysics for sets, this means that there must be the property of 'being one of these things', where this is a property instantiated by all and only the von Neumann numbers. This is intrinsically plausible under our construal of sets as plural essences.

It is good news for our metaphysics that it makes the infinity axiom intrinsically plausible. There are *various* things which the von Neumann numbers have in common, *several* universals which are instantiated by all and only the von Neumann numbers. There is a natural presumption, within such a theory, that there will be a plural essence for all the von Neumann numbers.

The axiom of infinity, then, gains plausibility on two scores under our metaphysics. Given the existence of an initial segment of ω, the existence of the next in the sequence is easy to justify. This entails that, given any initial segment of ω, all the members of ω are easily justified. Then, given all the members of ω, the existence of ω itself is easily justified.

Infinity, then, is plausible on our metaphysics, provided that we grant the existence of some initial segment of the von Neumann numbers. This sequence does need an initial seed, however, to get it started: the empty set. How plausible is the existence of the empty set on our metaphysics?

We could, in fact, use something else for an initial seed to generate something equivalent to the von Neumann numbers. We could use Descartes's *cogito* to give an *a priori* assurance of the existence of the thinker, who can be substituted for the empty set as a seed for an ω-sequence. Each thinker would generate her own, personalized ω-sequence. Alternatively, we could pool our resources and we could all appeal to a common Cartesian *cogitamus* ("*we* think"). Each knows with Cartesian certainty that at least one thinker exists, and with almost equal certainty, therefore, that the aggregate of all thinkers exists. This, then, is something whose existence they all can know with certainty. It can serve as a seed for a von Neumann sort of sequence: just replace the empty set, ϕ, by θ (for 'thinkers').

Thus, if we found reasons to believe that there is no empty set (or if we found insufficient reasons to believe that there is an empty set), this would not scuttle the whole set-theoretical project. However, it would be nice if the empty set could be taken realistically. Then the axiom of infinity, and the rest of set theory, could be taken at face value, with no subterfuge. It is not easy to see how this is to be justified, however, with sets construed as plural essences. It is not easy to see how the empty set could plausibly be construed as either a singular or a plural essence.

For this reason, among others, it is worth examining a rival theory about which universals constitute sets. This rival theory was alluded to earlier, in Section 2.2. It is the theory that sets arise out of the relations of *coextensiveness* among universals.

There are various advantages in thinking of a set as something that coextensive universals share. More generally, even when two universals are not coextensive, they may still have something in common, in virtue of the overlap of their extensions. This something that they have in common is the *set* of things which instantiate both of them.

This makes a set a property of a property. This is, on the face of things, quite different from what we have been calling a plural essence. A plural essence need not be a property of a property. It may be, and is most naturally conceived as, a universal which is instantiated by individuals. Of course, universals of any rank in the type hierarchy may have a plural essence; so a plural essence may be instantiated by universals as well as by individuals. But the fact that it *may* be instantiated by individuals does serve to distinguish the plural essence construal of sets from the rival which sees sets essentially as what coextensive universals have in common. Call the rival theory the higher-order theory.

One advantage of the higher-order theory is that it makes it easier to make sense of the empty set. Consider a pair of universals whose extensions are disjoint. Consider another pair of universals whose extensions are disjoint. These two pairs have something in common. What do all disjoint pairs of sets have in common? The empty set. There is justification for thinking disjoint sets do have something in common. If we construe the empty set as what disjoint sets have in common, there is justification for believing that the empty set does exist.

Although the empty set furnishes an advantage for the higher-order theory, there are also intuitive advantages for the plural essence theory. We have dealt with some of those advantages, implicitly, in showing how the plural essence idea fits snugly onto the axioms of ZF.

The higher-order theory, however, can strike back by neatly turning all the merits of its rival to its own advantage. The higher-order theory can accommodate plural essences as a derived consequence of the core, higher-order theory. The technique employed is precisely the same as one which featured prominently in our theory of quantities in Chapter 2.

Suppose there are some things, x, y, and so on, which instantiate a property F and that property F instantiates property G. Then this structure induces extra properties of the original things x, y, and so on, properties which, although they are instantiated by the lower-level things like x and y, nevertheless involve the higher-level property G. In general, when x has property F, which in turn has property G, then x not only has property F, but also has the property of having a G-type property.

We can approach this second-order property another way. Suppose x has a property F, which in turn has property G. Suppose something else, z say, has some other property, H say, which also, in turn, has property G. We may suppose that x lacks properties H and G. Conversely, z lacks properties F and G. It then follows that x and z do have something in common. What they have in common is neither property F, nor G, nor H. What they have in common is having a property which has property G.

This general strategy can be reapplied in the higher-order theory of sets. Suppose we initially think of a set as something that universals may have in common. We then have things, say x and so forth, which instantiate the universal, F say, which in turn instantiates a further universal, G, which we are provisionally calling a set. In virtue of this structure, the thing x will have the further property of having a G-type property.

This new property now emerges as a contender for the title 'set'. And there are various grounds for identifying sets with this induced property rather than with the property of properties. In the first place, this generates a new and more fully articulated version of plural essences. All the advantages of that theory thereby accrue also to this theory. Furthermore, this version of the theory ensures that the membership relation is simply instantiation. It is a great merit of the theory that it gives a straightforward account of the relationship between a set and its members.

The earlier version of the higher-order theory had the disadvantage of making the link between a set and its members very tenuous. If a set is a property *G* of a property *F* which has instances like *x*, then the link between set *G* and thing *x* is very indirect: they are separated by a layer in the type hierarchy. And yet *x* would have to be a *member* of set *G*. Hence, membership could not be simply instantiation.

On the revised higher-order theory, membership will be simply instantiation. Sets will be plural essences induced by properties of properties. There will be just one exception. The empty set will be construed as a property of properties or, more strictly, a relation between universals. It is what disjoint pairs of universals have in common. But in this one case, the universal one level up the type hierarchy does *not* induce a new property of the things two levels below. That is why the empty set alone cannot be construed as a plural essence.

Nevertheless, the empty set does exist. We can put this final piece of the jigsaw together with all the other pieces which together make ZF. A justification for the empty set was all we had left over to justify the axiom of infinity. That completes our justification for all the core axioms of set theory. There are a few further axioms, such as the axiom of choice. It would be interesting to explore ways in which our theory could interpret these, but we will not do so here. It suffices, for a defence of our metaphysics, to show that it legitimates the core axioms, the ones on which there is most nearly a consensus and the ones which generate those chunks of mathematics which are needed for science.

8.6 MATHEMATICAL REFLEXIVITY

Our primary focus, within mathematics, has been on the nature of numbers and sets. This is appropriate since numbers and sets pose a worse prima facie problem for our theory than do other branches of mathematics. If we can crack numbers and sets, the rest of mathematics will be a breeze, so to speak. The reason that numbers and sets are a special problem for us is that they do seem to be *objects,* abstract *individuals,* rather than universals. So our view, that numbers and sets are universals, runs counter to initial impressions. In other branches of mathematics, however, there is not so strong an initial impression that the subject-matter consists of objects of some special, otherworldly sort. It is often very natural to construe such branches of mathematics as dealing with properties and relations of one kind or another.

Our account of mathematics, we urge, supplies mathematics with precisely the right degrees of unity and diversity. Mathematics is locally tidy, but globally open-ended and impossible to compartmentalize. It is like an ancient city filled with immaculate houses and winding lanes, but having no simple, overall plan. In recent times there have been attempts to impose a totalitarian central authority on this happy anarchy. There have been attempts to subordinate all of mathematics to logic; and there have been attempts to rebuild the whole of mathematics with set theory. In our view, such attempts have been misguided.

Our theory construes mathematics as a study of rich and interconnected patterns among universals. There is thus a unity in mathematics: all branches study universals. Yet there is also an open-endedness about it. The universals on which mathematics is grounded have a variety of sources. Hence, different branches of mathematics are in some respects the same, but they are also very different from one another. The greatest divide, historically, has been that between arithmetic and geometry. This divide is bridged, at higher levels, in many ways. And yet the existence of such bridges should not obscure for us an underlying difference in the sources of geometry and arithmetic. There are universals which involve spatial properties and relations, such as straightness and being longer than. There are also universals which arise from the numbering of things which may or may not be spatial. The latter universals are distinct from the former. The two sorts interlock, but they are not the same.

Our metaphysical theory of mathematics thus explains the diversities in mathematics, as well as its underlying unity. And there is another very striking feature of mathematics which our metaphysical theory accounts for: what we might call the *reflexivity* of mathematics. Mathematics is a miraculous creature: it feeds on itself, and it grows bigger. First, some patterns are discovered in one area – say, a pattern in number theory, modular arithmetic perhaps. Then a pattern is discovered in, say, geometry – the pattern generated by rotations and reflections of a regular polygon perhaps. Then it is noticed that these two patterns have something in common. Relationships emerge between some such patterns and others. A vast range of patterns – geometrical, arithmetical, and others – begin to display a rich range of patterns of their own. Mathematicians begin to study those *patterns of patterns*. Thus emerges the Gallois theory of *groups* and other branches of abstract algebra and eventually, at a very high level, category theory.

Patterns display patterns. That is why mathematics feeds on itself. The properties and relations studied by mathematics are very often ones which are instantiated not only by hardware in the physical world, but also by "mathware", by *other* properties and relations studied elsewhere in mathematics. So pure mathematics becomes increasingly introverted. The ground of this introversion lies in the way properties and relations can instantiate one another. For instance, all sorts of things can be counted: your parents are two, your virtues are many, your vices are few. But among the things that can be counted, we must include numbers. The primes between nine and fourteen are two in number. *Numbers can be counted.* This epitomizes the reflexivity of mathematics. Similarly, lengths, masses, and other quantities can stand in proportions to one another: one may be $(\pi/4)$ times another, for instance. Yet numbers, too, may stand in proportions to one another: π is $(\pi/4)$ times 4. *Proportions stand in proportions.* Again, there is an inescapable reflexivity in the universals studied in mathematics.

This casts a different light on the paramount importance which set theory has in current mathematics. Sets are important because they *instantiate* all (or almost all) the universals – the structures, patterns, relations – studied in mathematics. For instance, no sets are numbers – or so we argue, against orthodoxy. But sets instantiate numbers. It is convenient to deal with sets rather than with countable contingent objects, like quarks, because this enables us to minimize distractions arising from inessential properties and relations.

In order to deal adequately with the reflexivity of mathematics, we have to rest heavily on a specific aspect of our metaphysics of universals. We construe universals as, roughly, *recurrences*. They are (or most of them are) capable of being in several places at once. They are also distinguished by the fact that there is somehow that individuals may be related to them – individuals may *instantiate* them – and universals cannot instantiate individuals. Yet aside from these distinguishing features, universals are *things in the world* like any others. In particular, universals are nameable. There is no essential link between universals and predicates: universals can be referred to in subject position. The syntactic hierarchy of higher-order logics, therefore, does not carry over into a hierarchy of universals. If our theory of universals were traced straight off language, universals would have to be partitioned hierarchically. That is what happened with Whitehead and Russell's theory of types. And a hierarchical partitioning of the entities of mathematics runs entirely against the grain. It fails to respect the essential reflexivity in mathematics.

Hence, it is of great importance to free the theory of universals from its domination by semantics. Universals, as we construe them, are not hierarchically generated out of language. They do fall into patterns which are to some degree hierarchical. Individuals are in some sense at the bottom of a hierarchy: they instantiate things, but nothing instantiates them. But once we rise up a few levels in this incipient hierarchy, we find many universals which straddle levels in such a way as to blur entirely its hierarchical appearance.

By permitting mathematical reflexivity, by permitting universals to instantiate one another, we run a high risk of generating self-referential antinomies, such as Russell's famous paradox.[15] If universals are allowed to instantiate themselves, they are like barbers who shave themselves: and what are we to say about the barber who shaves all and only those who do not shave themselves? Clearly there is no such barber. The supposition that there is such a barber leads to contradiction.

Yet why should this be a problem for us? The problems of self-reference arise precisely for those who try to extract *a priori* foundations for mathematics from logic alone. That is exactly what we do not try to do. If we were to suppose that every linguistic item, such as a description, determines a mathematical entity, such as a set, *then* we would run into contradictions. Yet that is to read ontology straight off language in a way which should be resisted. We can frame hypotheses about which universals there are, and if such hypotheses lead to contradictions, they must be revised. That is only to be expected. There is no threat here for *a posteriori* realism about universals.

In a sense, we are immune to the threat of self-referential paradox

15 For one of the earliest discussions of Russell's paradox, see Russell (1903, chap. 10).

only because we have been very inexplicit about which universals we claim exist. If we were formally to set out, axiomatically, a theory of universals, we would run risks of inconsistency. Yet we protest that it is perfectly legitimate to develop a theory gradually, under the constant danger of uncovering inconsistencies. We offer no cast-iron guarantee of consistency. There is no such guarantee to be had.

In summary, then, our metaphysical foundation for mathematics accounts for the unity, the diversity, the reflexivity, and the paradoxes of mathematics. It also accounts for the intimate relationships between pure and applied mathematics, and hence accounts for the central role of mathematics in science. To that topic we now turn.

8.7 ABSTRACTIONS AND EXPLANATIONS

It is sometimes thought that numbers are nonphysical "abstract objects", and hence to be repudiated by any down-to-earth, sensible scientific realist. And what applies to numbers will, we claim, apply equally to other mathematical objects. Part of the complaint about mathematical objects lies in the allegation that they are not observable. This complaint traces back to a deeper allegation: that numbers lack causal powers. They do not cause anything. Assuming some broadly causal account of perception, it follows from their causal impotence that numbers cannot be perceived. Empiricists are suspicious of anything they cannot see. And they are even more suspicious of idle entities, entities which do no work at all.[16]

Defenders of mathematical objects could lock horns with empiricists over their Calvinistic ontological "work ethic". It could be argued that we should believe in entities even when they do not do any work. Our strategy, however, is to deny that mathematical entities are idle. In fact, we believe that some mathematical entities are observable after all. But we do not need to establish observability here. That would take us far into the philosophy of perception, and in this book we have set such epistemological issues to one side. Of more central concern for us here is the more basic question of causation.

16 Our uneasiness about the ambiguities of the term 'abstract' owes much to Lewis; see, e.g., Lewis (1986a, Sec. 1.7). Armstrong (1978, Vol. 2, pp. 45–6) speaks sympathetically about causal efficacy as a mark of being; Oddie (1982) dubs this the *eleatic principle*.

Admittedly, it sounds odd to say that the number $\sqrt{5}$, say, caused something. Yet this is partly due to the fact that, in general, causes and effects are not primarily objects, or properties, or relations, but *events*. In the most basic cases a cause is, we have argued, a state of or change in a field. And this means that a mathematical object, like a number, cannot be a cause. But you cannot be a cause either, strictly speaking. Yet you exist, and you are, in a less strict sense, causally efficacious. That is because you are involved in causal processes.

Numbers, too, are involved in causal processes. If various objects had not instantiated certain quantities, and if these quantities had not stood in the proportions in which they did stand – then fields would not have been in the states they were or have undergone the changes they did undergo. Hence, causes and effects would have been different from what they actually were if different proportions had held between quantities. That shows that proportions, or numbers, are involved in causal processes.

Consider Archimedes' theory of the lever. Imagine a fulcrum under a long beam, with two different masses on the two ends of the beam. The beam will balance horizontally, in equilibrium, provided that there is a specific proportion between the length from the smaller weight to the fulcrum and the length from the larger weight to the fulcrum. The length from the smaller weight to the fulcrum should be *longer* than that from the larger weight to the fulcrum. Furthermore, the proportion between these lengths should be the same as that between the weights. If the proportion deviates in one way, the smaller weight will descend; if it deviates in the other way, the greater weight will descend. The different events, of the beam's tipping one way or the other or neither, are results of causal processes which involve proportions.

The case of the lever illustrates something that occurs again and again in science: the functional dependence of one quantity on another. Science, particularly physics, is replete with equations which link one quantity with another. In the case of the lever, there is an inverse proportion between length and weight. As the weight on one side increases, the length must decrease to maintain equilibrium.

Equations such as these establish a pattern of counterfactual dependences among quantities. Archimedes' law, for instance, ensures that, when a beam is in equilibrium, it would tip one way if weight were added to one side, and it would tip the other way if weight were added to the other side. Weight having been added to one side or the other, the beam would balance if the fulcrum were placed at another point. All these claims about what *would* happen are counterfactuals. There is an infinite battery of them, summed up in one simple equation.

Thus, what the equations in science do is to describe what we have called *tracking*.[17] One thing tracks another when changes in one are always matched by corresponding changes in the other. A thermometer, for instance, tracks the temperature. As the temperature increases, the mercury rises; as the temperature decreases, the mercury falls.

Proportions among quantities enter into very rich patterns of counterfactual dependence. For this reason, they enter into one of the important patterns of explanation. Whenever a chain of counterfactuals holds,

$$\begin{array}{c} p_1 \;\Box\!\!\rightarrow\; q_1 \\ p_2 \;\Box\!\!\rightarrow\; q_2 \\ \cdot \\ \cdot \\ \cdot \\ p_i \;\Box\!\!\rightarrow\; q_i \\ \cdot \\ \cdot \\ \cdot \end{array},$$

we may explain why one of the q's is true rather than the others. If, say, p_2 is true, we may then say that q_2 is true *because* p_2 is true. For instance, we can explain that the mercury is at this level rather than higher, *because* the temperature is, say, 60°C rather than some temperature above 60°C.

Such explanations, furthermore, are not merely explanations of *how* things occur, but also of *why*. The counterfactuals involved do give very rich causal information, at least in most cases. Admittedly, there may be some cases in which counterfactual dependences are known, and yet they give only very minimal information about causal connections. But in most cases of functional dependences between quantities, causal connections are not at all far below the surface. It follows, therefore, that mathematical ob-

17 Nozick (1981) brought this idea to widespread attention and used the term 'tracking' to great effect. We derived this idea, however, mainly from Lewis (1973b), as explained in Section 7.3.

jects, like numbers, do play an integral role in causal explanations. It is also easy to see why equations are so important in science, since they provide such a convenient way of describing a functional dependence of the sort which issues in an infinite battery of counterfactuals.

There is a commonly held view about the role of mathematics which, for all its popularity among scientists themselves, is altogether mistaken. This common view depicts the scientist as beginning with real physical things and translating these into the entirely separate, nonphysical, "abstract" realm of mathematics. Then manipulations occur in this otherworldly realm, yielding some abstract result. This result is then translated back into the physical world, and the scientist checks to see whether this result matches observations. On this view, mathematics is a useful shortcut, taking us from physical facts to physical facts, but through an intervening territory outside the physical world, a shortcut through a kind of abstract hyperspace.

On this view of mathematics, it is very tempting to think that there must be a way of getting from a physical starting point to a physical result *without* taking a detour through the abstract realm of mathematics – hence, the temptation of Hartry Field's project, in *Science without Numbers,* of reconstruing science nominalistically.[18] Field's project is unnecessary, however, if we abandon the notion that mathematics is any sort of a detour at all. Mathematics is just further description of the physical properties and relations of things in the world. Any technique which bypasses the mathematics thus also bypasses some of the physical reality.

8.8 MATHEMATICS AND MODALITY

The hardness of the mathematical 'must' is something that has to be reckoned with. One of the things science does is to uncover necessities in nature. Some of these necessities are only relative, or conditional, necessities. But some of them are unconditional ones. These are the hardest sorts of necessities. And most of these spring from mathematics.

Necessities impose constraints. But the modalities in science are

18 Field (1980).

not all so inhibitory. One of the things science does is to reveal possibilities which had never occurred to us. And here again, mathematics plays a key role. It generates a space of possibilities within which we can locate the world as we know it and within which we can choose among possible futures for our world. Mathematics uncovered for us the amazing possibilities of non-Euclidean geometries, for instance; the possibility of a one-sided surface, the Möbius strip; and so forth. Mathematics sets limits to possibilities; but it also opens up undreamed of possibilities. Mathematics lies at the core of modalities in science.

It is important not to lose sight of the centrality of mathematics for any theory of modality. Philosophers have a tendency to focus on fairly simple examples in order to get to the heart of the issues without the distracting complexities of real live cases. This is a good thing, within limits, and is analogous to the so-called thought experiments of many of the great scientists, from Aristotle to Galileo to Einstein. Following this kind of strategy, philosophers of modality have focussed on simple examples, such as the necessity of 'All bachelors are unmarried'. Yet a philosopher whose diet is too one-sided is in danger of producing a scurvy philosophy. In understanding modality, in particular, it is essential that we round out our diet with at least a few examples from mathematics, both pure and applied. We should remember that modality cannot be separated from mathematics, and mathematics cannot be separated from science.

A realist about science, therefore, should be a realist about mathematics, and about modality. The difficulty, for many realists, is to see how they can be realists about mathematics, or modality, without wallowing in otherworldly Platonic mysticism.

And indeed, if a scientific realist is also a nominalist, the prospects are bleak indeed for any hopes of realism about mathematics or modality. A nominalist must try to draw a line between the parts of science which can be cut free of mathematics and the parts which cannot be cleansed and must therefore be either consigned to the flames or tolerated only under heavy instrumentalist warnings. Since there is, we have argued, no such line to be drawn, there is a strong tendency for nominalist leanings to draw a philosopher away from realism altogether. Witness, for instance, the extreme antirealism which that staunch nominalist Nelson Goodman has voiced in his aptly titled book, *Ways of Worldmak-*

ing.[19] His early ally, Quine, fell away from the hardline nominalism of their early joint paper. It is interesting that the reason Quine gave for his lapse was the necessity for sets if we are to ground enough mathematics to meet the needs of science. It was scientific realism which drew Quine away from nominalism.[20]

Nominalism, we argue, often militates against mathematical and modal realism. Admittedly much of Lewis's modal realism was formulated within a thoroughly nominalistic framework. Lewis's later sympathies for theories of universals should not blind us to the fact that his central theories have rested on a framework including individuals and set theory, but not universals.[21] And yet his theories are whole-heartedly realist about both mathematics and modality. This shows that it is not nominalism alone which militates against mathematical and modal realism. It is only within the context of some sort of "this-worldly" presuppositions that nominalism leads to objectionably antirealist consequences.

Hence, we argue, a this-worldly scientific realist should not be a nominalist. Mathematics stands as the most powerful of all the arguments for metaphysical realism – that is, for realism about universals. One of the most distinctive features of our version of realism, which we call scientific Platonism, is its realist stand on the role of mathematics in science and its construal of mathematics as a multilayered theory of universals. Any full-blooded realist about science should be a scientific Platonist. The arguments which lead to scientific realism, to realism about electrons, frogs, and so forth, if carried to their logical conclusion, lead in the end to mathematical realism as well – that is, to scientific Platonism.

Once it is accepted that the world contains universals, including those studied in mathematics, the way is open for casting new light on the nature of modality. The theory we recommend is a broadly combinatorial one. The world contains a stock of particulars and universals. Any combination or permutation of these will generate what we have called a world book. Some world books, in turn, will correspond to a complex property that the world may or may not

19 Goodman (1978).

20 For Quine's earlier joint paper with Goodman, see Goodman and Quine (1947). The closest investigation of the need for sets is found in Quine (1969b).

21 Lewis (1986a) sums up the issues comprehensively. Lewis (1968) is an instance of modal realism within a broad sympathy for nominalism; Lewis (1983a) marks a shift towards somewhat greater sympathy for universals.

instantiate. Such world properties, when they are maximally specific, may be called possible worlds. Then, using possible worlds, we may define modalities of all sorts, including probabilities, by way of the orthodox mathematical theory of probability which rests on a so-called sample space. We take this sample space to be a set of possible worlds, that is, a set of world properties, of ways the world could be.

This theory of modality rests ultimately on the stock of universals which are available for the recombinations which generate possible worlds. We also suspect that there will be constraints on the allowable recombinations, the recombinations that yield possible worlds. These constraints will flow from the natures of the universals. There may, for instance, be part–whole relations among universals, and these may sustain constraints on allowable recombinations. Furthermore, a universal may contain "parts" in two distinct senses. It may contain a part, in one sense (the most basic sense), when *there is something* to which it is somehow related, something to which it stands in the part–whole relation. It may contain a part, in another, quite different sense, when *there is somehow* that its parts (in the first sense) stand to one another. These two different senses of parthood correspond to the two different levels of quantification – the first-order "something" and the second-order "somehow". Both senses of parthood are relevant to the way in which complex universals are generated by simpler universals. This feeds into a full theory of world properties and thereby into a full analysis of modality.

Combinatorialism rests on the stock of universals which the world contains. We have maintained that the world contains quantities and proportions, as well as other patterns, structures, properties, and relations studied in mathematics. The subject-matter for mathematics furnishes the stock of raw materials with which a combinatorial theory can be built. Mathematics thus enters into the ground floor of our theory of modality. There is much more to be discovered in detail about exactly how combinatorialism works and how mathematics is linked to modality. But at least the broad outlines are clear. Mathematics studies universals, and universals underpin modality. Hence, it is no surprise that mathematics is at the core of most modalities.

Coda: scientific Platonism

There are more things in heaven and earth than the middle-sized goods that people can buy and sell. There are also recurrences: properties and relations which can be instantiated by several different things in different places at the same time. And there are properties and relations which make individuals unique, ones which they could not lack and could not share – individual essences. There are, in other words, universals – universals instantiated by the individuals around us, and even universals that will never be instantiated.

A scientific realism which embraces the existence of universals can provide an account of the nature of science which does justice to its modal character. Realism about universals permits pure mathematics, or at least most of it, to be literally true. Arithmetic and geometry deal with patterns, universals, and many of these are instantiated in nature. Even set theory can be construed realistically once we accept the existence of individual essences along with other universals. Realism about universals, and hence mathematics, then permits an analysis of the quantities measured in science: not only so-called scalars, like mass and volume, but also vectors, like relative velocity, acceleration, and force.

Forces are vectors, and we analyse these in terms of universals. In the light of modern science, we argue, forces should be accepted as part of the furniture of the world. Once we accept the existence of forces, thus construed, we have available a new analysis of causation. Causation, we urge, amounts to the action of forces. Realism about universals thus permits realism about causation.

When one thing contains others, the properties and relations of the parts together constitute a property of the whole. This is true for all things great and small, and even for the whole world around us. There is a grand world property which is instantiated by the actual world. There are also world properties which this world around us does not instantiate. We suppose these world properties to be uninstantiated; but we conjecture, with Plato, that a property does not have to be instantiated in order to exist. Uninstantiated world properties constitute other ways that a world could be – or in other words, possible worlds. Realism about universals thus permits realism about possibilities, modal realism.

Realism about possibilities then permits realism about reason. Deductively valid arguments are ones for which there is no possible world in which the premisses are true and the conclusion false. Probabilistic reasoning, too, can be construed realistically once we accept the existence of possible worlds.

But not only does realism about possibilities permit the character of scientific reasoning to be analysed; it also casts light on many of the other striking features of the theories of science. We can realistically analyse the objective chances which feature so prominently in recent science. We can realistically analyse the natural necessity which is possessed by scientific laws and explains their intricate tie-up with counterfactuals. And finally, explanation, one of the central goals of science, can then be construed realistically. It is even possible to justify realism about function and purpose in nature.

Scientific Platonism, the conjecture that universals as well as individuals exist, opens the door to both mathematical and modal realism and to the gardens of delight that these afford both the intellect and the emotions. Science is not just a cold instrumentalist machine with a technological output. Nor is science just a store of experimental data, with no rhyme or reason behind it all. Any view of science which leaves out its modal and explanatory dimensions will rob it of what matters most to us, the provision of an understanding of the world we live in and our place in it. Science does, fallibly yet with increasing success, represent the ways of the world. For scientific Platonism, or *metaphysicalism* as we might also call it, the theories of science, and even the logic and the mathematics in science, correspond to realities.

Bibliography

Adams, R. M. (1979). Primitive thisness and primitive identity. *Journal of Philosophy 76,* pp. 5–26.

Adjukiewicz, K. (1967). Syntactic connection. In *Polish Logic* (ed. S. McCall). Oxford University Press, pp. 207–31. (English translation of Die syntaktische konnexitat. *Studia Philosophia 1* [1935], pp. 1–27.)

Armstrong, D. M. (1961). *Perception and the physical world.* London, Routledge & Kegan Paul.

(1978). *Universals and scientific realism,* Vols. 1 and 2. Cambridge University Press.

(1983). *What is a law of nature?* Cambridge University Press.

(1986). In defence of structural universals. *Australasian Journal of Philosophy 64,* pp. 85–8.

(1988). Are quantities relations? A reply to Bigelow and Pargetter. *Philosophical Studies 44,* pp. 305–16.

(1989). *A combinatorial theory of possibility.* Cambridge University Press.

Aronson, J. L. (1982). Untangling ontology from epistemology in causation. *Erkenntnis 18,* pp. 293–305.

Ayer, A. J. (1956). What is a law of nature? *Revue Internationale de Philosophie 2,* pp. 144–65. (Reprinted in *The concept of a person.* London, Macmillan Press, 1963.)

Barker, A. O. (1969). An approach to the theory of natural selection. *Philosophy 44,* pp. 271–90.

Barrow, J. D. and Tipler, F. T. (1985). *The anthropic cosmological principle.* Oxford University Press.

Barwise, J., and Perry, J. (1983). *Situations and attitudes.* Cambridge Mass., MIT Press.

Bennett, J. (1974). Counterfactuals and possible worlds. *Canadian Journal of Philosophy 4,* pp. 391–3.

(1976). *Linguistic Behaviour.* Cambridge University Press.

Berkeley, G. (1965). *A treatise concerning the principles of human knowledge,* (1710) (ed. D. M. Armstrong). New York, Collier.

Bigelow, J. C. (1975). Contexts and quotation, Parts 1 and 2. *Linguistische Berichte 38,* pp. 1–21; *39,* pp. 1–21.

(1976). Possible worlds foundations for probability. *Journal of Philosophical Logic 5,* pp. 299–320.

(1977). Semantics of probability. *Synthese 36,* pp. 459–72.
(1979). Quantum probability in logical space. *Philosophy of Science 46,* pp. 223–43.
(1981). Semantic nominalism. *Australasian Journal of Philosophy 61,* pp. 403–21.
(1988a). *The reality of numbers: A physicalist's philosophy of mathematics.* Oxford, Clarendon Press.
(1988b). Real possibilities. *Philosophical Studies 53,* pp. 37–64.
(1989). Sets are universals. In *Physicalism in mathematics* (ed. A. D. Irvine), pp. 291–305. Dordrecht, Kluwer Academic Publishers.
Bigelow, J., and Pargetter, R. (1987a). Functions. *Journal of Philosophy 84,* pp. 181–96.
(1987b). Beyond the blank stare. *Theoria 53,* pp. 97–114.
(1988). Quantities. *Philosophical Studies 54,* pp. 287–304.
(1989a). From extroverted realism to correspondence: A modest proposal. *Philosophy and Phenomenological Research 50,* pp. 435–60.
(1989b). Vectors and change. *British Journal for the Philosophy of Science 40,* pp. 289–306.
(1989c). A theory of structural universals. *Australasian Journal of Philosophy 67,* pp. 1–11.
(1990). The metaphysics of causation. *Erkenntnis 33,* pp. 89–119.
Bigelow, J., Ellis, B., and Pargetter, R. (1988). Forces. *Philosophy of Science 55,* pp. 614–30.
Boolos, G. S. (1975). On second-order logic. *Journal of Philosophy 72,* pp. 509–27.
Boolos, G. S., and Jeffrey, R. C. (1980). *Computability and logic,* 2d ed. Cambridge University Press.
Boorse, C. (1976). Wright on functions. *Philosophical Review 85,* pp. 70–86.
(1977). Health as a theoretical concept. *Philosophy of Science 44,* pp. 542–73.
Bradley, F. H. (1914). *Essays on truth and reality.* Oxford, Clarendon Press.
Braithwaite, R. B. (1968). *Scientific explanation.* Cambridge University Press.
Cannon, H. G. (1958). *The evolution of living things.* Manchester, Manchester University Press.
Cantor, G. (1955). *Contribution to the founding of the theory of transfinite numbers* (1915) (Trans. P. E. B. Jourdain). New York, Dover.
Carnap, R. (1947). *Meaning and necessity: A study in semantics and modal logic.* Chicago, University of Chicago Press.
Cartwright, N. (1979). Causal laws and effective strategies. *Nous 13,* pp. 419–37.
(1980). Do the laws of physics state the facts? *Pacific Philosophical Quarterly 61,* pp. 343–77.
(1983). *How the laws of physics lie.* Oxford, Clarendon Press.
Churchland, P. M. (1979). *Scientific realism and the plasticity of mind.* Cambridge University Press.

Cresswell, M. J. (1972). The world is everything that is the case. *Australasian Journal of Philosophy 50,* pp. 1–13.
(1973). *Logics and languages.* London, Methuen.
(1985a). *Adverbial modification.* Dordrecht, Reidel.
(1985b). *Structured meanings: The semantics of propositional attitudes.* Cambridge, Mass., MIT Press.
Crew, H., and De Salvio, A., eds. and trans. (1954) *Dialogues concerning two new sciences* (by Galileo Galilei). Dover, New York.
Cummins, R. (1975). Functional analysis. *Journal of Philosophy 72,* pp. 741–65.
Davidson, B., and Pargetter, R. (1980). Possible worlds and a theory of meaning for modal languages. *Australasian Journal of Philosophy 58,* pp. 388–94.
Davidson, D. (1980). *Essays on actions and events.* Oxford University Press.
Dedekind, R. (1963). Was sind und was sollen die zahlen? (1888). *Essays on the theory of numbers* (trans. W. W. Beman). Dover, New York.
Dennett, D. G. (1978). *Brainstorms.* Montgomery, Vt., Bradford Books.
Devitt, M. (1984). *Realism and truth.* Princeton, N.J., Princeton University Press.
Devitt, M., and Sterelny, K. (1987). *Language and reality.* Oxford, Blackwell Publisher.
Dijksterhuis, E. J. (1961). *The mechanization of the world picture.* Oxford University Press.
Dretske, F. I. (1973). Contrastive statements. *Philosophical Review 81,* pp. 411–37.
(1977). Laws of nature. *Philosophy of Science 44,* pp. 248–68.
Dreyer, J. L. E. (1953). *A history of astronomy from Thales to Kepler* (1906). New York, Dover.
Ducasse, C. J. (1925). Explanation, mechanism and teleology. *Journal of Philosophy 22,* pp. 150–5.
Dudman, V. R. (1983). Tense and time in English verb clusters of the primary pattern. *Australian Journal of Linguistics 3,* pp. 25–44.
(1984). Conditional interpretations of if-sentences. *Australian Journal of Linguistics 4,* pp. 143–204.
Dummett, M. (1978). *Truth and other enigmas.* London, Duckworth.
(1982). Realism. *Synthese 52,* pp. 55–112.
Earman, J. (1976). Causation: A matter of life and death. *Journal of Philosophy 73,* pp. 23–5.
Ellis, B. D. (1970). Explanation and the logic of support. *Australasian Journal of Philosophy 48,* pp. 177–89.
(1979). *Rational belief systems.* Oxford, Blackwell Publisher.
(1987). The ontology of scientific realism. In *Metaphysics and morality: Essays in honour of J. J. C. Smart* (ed. P. Pettit, R. Sylvan, and J. Norman), pp. 50–70. Oxford, Blackwell Publisher.
Ellis, B. D., Bigelow J., and Lierse, C. (1989). The world is one of a kind. Paper read at the annual conference of the Australasian Association of Philosophy, Canberra.

Ellis, B. D., Jackson, F., and Pargetter, R. (1977). An objection to possible-worlds semantics for counterfactual logics. *Journal of Philosophical Logic 6,* pp. 355–7.

Euler, L. (1822). *Elements of algebra,* 3d ed. (trans. J. Hewlett). London, Johnson.

Fair, D. (1979). Causation and the flow of energy. *Erkenntnis 14,* pp. 219–50.

Feigl, H. (1945). Operationism and scientific method. *Psychological Review 52,* pp. 250–9, 284–8.

Field, H. H. (1972). Tarski's theory of truth. *Journal of Philosophy 69,* pp. 347–75.

(1980). *Science without numbers: A defense of nominalism.* Oxford, Blackwell Publisher.

Forge, J. (1986). The instance theory of explanation. *Australasian Journal of Philosophy 64,* pp. 131–42.

Forrest, P. (1986a). Ways worlds could be. *Australasian Journal of Philosophy 64,* pp. 15–24.

(1986b). Neither magic nor mereology. *Australasian Journal of Philosophy 64,* pp. 89–91.

Frege, G. (1879). *Begriffsschrift.* Halle, Louis Nebert. (Translated by J. van Heijenoort in *From Frege to Gödel: A source book in mathematical logic, 1879–1931.* Cambridge, Mass., Harvard University Press, 1967.)

(1884). *Die Grundlagen der Arithmetik.* Breslav, Verlag von Wilhelm Koebner. (Translated by J. L. Austin as *The foundations of arithmetic: A logico-mathematical enquiry into the concept of number,* 2d rev. ed. Oxford, Blackwell Publisher, 1959.)

(1893). *Grundgesetze der Arithmetik,* 2 vols. Jena, Hermann Pohle, 1893–1903.

(1980). *Philosophical and mathematical correspondence* (ed. G. Gabriel, H. Hermes, K. Kambartel, C. Thiel, and A. Veraart; abridged B. McGuinnes and trans. H. Kaal). Chicago, University of Chicago Press.

Friedman, A. (1970). *Foundations of modern analysis.* New York, Holt, Rinehart & Winston.

Gasking, E. (1967). *Investigations into generation, 1651–1828.* London, Hutchinson.

Goodman, N. (1951). *The structure of appearance.* New York, Bobbs-Merrill.

(1955). *Fact, fiction and forecast.* New York, Bobbs-Merrill.

(1978). *Ways of worldmaking.* Indianapolis, Ind., Hackett.

Goodman, N., and Quine, W. V. (1947). Steps toward a constructive nominalism. *Journal of Symbolic Logic 12,* pp. 105–22.

Hacking, I. (1983). *Representing and intervening: Introductory topics in the philosophy of natural science.* Cambridge University Press.

Halmos, P. R. (1960). *Naive set theory.* New York, Van Nostrand.

Harper, W. L., Stalnaker, R., and Pearce, G., eds. (1980). *Ifs.* Dordrecht, Reidel.

Heath, T. L. (1921). *A history of Greek mathematics,* Vol. 1: *From Thales to Euclid.* Oxford University Press.

Heath, T. L., ed. and trans. (1956). *The thirteen books of the elements.* New York, Dover.
Heathcote, A. (1989). A theory of causality: Causality = interaction (as defined by a suitable quantum field theory). *Erkenntnis 31,* pp. 77–108.
Hempel, C. G. (1965). *Aspects of scientific explanation.* New York, Free Press; London, Collier-Macmillan.
(1966). *Philosophy of natural science.* Englewood Cliffs, N.J., Prentice-Hall.
Hempel, C. G., and Oppenheim, P. (1948). Studies in the logic of explanation. *Philosophy of Science 15,* pp. 135–75.
Hirst, R. J. (1959). *The problems of perception.* London, Allen & Unwin.
Hofstadter, D. R., and Dennett, D. C. (1982). *The mind's I.* Harmondsworth, Penguin Books.
Hospers, J. (1946). On explanation. *Journal of Philosophy 43,* pp. 337–56.
Hughes, G. E., and Cresswell, M. J. (1968). *An introduction to modal logic.* London, Methuen.
(1984). *A companion to modal logic.* London, Methuen.
Hume, D. (1960). *A treatise of human nature: Being an attempt to introduce the experimental method of reasoning into moral subjects* (1739) (ed. L. A. Selby-Bigge). Oxford, Clarendon Press.
Jackson, F. (1977a). A causal theory of counterfactuals. *Australasian Journal of Philosophy 55,* pp. 3–21.
(1977b). *Perception: A representative theory.* Cambridge University Press.
(1987). *Conditionals.* Oxford, Blackwell Publisher.
Jackson, F., and Pargetter, R. (1989). Causal statements. *Philosophical Topics 16,* pp. 109–27.
Jackson, F., Pargetter, R., and Prior, E. W. (1982). Functionalism and type–type identity theories. *Philosophical Studies 42,* pp. 209–25.
James, W. (1907). *Pragmatism,* Lectures 2 and 6. New York, McKay.
Jeffrey, R. C. (1965). *The logic of decision.* Chicago, University of Chicago Press.
Jeffrey, R. C., ed. (1980). *Studies in inductive logic and probability,* Vol. 2. Berkeley and Los Angeles, University of California Press.
Jevons, W. S. (1877). *The principles of science: A treatise on logic and scientific method,* 2d ed. London, Macmillan Press.
Johnson, W. E. (1921). *Logic,* Part I. Cambridge University Press.
Kaplan, D. (1975). How to Russell a Frege-Church. *Journal of Philosophy 22,* pp. 716–29. (Reprinted in *The possible and the actual* [ed. M. J. Loux], Ithaca, N.Y., Cornell University Press, 1979.)
Kim, J. (1973). Causes and counterfactuals. *Journal of Philosophy 70,* pp. 570–2. (Reprinted in *Causation and conditionals* [ed. E. Sosa], Oxford University Press, 1975.)
Kline, A. D. (1985). Transference and the direction of causation. *Erkenntnis 23,* pp. 51–4.
Kolmogorov, A. N. (1956). *Foundations of the theory of probability.* New York, Chelsea.
Krabbe, E. C. W. (1978). Note on a completeness theorem in the theory of counterfactuals. *Journal of Philosophical Logic 7,* pp. 91–3.

Kripke, S. A. (1959). A completeness theorem in modal logic. *Journal of Symbolic Logic 24,* pp. 1–14.
(1962). Semantical considerations on modal logic. *Acta Philosophica Fennica 16,* pp. 83–94.
(1980). *Naming and necessity.* Cambridge, Mass., Harvard University Press.
Leeds, S. (1978). Theories of reference and truth. *Erkenntnis 13,* pp. 111–29.
Lemmon, E. J. (1969). *Introduction to axiomatic set theory.* London, Routledge & Kegan Paul.
Lesley, J. (1982). Anthropic principle, world ensemble, design. *American Philosophical Quarterly 19,* pp. 141–51.
Lewis, D. K. (1968). Counterpart theory and quantified modal logic. *Journal of Philosophy 65,* pp. 113–26.
(1973a). *Counterfactuals.* Oxford, Blackwell Publisher.
(1973b). Causation. *Journal of Philosophy 70,* pp. 556–67.
(1979). Counterfactual dependence and time's arrow. *Nous 13,* pp. 455–76. (Reprinted with notes in his *Philosophical Papers II,* Oxford University Press, 1986.)
(1980). A subjectivist's guide to objective chance. In *Studies in inductive logic and probability,* Vol. 2 (ed. R. C. Jeffrey), pp. 263–93. Berkeley and Los Angeles, University of California Press.
(1983a). New work for a theory of universals. *Australasian Journal of Philosophy 61,* pp. 343–77.
(1983b). *Philosophical papers,* Vol. 1. New York, Oxford University Press.
(1986a). *On the plurality of worlds.* Oxford, Blackwell Publisher.
(1986b). Against structural universals. *Australasian Journal of Philosophy 64,* pp. 25–46.
(1986c). Comment on Armstrong and Forrest. *Australasian Journal of Philosophy 64,* pp. 92–3.
(1986d). *Philosophical papers,* Vol. 2. New York, Oxford University Press.
(1986e). Causal explanation. In *Philosophical Papers,* Vol. 2, pp. 214–40. New York, Oxford University Press.
Lycan, W. G. (1979). The trouble with possible worlds. In *The possible and the actual* (ed. M. J. Loux), pp. 274–316. Ithaca, N. Y., Cornell University Press.
Mackie, J. L. (1965). Causes and conditions. *American Philosophical Quarterly 2,* pp. 245–55, 261–4. (Reprinted in *Causation and conditionals* [ed. E. Sosa], New York, Oxford University Press, 1975.)
(1973). *Truth, probability and paradox.* Oxford, Clarendon Press.
Malebranche. (1923). *Dialogues on metaphysics and on religion* (1688) (trans. M. Ginsberg). London, Allen & Unwin.
Manier, E. (1969). 'Fitness' and some explanatory patterns in biology. *Synthese 20,* pp. 206–18.
Mellor, D. H. (1976). Probable explanation. *Australasian Journal of Philosophy 54,* pp. 231–41.

(1980). Necessities and universals in natural laws. In *Science belief and behaviour* (ed. D. H. Mellor), pp. 105–25. Cambridge University Press.
Mill, J. S. (1843). *A system of logic*. (Reprinted in *John Stuart Mill's philosophy of scientific method* [ed. E. Nagel], New York, Hafner, 1950.)
Millikan, R. (1984). *Language, thought and other biological categories: New foundations for realism*. Cambridge, Mass., MIT Press.
Montague, R. (1974). *Formal philosophy: Selected papers of Richard Montague* (ed. R. H. Thomason). New Haven, Conn., Yale University Press.
Munroe, M. E. (1953). *Introduction to measure and integration*. Reading, Mass., Addison-Wesley.
Nagel, E. (1961). *The structure of science*. New York, Harcourt Brace; London, Routledge & Kegan Paul.
Newton, I. (1728). *Universal arithmetik*, 2d ed. (trans. Ralphson). London, Longman.
Nozick, R. (1981). *Philosophical explanations*. Cambridge, Mass., Harvard University Press.
Oddie, G. (1982). Armstrong on the eleatic principle and abstract entities. *Philosophical Studies 41*, pp. 285–95.
Pap, A. (1963). *An introduction to the philosophy of science*. London, Eyre & Spottiswoode.
Pargetter, R. (1984). Laws and modal realism. *Philosophical Studies 46*, pp. 335–47.
(1987). Fitness. *Pacific Philosophical Quarterly 68*, pp. 44–56.
Pargetter, R., and Prior, E. W. (1982). The dispositional and the categorical. *Pacific Philosophical Quarterly 63*, pp. 366–70.
Peacocke, C. (1979). *Holistic explanations*. Oxford University Press.
Plantinga, A. (1974). *The nature of necessity*. Oxford, Clarendon Press.
(1976). Actualism and possible worlds. *Theoria 42*, pp. 139–60. (Reprinted in *The possible and the actual* [ed. M. J. Loux] Ithaca, N.Y., Cornell University Press, 1979.)
(1987). Two concepts of modality: Modal realism and modal reductionism. In *Philosophical perspectives*, Vol. 1: *Metaphysics* (ed. J. E. Tomberlin), pp. 189–231. Atascadero, Calif., Ridgeview.
Pollock, J. L. (1974). *Knowledge and justification*. Princeton, N.J., Princeton University Press.
(1976). *Subjunctive reasoning*. Dordrecht, Reidel.
Presley, C. F., ed. (1967). *The identity theory of mind*. St. Lucia, University of Queensland Press.
Priest, G. (1979). The logic of paradoxes. *Journal of Philosophical Logic 8*, pp. 219–41.
Prior, A. (1949). Determinables, determinates and determinants. *Mind 58*, Part 1, pp. 1–20, Part 2, pp. 178–94.
Prior, E. W. (1985a). What is wrong with etiological accounts of biological function? *Pacific Philosophical Quarterly 64*, pp. 310–28.
(1985b). *Dispositions*. Aberdeen, Aberdeen University Press.
Prior, E. W., Pargetter, R., and Jackson, F. (1982). Three theses about dispositions. *American Philosophical Quarterly 19*, pp. 251–7.

Putnam, H. (1962). It ain't necessarily so. *Journal of Philosophy 59,* pp. 658–71. (Reprinted in *Philosophical Papers,* Vol. 1. Cambridge University Press, 1975, pp. 237–49.)

(1975). On properties. In *Philosophical Papers,* Vol. 1. Cambridge University Press.

(1978). *Meaning and the moral sciences.* London, Henley; Boston, Routledge & Kegan Paul.

(1981). *Reason, truth and history.* Cambridge University Press.

Quine, W. V. (1941). Whitehead and the rise of modern logic. In *The philosophy of Alfred North Whitehead* (ed. P. A. Schilpp), pp. 125–63. La Salle, Ill., Open Court.

(1960). *Word and object.* Cambridge, Mass., MIT Press.

(1961). *From a logical point of view: Logico-philosophical essays,* 2d ed. New York, Harper & Row.

(1966). *Selected logic papers.* New York, Random House.

(1969a). *Ontological relativity and other essays.* New York, Columbia University Press.

(1969b). *Set theory and its logic,* rev. ed. Cambridge, Mass., Harvard University Press.

Railton, P. (1978). A deductive-nomological model of probabilistic explanation. *Philosophy of Science 45,* pp. 206–26.

Ramsey, F. P. (1929). General propositions and causality. In *Foundations* (ed. D. H. Mellor), London, Routledge & Kegan Paul.

(1931). *The foundations of mathematics and other logical essays* (ed. R. B. Braithwaite). London, Routledge & Kegan Paul.

Reichenbach, H. (1958). *The philosophy of space and time.* New York, Dover.

Robinson, D. (1982). Re-identifying matter. *Philosophical Review 91,* pp. 317–41.

Ross, W. D., ed. (1928). *The works of Aristotle,* Vol. 8: *Metaphysica,* 2d ed. Oxford, Clarendon Press.

Routley, R. (1980). *Exploring Meinong's jungle and beyond: An investigation of noneism and the theory of items,* interim ed., Departmental Monograph. Canberra, A.C.T., Australian National University.

Ruse, M. (1971). Natural selection in the *Origin of Species. Studies in the History and Philosophy of Science 1,* pp. 311–51.

Russell, B. (1900). *A critical exposition of the philosophy of Leibniz.* London, Allen & Unwin.

(1903). *The principles of mathematics.* New York, Norton.

(1912). *The problems of philosophy.* New York, Oxford University Press.

Salmon, W. C. (1982). Comets, pollen and dreams: Some reflections on scientific explanation. In *What? Where? When? Why? Essays on induction, space, time, and explanation* (ed. R. McLaughlin), pp. 155–78. Dordrecht, Reidel.

Skyrms, B. (1980). *Causal necessity.* New Haven, Conn., Yale University Press.

(1981). Tractarian nominalism. *Philosophical Studies 40,* pp. 199–206.

Smart, J. J. C. (1968). *Between science and philosophy.* New York, Random House.

Smith, D. E. and Latham, M. L., eds. and trans. (1954). *The geometry of René Descartes*. Dover, New York.
Sobel, J. H. (1985). Circumstances and dominance in a causal decision theory. *Synthese 63,* pp. 167–202.
Sober, E. (1986). Explanatory presupposition. *Australasian Journal of Philosophy 64,* pp. 143–9.
Sosa, E. (ed.) (1975). *Causation and conditionals*. Oxford University Press.
Stalnaker, R. C. (1968). A theory of conditionals. In *Studies in logical theory* (ed. N. Rescher), pp. 98–112. Oxford, Blackwell Publishers.
(1976). Possible worlds. *Nous 10,* pp. 65–75. (Reprinted in *The possible and the actual* [ed. M. J. Loux], Ithaca, N.Y., Cornell University Press, 1979.)
(1981). A defense of conditional excluded middle. In *Ifs* (ed. W. L. Harper, R. Stalnaker, and G. Pearce), pp. 87–104. Dordrecht, Reidel.
(1984). *Inquiry*. Cambridge, Mass., MIT Press.
Stich, S. P. (1983). *From folk psychology to cognitive science: The case against belief*. Cambridge, Mass., MIT Press.
Strawson, P. F. (1959). *Individuals: An essay in descriptive metaphysics*. London, Methuen.
Suppes, P. (1970). *A probabilistic theory of causality*. Amsterdam, North Holland.
Tarski, A. (1956). *Logic, semantics, metamathematics* (trans. J. H. Woodger). Oxford University Press.
Taylor, B. (1987). The truth in realism. *Revue Internationale de Philosophie 41ᵉ année 160,* pp. 45–63.
Tichý, P. (1971). An approach to intentional analysis. *Nous 5,* pp. 273–97.
(1976). A counterexample to the Stalnaker–Lewis analysis of counterfactuals. *Philosophical Studies 29,* pp. 271–3.
Tooley, M. (1977). The nature of laws. *Canadian Journal of Philosophy 7,* pp. 667–98.
(1987). *Causation: A realist approach*. Oxford, Clarendon Press.
van Fraassen, B. (1980). *The scientific image*. Oxford, Clarendon Press.
van Inwagen, P. (1985). Two concepts of possible worlds. *Midwest Studies in Philosophy 9,* pp. 185–92.
Waddington, C. H. (1957). *The strategy of the genes*. London, Allen & Unwin.
Wiener, N. (1912). A simplification of the logic of relations. *Proceedings of the Cambridge Philosophical Society* 17 (1912–14), pp. 387–90. (Reprinted in J. van Heijenoort, *From Frege to Gödel: A source book in mathematical logic, 1879–1931*. Cambridge, Mass., Harvard University Press, 1967.)
Whitehead, A. N., and Russell, B. (1910). *Principia mathematica,* Vol. 1. Cambridge University Press.
Wimsatt, W. (1972). Teleology and the logical structure of function statements. *Studies in the History and Philosophy of Science 3,* pp. 1–80.
Wittgenstein, L. (1922). *Tractatus logico-philosophicus*. London, Routledge & Kegan Paul; New York, Humanities Press.
(1929). Some remarks on logical form. *Aristotelian Society,* suppl. vol. 9, pp. 162–71.

(1953). *Philosophical investigations* (trans. G. E. M. Anscombe). Oxford, Blackwell Publisher.
Woodfield, A. (1976). *Teleology.* Cambridge University Press.
Wright, L. (1973). Functions. *Philosophical Review 82,* pp. 139–68.
(1976). *Teleological explanations.* Berkeley and Los Angeles, University of California Press.

General index

Index of examples, illustrations, and stories